Integrating Food Science and Engineering Knowledge Into the Food Chain

Series Editor
Kristberg Kristbergsson

More information about this series at http://www.springer.com/series/7288

ISEKI-Food Series

Series editor: Kristberg Kristbergsson, *University of Iceland, Reykjavík, Iceland*

FOOD SAFETY: A Practical and Case Study Approach
Edited by Anna McElhatton and Richard J. Marshall

ODORS IN THE FOOD INDUSTRY
Edited by Xavier Nicolay

UTILIZATION OF BY-PRODUCTS AND TREATMENT OF WASTE IN THE FOOD INDUSTRY
Edited by Vasso Oreopoulou and Winfried Russ

PREDICTIVE MODELING AND RISK ASSESSMENT
Edited by Rui Costa and Kristberg Kristbergsson

EXPERIMENTS IN UNIT OPERATIONS AND PROCESSING OF FOODS
Edited by Maria Margarida Vieira Cortez and Peter Ho

CASE STUDIES IN FOOD SAFETY AND ENVIRONMENTAL HEALTH
Edited by Maria Margarida Vieira Cortez and Peter Ho

NOVEL TECHNOLOGIES IN FOOD SCIENCE: Their Impact on Products, Consumer Trends, and the Environment
Edited by Anna McElhatton and Paulo José do Amaral Sobral

TRADITIONAL FOODS: General and Consumer Aspects
Edited by Kristberg Kristbergsson and Jorge Oliveira

MODERNIZATION OF TRADITIONAL FOOD PROCESSES AND PRODUCTS
Edited by Anna McElhatton and Mustapha Missbah El Idrissi

FUNCTIONAL PROPERTIES OF TRADITIONAL FOODS
Edited by Kristberg Kristbergsson and Semih Ötles

FOOD PROCESSING
Edited by Kristberg Kristbergsson and Semih Ötles

APPLIED STATISTICS FOR FOOD AND BIOTECHNOLOGY
Edited by Gerhard Schleining, Peter Ho and Severio Mannino

PHYSICAL CHEMISTRY FOR FOOD SCIENTISTS
Edited by Stephan Drusch and Kirsi Jouppila

PROCESS ENERGY IN FOOD PRODUCTION
Edited by Winfried Russ, Barbara Sturm and Kristberg Kristbergsson

CONSUMER DRIVEN DEVELOPMENT OF FOOD FOR HEALTH AND WELL BEING
Edited by Kristberg Kristbergsson, Paola Pittia, Margarida Vieira and Howard R. Moskowitz

BOOK ON ETHICS IN FOOD PRODUCTION AND SCIENCE
Edited by Rui Costa and Paola Pittia

Kristberg Kristbergsson • Jorge Oliveira
Editors

Traditional Foods

General and Consumer Aspects

Springer

Editors
Kristberg Kristbergsson
Faculty of Food Science and Nutrition
University of Iceland
Reykjavík, Iceland

Jorge Oliveira
School of Engineering
University College Cork
Cork, Ireland

Integrating Food Science and Engineering Knowledge Into the Food Chain
ISBN 978-1-4899-7646-8   ISBN 978-1-4899-7648-2   (eBook)
DOI 10.1007/978-1-4899-7648-2

Library of Congress Control Number: 2015958898

Springer New York Heidelberg Dordrecht London

Printed on acid-free paper

Springer Science+Business Media LLC New York is part of Springer Science+Business Media (www.springer.com)

Series Preface

The ISEKI-Food series was originally planned to consist of six volumes of texts suitable for food science students and professionals interested in food safety and environmental issues related to sustainability of the food chain and the well-being of the consumer. As the work progressed it soon became apparent that the interest and need for texts of this type exceeded the topics covered by the first six volumes published by Springer in 2006–2009. The series originate in work conducted by the European thematic network "ISEKI-Food," an acronym for "Integrating Food Science and Engineering Knowledge Into the Food Chain." Participants in the ISEKI-Food network come from most countries in Europe, and most of the institutes and universities involved with food science education at the university level are represented. The network was expanded in 2008 with the ISEKI Mundus program with 37 partners from 23 countries outside of Europe joining the consortium, and it continues to grow with approximately 200 partner institutions from 60 countries from all over the world in 2011. Some international companies and nonteaching institutions have also participated in the program. The network was funded by the ERASMUS program of the EU from 1998 to 2014 first as FoodNet coordinated by Professor Elisabeth Dumoulin at AgroParisTech-site de MASSY in France. The network then became known as "ISEKI-Food" and was coordinated by Professor Cristina Silva at the Catholic University of Portugal, College of Biotechnology (Escola) in Porto from 2002 to 2011 when Professor Paola Pittia at the University of Teramo in Italy became coordinator of ISEKI Food 4.

The main objectives of ISEKI-Food network have been to improve the harmonization of studies in food science and engineering in Europe and to develop and adapt food science curricula emphasizing the inclusion of safety and environmental topics. The program has been further expanded into the ISEKI-Food Association (https://www.iseki-food.net/), an independent organization devoted to the objectives of the ISEKI consortium to further the safety and sustainability of the food chain through education. The motto of the association is "Integrating Food Science and Engineering Knowledge into the Food Chain." The association will continue work on the ISEKI-Food series with several new volumes to be published in the near future. The series continued with the publication in 2012 of *Novel Technologies in Food Science: Their Impact on Products, Consumer Trends and The Environment*, edited by Anna McElhatton and Paolo J. Sobral. The book is intended for food scientists, engineers, and readers interested in the new and emerging food processing technologies developed to produce safe foods that maintain most of their original freshness. All 13 chapters are written from a safety and environmental standpoint with respect to the emerging technologies.

We now see the publication of the Trilogy of Traditional Foods written for food science professionals as well as for the interested general public. The trilogy is in line with the internationalization of the ISEKI consortium and will offer 74 chapters dedicated to different traditional foods from all over the world. The trilogy starts with *Traditional Foods; General and Consumer Aspects* edited by the undersigned and Jorge Oliveira. The book offers general descriptions of different traditional foods and topics related to consumers and sensory aspects The second book in the trilogy is *Modernization of Traditional Food Processes and Products* edited by Anna McElhatton and Mustapha Missbah El Idrissi. The chapters are devoted to recent changes and modernizations of specific traditional foods with a focus on processing and engineering aspects. The third volume in the trilogy, *Functional Properties of Traditional Foods*, is devoted to functional and biochemical aspects of traditional foods and the beneficial effects of bioactive components found in some traditional foods.

The series will continue with several books including the textbook *Food Processing* intended for senior level undergraduates and junior graduate students. This textbook, edited by the undersigned and Semih Ötles, will provide a comprehensive introduction to food processing. The book should also be useful to professionals and scientists interested in food processing both from the equipment and process approach and in the physicochemical aspect of food processing. The book will contain five sections starting with chapters on the basic principles and physicochemical properties of foods, followed by sections with chapters on conversion operations, preservation operations, and food processing operations with separate chapters on most common food commodities. The final section will be devoted to post processing operations.

Applied Statistics for Food Technology and Biotechnology, edited by Gerhard Schleining, Peter Ho, and Saverio Mannino, will be intended for graduate students and industry personnel who need a guide for setting up experiments so that the results will be statistically valid. The book will provide numerous samples and case studies on how to use statistics in food and biotechnology research and testing. It will contain chapters on data collection, data analysis and presentation, handling of multivariate data, statistical process control, and experimental design.

The book *Process Energy in Food Production*, edited by Winfried Russ, Barbara Sturm, and the undersigned, is being prepared for publication. The book will offer an introduction section on basic thermodynamics and an overview of energy as a global element. It will also cover environmental effects of energy provision and usage. This will be followed by chapters on the use of energy in various food processes such as flour production, bakery, fish processing, meat processing, brewery and beverage production, direct and indirect heat integration in breweries, fruit juice, spray drying systems (milk powder), chilling, and storage of fresh horticultural products. There will also be chapters related to energy supply (thermal, solar, and hydroelectric), energy distribution, insulation for energy saving, storage systems for heat and coldness, waste heat recovery, and energy management systems.

Three more books are being prepared. The textbook *Physical Chemistry for Food Scientists*, edited by Stephan Drusch and Kirsi Jouppila, will provide senior undergraduate and beginning graduate level students an overview of the basic principles of physical chemistry of foods. The first part of the book will be devoted to fundamental principles of physical chemistry. The second part of the book will be devoted to the physical chemistry of food systems. The book *Consumer Driven Development of Food for Health and Well Being*, edited by the undersigned, Paola Pittia, Margarida Vieira, and Howard R. Moskowitz, is in preparation with chapters on general aspects of food development, the house of quality and Stage-Gate® process, consumer aspects of food development, mind genomics, conceptualization of well-being in the framework of food consumption, formulation of foods in the development of food for health and well-being, ingredients' contribution for health and well-being, new trends on the extension of shelf life, nutritional aspects of development of foods focusing on health and well-being, regulatory and policy aspects, and several case studies on product development with special emphasis on health and well-being. Finally, there is the book *Ethics in Food Production and Science* that will be edited by Rui Costa and Paola Pittia.

The ISEKI-Food series draws on expertise from universities and research institutions all over the world, and we sincerely hope that it may offer interesting topics to students, researchers, professionals, as well as the general public.

Reykjavík, Iceland   Kristberg Kristbergsson

Series Preface Volumes 1–6

The single most important task of food scientists and the food industry as a whole is to ensure the safety of foods supplied to consumers. Recent trends in global food production, distribution, and preparation call for increased emphasis on hygienic practices at all levels and for increased research in food safety in order to ensure a safer global food supply. The ISEKI-Food book series is a collection of books where various aspects of food safety and environmental issues are introduced and reviewed by scientists specializing in the field. In all of the books, a special emphasis was placed on including case studies applicable to each specific topic. The books are intended for graduate students and senior level undergraduate students as well as professionals and researchers interested in food safety and environmental issues applicable to food safety.

The idea and planning of the books originates from two working groups in the European thematic network "ISEKI-Food," an acronym for "Integrating Food Science and Engineering Knowledge Into the Food Chain." Participants in the ISEKI-Food network come from 29 countries in Europe, and most of the institutes and universities involved with food science education at the university level are represented. Some international companies and nonteaching institutions have also participated in the program. The ISEKI-Food network is coordinated by Professor Cristina Silva at the Catholic University of Portugal, College of Biotechnology (Escola) in Porto. The program has a website at: http://www.esb.ucp.pt/iseki/. The main objectives of ISEKI-Food network have been to improve the harmonization of studies in food science and engineering in Europe and to develop and adapt food science curricula emphasizing the inclusion of safety and environmental topics. The ISEKI-Food network started on October 1st in 2002 and has recently been approved for funding by the EU for renewal as ISEKI Food 2 for another 3 years. ISEKI has its roots in an EU-funded network formed in 1998 called FoodNet where the emphasis was on casting a light on the different food science programs available at the various universities and technical institutions throughout Europe. The work of the ISEKI-Food network was organized into five different working groups with specific task all aiming to fulfill the main objectives of the network.

The first four volumes in the ISEKI-Food book series come from WG2 coordinated by Gerhard Schleining at Boku University in Austria and the undersigned. The main task of the WG2 was to develop and collect materials and methods for teaching of safety and environmental topics in the food science and engineering curricula. The first volume is devoted to food safety in general with a practical and a case study approach. The book is composed of 14 chapters which were organized into three sections on preservation and protection, benefits and risk of microorganisms, and process safety. All of these issues have received high public interest in recent years and will continue to be in the focus of consumers and regulatory personnel for years to come. The second volume in the series is devoted to the control of air pollution and treatment of odors in the food industry. The book is divided into eight chapters devoted to defining the problem, recent advances in analysis and methods for prevention, and treatment of odors. The topic should be of special interest to industry personnel and researchers due to recent and upcoming regulations by the European Union on air pollution from food processes. Other countries will likely follow suit with more strict regulations on the level of odors permitted to enter the environment from food processing operations. The third volume in the series is devoted to utilization and treatment of waste in the food industry. Emphasis is placed on sustainability of food sources and how waste can be turned into by-products rather than pollution or landfills. The book is composed of 15 chapters starting off with an introduction of problems related to the treatment of waste and an introduction to the ISO 14001 standard used for improving and maintaining environmental management systems. The book then continues to describe the treatment and utilization of both liquid and solid waste with case studies from many different food processes. The last book from WG2 is on predictive modeling and risk assessment in food products and processes. Mathematical modeling of heat and mass transfer as well as reaction kinetics is introduced. This is followed by a discussion of the stoichiometry of migration in food packaging, as well as the fate of antibiotics and environmental pollutants in the food chain using mathematical modeling and case study samples for clarification.

Volumes 5 and 6 come from work in WG5 coordinated by Margarida Vieira at the University of Algarve in Portugal and Roland Verhé at Gent University in Belgium. The main objective of the group was to collect and develop materials for teaching food safety-related topics at the laboratory and pilot plant level using practical experimentation. Volume 5 is a practical guide to experiments in unit operations and processing of foods. It is composed of 20 concise chapters each describing different food processing experiments outlining theory, equipment, procedures, applicable calculations, and questions for the students or trainee followed by references. The book is intended to be a practical guide for the teaching of food processing and engineering principles. The final volume in the ISEKI-Food book series is a collection of case studies in food safety and environmental health. It is intended to be a reference for introducing case studies into traditional lecture-based safety courses as well as being a basis for problem-based learning. The book consists of 13 chapters containing case studies that may be used, individually or in a series, to discuss a range of food safety issues. For convenience the book was divided into three main sections with the first devoted to case studies, in a more general framework with a number of specific issues in safety and health ranging from acrylamide and nitrates to botulism and listeriosis. The second section is devoted to some well-known outbreaks related to food intake in different countries. The final section of the book takes on food safety from the perspective of the researcher. Cases are based around experimental data and examine the importance of experimental planning, design, and analysis.

The ISEKI-Food books series draws on expertise from close to a hundred universities and research institutions all over Europe. It is the hope of the authors, editors, coordinators, and participants in the ISEKI-Food network that the books will be useful to students and colleagues to further their understanding of food safety and environmental issues.

Reykjavík, Iceland   Kristberg Kristbergsson

Acknowledgments

ISEKI_Food 3 and ISEKI_Mundus 2 were thematic networks on food studies, funded by the European Union through the Lifelong Learning and Erasmus Mundus programs as projects N° 142822-LLP-1-2008-PT-ERASMUS-ENW and 145585-PT-2008-ERA MUNDUS—EM4EATN, respectively. The ISEKI_Mundus 2 project was established to contribute for the internationalization and enhancement of the quality of the European higher education in food studies and work toward the network sustainability, by extending the developments undergoing through the Erasmus Academic Network ISEKI Food 3 to other countries and developing new activities toward the promotion of good communication and understanding between European countries and the rest of the world.

Lifelong Learning Programme

ERASMUS MUNDUS

Preface

Traditional Foods; General and Consumer Aspects is the first book in the Trilogy of Traditional Foods and the eighth book in the ISEKI-Food series. The books in the trilogy are devoted to different characteristics of traditional foods. The trilogy covers general and consumer aspects, modernization of traditional foods, and functional properties of traditional foods in a total of 74 chapters written by authors from all over the world. The list of contributors in this first book of the trilogy, which includes authors from China, Bulgaria, Portugal, France, Norway, Romania, and Slovakia, to name a few, shows its truly international perspective. *Traditional Foods; General and Consumer Aspects* is divided into six parts, with the first part focusing on general aspects. The two first chapters are devoted to consumers' definition and perception of traditional foods. This is followed by chapters with general descriptions of the available traditional foods in four different countries. The following parts cover traditional dairy products, traditional cereal-based products, traditional meat and fish products, traditional beverages, and traditional deserts, side dishes, and oil products from various countries. All the chapters are written by practicing food scientists or engineers but are written with the interested general public in mind. The book should cater to the practicing food professional as well as the interested reader. The volume offers numerous recipes of traditional foods from across the world and sometimes detailed descriptions on how to proceed to mix, cook, bake, or store a particular food item in order to produce the desired effect.

Reykjavík, Iceland   Kristberg Kristbergsson
Cork, Ireland   Jorge Oliveira

Contents

Part I   General Aspects of Traditional Foods

**1   European Consumers' Definition and Perception
of Traditional Foods**...   3
Wim Verbeke, Luis Guerrero, Valerie Lengard Almli,
Filiep Vanhonacker, and Margrethe Hersleth

**2   Consumer's Valuation and Quality Perception
of Kid's Meat from Traditional "Cabrito da Gralheira":
Protected Geographical Indication**.......................................   17
António Lopes Ribeiro, Ana Pinto de Moura, and Luís Miguel Cunha

3   Traditional Fermented Foods in Thailand............................   31
Busaba Yongsmith and Wanna Malaphan

**4   Traditional Food in Romania Integrated in a Protected
Geographical Designations System**   61
Gabriela Nedita, Nastasia Belc, Lucia Romanescu,
and Roxana Cristina Gradinariu

5   Traditional Foods in Slovakia   71
Jan Brindza, Dezider Toth, Radovan Ostrovsky,
and Lucia Kucelova

6   Traditional Foods in Turkey: General and Consumer Aspects   85
Semih Ötleş, Beraat Özçelik, Fahrettin Göğüş,
and Ferruh Erdoğdu

Part II   Traditional Dairy Products

7   Indian Traditional Fermented Dairy Products   101
Narender Raju Panjagari, Ram Ran Bijoy Singh,
and Ashish Kumar Singh

8   **Traditional Bulgarian Dairy Food**.. 115
     Anna Aladjadjiyan, Ivanka Zheleva, and Yordanka Kartalska

9   *Dulce de Leche*: **Technology, Quality, and Consumer Aspects
     of the Traditional Milk Caramel of South America**........................... 123
     María Cecilia Penci and María Andrea Marín

Part III   Traditional Cereal Based Products

10   **Austrian Dumplings**.. 139
     Josefa Friedel, Helmut Glattes, and Gerhard Schleining

11   **French Bread Baking**... 157
     Alain Sommier, Yannick Anguy, Imen Douiri,
     and Elisabeth Dumoulin

12   **Traditional Rye Sourdough Bread in the Baltic Region**.................. 173
     Grazina Juodeikiene

13   **The Legume Grains: When Tradition Goes Hand
     in Hand with Nutrition**.. 189
     Marta Wilton Vasconcelos and Ana Maria Gomes

14   **Traditional Food Products from *Prosopis* sp. Flour**...................... 209
     Leonardo Pablo Sciammaro, Daniel Pablo Ribotta,
     and Maria Cecilia Puppo

15   **Amaranth: An Andean Crop with History, Its Feeding
     Reassessment in America**.. 217
     Myriam Villarreal and Laura Beatriz Iturriaga

Part IV   Traditional Meat and Fish Products

16   **Yunnan Fermented Meat: Xuanwei Ham, Huotui**......................... 235
     Huang Aixiang, Sarote Sirisansaneeyakul, and Yusuf Chisti

17   **Selected Viennese Meat Specialties**.. 251
     Elias Gmeiner, Helmut Glattes, and Gerhard Schleining

18   **Dried Norse Fish**... 259
     Trude Wicklund and Odd Ivar Lekang

19   **Utilization of Different Raw Materials from Sheep
     and Lamb in Norway**... 265
     Trude Wicklund

20   *Muxama* and *Estupeta*: **Traditional Food Products
     Obtained from Tuna Loins in South Portugal and Spain**................. 271
     Jaime Aníbal and Eduardo Esteves

21   Salting and Drying of Cod.. 275
    Helena Oliveira, Maria Leonor Nunes, Paulo Vaz-Pires,
    and Rui Costa

22   Brazilian Charqui Meats.. 291
    Massami Shimokomaki, Carlos Eduardo Rocha Garcia,
    Mayka Reghiany Pedrão, and Fabio Augusto Garcia Coró

Part V   Traditional Beverages

23   German Beer 297
    Martin Zarnkow, Christopher McGreger, and Nancy McGreger

24   "Pálinka": Hungarian Distilled Fruit 313
    Zsuzsanna László, Cecila Hodúr, and József Csanádi

25   Tokaji Aszú: "The Wine of Kings, the King of Wines" 319
    Cecilia Hodúr, József Csanádi, and Zsuzsa László

26   Mead: The Oldest Alcoholic Beverage 325
    Rajko Vidrih and Janez Hribar

27   Midus: A Traditional Lithuanian Mead 339
    Rimantas Kublickas

28   Horchata 345
    Idoia Codina, Antonio José Trujillo, and Victoria Ferragut

Part VI   Traditional Deserts, Side Dishes and Oil

29   "Dobos": A Super-cake from Hungary 359
    Elisabeth T. Kovács

30   Traditional Green Table Olives from the South of Portugal 367
    Maria Alves and Célia Quintas

31   Extra Virgin Olive Oil and Table Olives from Slovenian Istria 377
    Milena Bučar-Miklavčič, Bojan Butinar, Vasilij Valenčič,
    Erika Bešter, Mojca Korošec, Terezija Golob,
    and Sonja Smole Možina

**32   A Perspective on Production and Quality
    of Argentinian Nut Oils** 387
    Marcela Lilian Martínez and Damián Modesto Maestri

33   Pupunha (Bactris gasipaes): General and Consumption Aspects 399
    Carolina Vieira Bezerra and Luiza Helena Meller da Silva

Index.. 407

About the Editors

Kristberg Kristbergsson is professor of food science at the Faculty of Food Science and Nutrition at the University of Iceland. Dr. Kristbergsson has a B.S. in food science from the Department of Chemistry at the University of Iceland and earned his M.S., M.Phil., and Ph.D. in food science from Rutgers University. His research interests include physicochemical properties of foods, new processing methods for seafoods, modernization of traditional food processes, safety and environmental aspects of food processing, extrusion cooking, shelf life and packaging, biopolymers in foods, and delivery systems for bioactive compounds.

Jorge Oliveira is senior lecturer in chemical engineering at University College Cork. He is a chartered chemical engineer from the University of Porto, with a Ph.D. from the University of Leeds. His research interests include product design engineering with an emphasis on new methods to bridge cognitive science and neurophysiology, the study of food perception and choice using emotional space as a target for new product development, new methodologies, experimental design and data analysis methods for consumer assessment of foods and beverages, and kinetics of quality and safety factors in bioprocesses and process modeling.

Contributors

Huang Aixiang Faculty of Food Science and Technology, Yunnan Agricultural University, Kunming, P.R. China

Anna Aladjadjiyan Agricultural University, Plovdiv, Bulgaria

Valerie Lengard Almli Nofima, Ås, Norway

Maria Alves Departamento de Engenharia Alimentar, Campus de Gambelas, Instituto Superior de Engenharia, Universidade do Algarve, Faro, Portugal

Center for Mediterranean Bioresorces and Food (Meditbio), Campus de Gambelas, Faro, Portugal

Yannick Anguy I2M-TREFLE UMR CNRS 5295, Esplanade des Arts et Métiers, Talence Cedex, France

Jaime Aníbal CIMA—Centro de Investigação Marinha e Ambiental and Departamento de Engenharia Alimentar, Instituto Superior de Engenharia, Universidade do Algarve, Faro, Portugal

Nastasia Belc Institute of Food Bioresources, Bucharest, Romania

Erika Bešter Science and Research Centre of Koper, University of Primorska, Koper, Slovenia

Carolina Vieira Bezerra Food Engineering Faculty, Physical Measurements Laboratory, Federal University of Para, Para, Brazil

Jan Brindza Slovak University of Agriculture, Nitra, Slovak Republic

Milena Bučar-Miklavčič LABS LLC, Institute for Ecology, Olive Oil and Control, Izola, Slovenia

Science and Research Centre of Koper, University of Primorska, Koper, Slovenia

Faculty of Health Science, University of Primorska, Izola, Slovenia

Bojan Butinar Science and Research Centre of Koper, University of Primorska, Koper, Slovenia

Yusuf Chisti School of Engineering, Massey University, Palmerston North, New Zealand

Idoia Codina Veterinary Faculty, Autonomous University of Barcelona, Bellaterra, Spain

Fabio Augusto Garcia Coró Federal University of Technology—Paraná, Londrina, Paraná, Brazil

Rui Costa CERNAS, College of Agriculture of the Polytechnic Institute of Coimbra, Coimbra, Portugal

József Csanádi Faculty of Engineering, Institute of Food Engineering, University of Szeged, Szeged, Hungary

Luís Miguel Cunha DGAOT, Faculty of Sciences, University of Porto, Vairão, Portugal

LAQV, Rede de Química e Tecnologia (REQUIMTE), University of Porto, Porto, Portugal

Imen Douiri AgroParisTech-site de MASSY, Massy Cedex, France

Elisabeth Dumoulin AgroParisTech-site de MASSY, Massy Cedex, France

Ferruh Erdoğdu Department of Food Engineering, Ankara University, Ankara, Turkey

Eduardo Esteves Departamento de Engenharia Alimentar, Instituto Superior de Engenharia, Universidade do Algarve and Centro de Ciências do Mar CCMar—CIMAR Laboratório Associado, Faro, Portugal

Victoria Ferragut Veterinary Faculty, Autonomous University of Barcelona, Bellaterra, Spain

Josefa Friedel Department of Food Science and Technology, BOKU, Vienna, Austria

Carlos Eduardo Rocha Garcia Department of Pharmacy, Paraná Federal University, Curitiba, Paraná, Brazil

Helmut Glattes Department of Food Science and Technology, BOKU, Vienna, Austria

Elias Gmeiner Department of Food Sciences and Technology, BOKU, Wien, Austria

Fahrettin Göğüş Department of Food Engineering, Gaziantep University, Gaziantep, Turkey

Terezija Golob Department of Food Science and Technology, Biotechnical Faculty, University of Ljubljana, Ljubljana, Slovenia

Ana Maria Gomes Centro de Biotecnologia e Quimica Fina – Laboratorio Associado, Escola Superior de Biotecnologia, Universidade Catolica Portuguesa, Rua Arquiteto Lobao Vital, Apartado, Porto, Portugal

Roxana Cristina Gradinariu Ministry of Agriculture, Forests and Rural Development, Bucharest, Romania

Luis Guerrero IRTA Monells, Monells, Spain

Margrethe Hersleth Nofima, Ås, Norway

Cecila Hodúr Faculty of Engineering, Institute of Process Engineering, University of Szeged, Szeged, Hungary

Janez Hribar Biotechnical Faculty, University of Ljubljana, Ljubljana, Slovenia

Laura Beatriz Iturriaga Facultad de Agronomía, y Agroindustrias, Instituto de Ciencia y Tecnología de Alimentos, Universidad Nacional de Santiago del Estero, Santiago del Estero, Argentina

Grazina Juodeikiene Kaunas University of Technology, Kaunas, Lithuania

Yordanka Kartalska Agricultural University, Plovdiv, Bulgaria

Mojca Korošec Department of Food Science and Technology, Biotechnical Faculty, University of Ljubljana, Ljubljana, Slovenia

Elisabeth T. Kovács Faculty of Engineering, Institute of Process Engineering, University of Szeged, Szeged, Hungary

Rimantas Kublickas Department of Food Science and Technology, Kaunas University of Technology, Kaunas, Lithuania

Lucia Kucelova Slovak University of Agriculture, Nitra, Slovak Republic

Zsuzsanna László Faculty of Engineering, Institute of Process Engineering, University of Szeged, Szeged, Hungary

Damián Modesto Maestri Facultad de Ciencias Exactas, Físicas y Naturales, Instituto Multidisciplinario de Biología Vegetal (IMBIV. CONICET-UNC)— Instituto de Ciencia y Tecnología de los Alimentos (ICTA), Universidad Nacional de Córdoba, Córdoba, Argentina

Wanna Malaphan Department of Microbiology, Faculty of Science, Kasetsart University, Bangkok, Thailand

María Andrea Marín Departamento de Química Industrial y Aplicada, Facultad de Ciencias Exactas, Instituto de Ciencia y Tecnología de los Alimentos, Físicas y Naturales Universidad Nacional de Córdoba, Córdoba, Argentina

Marcela Lilian Martínez Facultad de Ciencias Exactas, Físicas y Naturales, Instituto Multidisciplinario de Biología Vegetal (IMBIV. CONICET-UNC)— Instituto de Ciencia y Tecnología de los Alimentos (ICTA), Universidad Nacional de Córdoba, Córdoba, Argentina

Christopher McGreger McGreger Translations and Consulting, Munich, Germany

Nancy McGreger McGreger Translations and Consulting, Munich, Germany

Ana Pinto de Moura DCeT, Universidade Aberta, Porto, Portugal

REQUIMTE, Universidade do Porto, Porto, Portugal

Sonja Smole Možina Department of Food Science and Technology, Biotechnical Faculty, University of Ljubljana, Ljubljana, Slovenia

Gabriela Nedita Institute of Food Bioresources, Bucharest, Romania

Maria Leonor Nunes Division of Aquaculture and Upgrading, Portuguese Institute for the Sea and Atmosphere, I.P. (IPMA, I.P.), Lisbon, Portugal

CIIMAR/CIMAR, Interdisciplinary Center of Marine and Environmental Research, University of Porto, Porto, Portugal

Helena Oliveira Division of Aquaculture and Upgrading, Portuguese Institute for the Sea and Atmosphere, I.P. (IPMA, I.P.), Lisbon, Portugal

ICBAS-UP, Abel Salazar Institute for the Biomedical Sciences, University of Porto, Porto, Portugal

CIIMAR/CIMAR, Interdisciplinary Center of Marine and Environmental Research, University of Porto, Porto, Portugal

CERNAS, College of Agriculture of the Polytechnic Institute of Coimbra, Coimbra, Portugal

Odd Ivar Lekang Department of Mathematical Sciences and Technology, Norwegian University of Life Sciences, Aas, Norway

Radovan Ostrovsky Slovak University of Agriculture, Nitra, Slovak Republic

Semih Ötleş Department of Food Engineering, Ege University, Izmir, Turkey

Beraat Özçelik Department of Food Engineering, Istanbul Technical University, Istanbul, Turkey

Narender Raju Panjagari Dairy Technology Division, National Dairy Research Institute, Karnal, Haryana, India

Mayka Reghiany Pedrão Federal University of Technology—Paraná, Londrina, Paraná, Brazil

María Cecilia Penci Departamento de Química Industrial y Aplicada, Facultad de Ciencias Exactas, Físicas y Naturales, Universidad Nacional de Córdoba, Córdoba, Argentina

Maria Cecilia Puppo Centro de Investigación y Desarrollo en Criotecnología de Alimentos (CONICET, UNLP), La Plata, Argentina

Célia Quintas Departamento de Engenharia Alimentar, Campus de Gambelas, Instituto Superior de Engenharia, Universidade do Algarve, Faro, Portugal

Center for Mediterranean Bioresorces and Food (Meditbio), Campus de Gambelas, Faro, Portugal

António Lopes Ribeiro DCeT, Universidade Aberta, Porto, Portugal

Daniel Pablo Ribotta CONICET-Universidad Nacional de Córdoba, Córdoba, Argentina

Lucia Romanescu National Office of the Romanian Traditional and Ecological Products, Brasov, Romania

Gerhard Schleining Department of Food Science and Technology, BOKU, Vienna, Austria

Leonardo Pablo Sciammaro Centro de Investigación y Desarrollo en Criotecnología de Alimentos (CONICET, UNLP), La Plata, Argentina

Massami Shimokomaki Federal University of Technology—Paraná, Londrina, Paraná, Brazil

Graduate Program in Animal Science, Department of Veterinary Medicine Preventive, Londrina State University, Londrina, Paraná, Brazil

Luiza Helena Meller da Silva Federal University of Para, Food Engineering Faculty, Physical Measurements Laboratory, Para, Brazil

Ashish Kumar Singh Dairy Technology Division, National Dairy Research Institute, Karnal, Haryana, India

Ram Ran Bijoy Singh Dairy Technology Division, National Dairy Research Institute, Karnal, Haryana, India

Faculty of Dairy Technology, Sanjay Gandhi Institute of Dairy Technology, Patna, Bihar, India

Sarote Sirisansaneeyakul Department of Biotechnology, Faculty of Agro-Industry, Kasetsart University, Bangkok, Thailand

Alain Sommier I2M-TREFLE UMR CNRS 5295, Esplanade des Arts et Métiers, Talence Cedex, France

Dezider Toth Slovak University of Agriculture, Nitra, Slovak Republic

Antonio José Trujillo Veterinary Faculty, Autonomous University of Barcelona, Bellaterra, Spain

Vasilij Valenčič Science and Research Centre of Koper, University of Primorska, Koper, Slovenia

Filiep Vanhonacker Department of Agricultural Economics, Ghent University, Ghent, Belgium

Marta Wilton Vasconcelos Centro de Biotecnologia e Quimica Fina – Laboratorio Associado, Escola Superior de Biotecnologia, Universidade Catolica Portuguesa, Rua Arquiteto Lobao Vital, Apartado, Porto, Portugal

Paulo Vaz-Pires ICBAS-UP, Abel Salazar Institute for the Biomedical Sciences, University of Porto, Porto, Portugal

CIIMAR/CIMAR, Interdisciplinary Center of Marine and Environmental Research, University of Porto, Porto, Portugal

Wim Verbeke Department of Agricultural Economics, Ghent University, Ghent, Belgium

Rajko Vidrih Biotechnical Faculty, University of Ljubljana, Ljubljana, Slovenia

Myriam Villarreal Facultad de Agronomía, y Agroindustrias, Instituto de Ciencia y Tecnología de Alimentos, Universidad Nacional de Santiago del Estero, Santiago del Estero, Argentina

Trude Wicklund Department of Chemistry, Biotechnology and Food Sciences, Norwegian University of Life Sciences, Aas, Norway

Busaba Yongsmith Department of Microbiology, Faculty of Science, Kasetsart University, Bangkok, Thailand

Martin Zarnkow Research Center Weihenstephan for Beer and Food Quality, Technische Universität München, Munich, Germany

Ivanka Zheleva Rousse University, Rousse, Bulgaria

Part I
General Aspects of Traditional Foods

Chapter 1
European Consumers' Definition and Perception of Traditional Foods

Wim Verbeke, Luis Guerrero, Valerie Lengard Almli, Filiep Vanhonacker, and Margrethe Hersleth

1.1 Introduction

Traditional food products (TFPs) constitute an important element of European culture, identity, and culinary heritage (European Commission 2007). They contribute to the development and sustainability of rural areas; protect them from depopulation; entail substantial product differentiation potential for food producers, processors, and retailers (Avermaete et al. 2004); and provide variety and choice for food consumers. The production of TFPs in Europe is mainly realized by small- and medium-sized enterprises (SMEs), and these products are for an important part sold under collective trademarks. For SMEs, which in general are not equipped with a separate and specialized marketing and communication business unit, a good understanding of consumers' beliefs, perceptions, and expectations is essential for further product development and innovation, for the implementation of successful marketing actions, and for the communication of targeted and tailor-made messages. The tendency of growing consumer opposition to an increasing globalization and industrialization of the food sector has also fuelled consumer interest in TFPs as a food product category (Jordana 2000).

To date, existing definitions for TFPs have been designed mainly from a food scientist or food technologist perspective (Table 1.1). These efforts have proven to be challengeable for some reasons. First, traditional food is a relative rather than

W. Verbeke (✉) • F. Vanhonacker
Department of Agricultural Economics, Ghent University,
Coupure links 653, Ghent 9000, Belgium
e-mail: wim.verbeke@ugent.be

L. Guerrero
IRTA Monells, Finca Camps i Armet, Monells 17121, Spain

V.L. Almli • M. Hersleth
Nofima, Osloveien 1, Ås 1430, Norway

© Springer Science+Business Media New York 2016
K. Kristbergsson, J. Oliveira (eds.), *Traditional Foods*, Integrating Food
Science and Engineering Knowledge Into the Food Chain 10,
DOI 10.1007/978-1-4899-7648-2_1

Table 1.1 Overview of published definitions of the concept of traditional food products (TFPs) prior to the Truefood consumer studies of 2006–2007

Definition of the concept of TFPs	Source
A TFP is a representation of a group, it belongs to a defined space, and it is part of a culture that implies the cooperation of the individuals operating in that territory	Bertozzi (1998)
In order to be traditional, a product must be linked to a territory, and it must also be part of a set of traditions, which will necessarily ensure its continuity over time	Jordana (2000)
Traditional means proven usage in the community market for a time period showing transmission between generations; this time period should be the one generally ascribed as one human generation, at least 25 years	EU (2006)
Traditional food is a food of a specific feature or features, which distinguish it clearly from other similar products of the same category in terms of the use of traditional ingredients (raw materials or primary products) or traditional composition or traditional type of production and/ or processing method	EuroFIR (2007)
TFPs are agri-food products whose methods of processing, storage, and ripening are consolidated with time according to uniform and constant local use	Ministero Agricoltura (1999)

absolute concept (Nosi and Zanni 2004), whose spectrum of foods continually evolves and grows. Concomitantly, the boundaries of what is or what is not a TFP are rather faint (Wycherley et al. 2008). This is illustrated by the diversity of food product examples of traditional foods discussed in literature. Kuznesof and others (1997) referred to regional foods as "products with a protected designation of origin (PDO)" as well as to "poorer people's food" and "old-fashioned food" as examples of traditional foods. In a similar vein, Cayot (2007) mentioned bread, cheese, and wine as typical examples of TFPs.

Second, a wide variety of terms or designations are applied for specific food product categories, which show clear interfaces with characteristics of traditional food. Examples pertain to "local food" (e.g., Chambers et al. 2007; Lobb and Mazzocchi 2007; Roininen et al. 2006), "original food" (Cembalo et al. 2008), "regional food" (e.g., Kuznesof et al. 1997), "typical food" (e.g., Caporale et al. 2006; Iaccarino et al. 2006; Nosi and Zanni 2004; Platania and Privitera 2006), "specialty food" (e.g., Guinard et al. 1999; Schamel 2007; Stefani et al. 2006; Wycherley et al. 2008), and "traditional (agri-)food" (e.g., Cayot 2007; Jordana 2000; Sanzo et al. 2003).

Third, it can be reasonably expected that different motivations to purchase and consume TFPs will exist among consumers (Platania and Privitera 2006). Differences in motivations may associate with different perceptions and conceptions of what consumers perceive to be traditional foods. As a consequence, insights in consumers' perceptions and conceptualizations of TFPs are crucial for future product development, market positioning, and marketing communication related to this food category.

Finally, Europe cannot be regarded as a homogeneous food consumption area (Askegaard and Madsen 1998). According to Jordana (2000), southern European countries have a more traditional food character due to a greater market share of small-sized companies and a warmer climate, which supports a more widespread availability of TFPs. Differences are apparent not only at the national level but also at a more regional level in terms of food-related preferences, food purchasing and eating habits, and food-related behavior and attitudes. Montanari (1994), for example, indicated that urban consumers might be more prone to reconnect with rural roots, while according to Weatherell et al. (2003), rural-based consumers tend to give a higher priority to "civic" issues (i.e., a set of collective principles and collective commitments) in food choice, to exhibit higher levels of concern over food provisioning issues, and to show a greater interest in local foods. This type of variability is particularly pronounced when dealing with TFPs and traditional cuisine that are based mainly on the natural resources available in the specific area.

This chapter will present findings with respect to the consumers' definition and perception of TFPs, based on consumer research performed within the European Union (EU)-funded project Truefood (Traditional United Europe Food). Truefood was an integrated research project financed by the European Commission as part of its sixth Framework Program for Research, Technology Development, and Demonstration. In order to account for expected cross-cultural differences, several European countries were included in the study. These countries vary in their geographical location, market presence of TFPs, and familiarity with EU food quality certification labels, like geographic origin or traditional specialty labels. With the selection of Belgium, France, Italy, Norway, Poland, and Spain as countries covered by the study, this research covered both the North–South and the East–West axis in Europe. Moreover these countries differ substantially in terms of the presence of EU certification labels and in food quality orientations (Becker 2009; Verbeke et al. 2012). Italy, France, and Spain had 227, 183, and 148 European Commission (EC)-registered products with geographical indications (protected designation of origin [PDO], protected geographical indication [PGI], and traditional specialty guaranteed [TSG]), respectively, while this number was no more than a dozen in Belgium and Poland in May 2011 (EU DOOR 2011). By May 2015, these numbers had increased to 273, 220, and 182 in Italy, France, and Spain, respectively, versus 17 in Belgium and 36 in Poland (EU DOOR 2015), herewith illustrating the persisting divide between these countries' food quality policy orientation. The chapter is organized as follows: First, a brief overview of the research methods is presented. Second, key findings are reported. Finally, conclusions and implications are set forth.

1.2 Research on Consumers and Traditional Foods

The Truefood consumer research consisted of both a qualitative and a quantitative research phase. The qualitative research phase combined focus group discussions with free word association tests and served as input for the development of a formal questionnaire that was used and analyzed in the cross-sectional quantitative research phase.

1.2.1 Focus Group Discussions

Focus group discussion is a method in which a small number of individuals are selected and interviewed in group with the goal to obtain information about their reaction to products and/or concepts (Resurreccion 1998). It is an efficient way to obtain preliminary insights into the concept of TFPs, for example (Krueger 1988). In order to safeguard the objective interpretation of the results, textual statistical analyses using the software ALCESTE were applied. A total of 12 focus group discussions (with 7 ± 2 participants in each group) were carried out between June and September 2006, i.e., two group discussions in each of the six countries involved in the study. In each country, one group discussion was held with rural consumers and the other with urban consumers. All selected participants were involved in deciding what food to buy and in food preparation at home. Participants' age ranged from 29 to 55 years. The focus group discussions had the objective to obtain a qualitative exploratory consumer-driven definition for the concept of TFPs and to compare the components of this definition across six different European countries.

1.2.2 Free Word Associations

Focus group discussions provide a rational and cognitive approach to a specific topic and can be affected, in some cases, by stereotype response behavior. The use of projective techniques might provide complementary information since these techniques allow to reveal the internal thoughts and feelings of a person and to record more spontaneous and affective responses. Free word association is one such projective technique, in which the participant projects his or her personality, attitudes, and opinions as a response to a keyword, in this case the word "traditional" in a food-related context. In this test, an interviewer verbally presents the word "traditional" to the participants and registers the verbal responses that were mentioned. Participants were asked to elicit up to three different words, within a maximum time interval of 30 s for each valid association. The analyses were performed using the software XLSTAT 2006 v.4. The goal was to identify European consumers' associations to the concept of "traditional" in a food context. The findings will be compared with the qualitative definition of TFPs obtained via the focus group interviews. About 120 participants were recruited in each of the six countries (total $n=721$). Participants were involved in deciding about food shopping and preparation of food at home. Further selection criteria were age (a minimum of 15 % of participants in each decade from 20 to 60 years old) and gender (a minimum of 25 % of participants of each gender within each age group).

1.2.3 Cross-Sectional Consumer Survey

A questionnaire was developed based on the findings from the focus group discussions and the free word association tests. Cross-sectional data were collected from consumer samples representative for age, gender, and regions in Belgium, France,

Italy, Norway, Poland, and Spain. Participants were recruited from the TNS European Online Access Panel, which is a large-scale and representative panel of individual consumers who agreed to take part in market research. Participants were selected from this panel using stratified random sampling and proportionate stratification in line with the national population distributions for age and region. All contact and questionnaire administration procedures were electronic. Data were collected during October–November 2007. The total sample size was 4828 participants, i.e., around 800 participants in each of the six countries involved in the study. The questionnaire was developed to allow constructing a consumer-driven definition for the concept of TFPs, to elaborate on cross-cultural differences and country specificities, and to map consumers' image, perceptions, and overall evaluation of TFPs on the European food market.

1.3 How Consumers Define and Perceive TFP

1.3.1 Exploratory Definition of TFPs Along Four Dimensions

Four main dimensions were distinguished in the way consumers define the concept of TFPs based on the focus group discussions (Guerrero et al. 2009). Despite obvious cultural differences between the countries, the overall results were very similar. The first dimension of the definition of TFPs was defined as "*habits and natural*." TFPs were perceived as food products that are eaten every day or quite frequently and as foods that are part of daily life and that are commonly used. It seems that most consumers associate TFPs with habits. Some TFPs were also defined as seasonal or consumed at special occasions such as Christmas and Easter. The concept of being a traditional food was associated with something anchored in the past to the present, transmitted from one generation to another, that has been consumed and is consumed from the past, has existed for a long time, and has "always" been part of the consumers' life. The TFP concept included aspects related to health, to naturalness, to homemade or made on the farm, to an artisan production method, without excessive industrial processing or handling and without additives.

A second dimension was called "*origin and locality*." Tradition in relation to food was linked to food origin, and in this sense all the country samples agreed that traditions cannot be exported or transferred to other regions. Local products outside their area of influence, outside their locality, region, or country, will simply be perceived as regular products, thus losing all or at least an important part of the additional values and feelings that may be conferred on consumers in their original place of manufacturing and/or distribution. However, some consumers participating in focus groups stated that in certain cases traditions may be created or taken over from other regions or countries (e.g., couscous in France), because information, fashions, or globalization may spread traditions and TFPs all over the world and may convert even a seemingly nontraditional product into a traditional one over time.

A third dimension pertained to *processing and elaboration*. There was general agreement across countries regarding the importance of the elaboration of the food. It seemed more appropriate to talk about traditional cuisine than to talk about TFPs.

Normally, it is the elaboration that makes the difference between a traditional and a nontraditional food product. In this context, the gastronomic heritage and artisan character of the elaboration method received great importance. When dealing with food, the transfer of the know-how or culinary arts among generations constitutes the gastronomic heritage. To be traditional a food product not only has to contain traditional ingredients, but it also has to be processed in a traditional way, according to traditional recipes. TFPs were perceived, in general, as relatively simple products, with a rather low complexity. TFPs tend to be basic, natural, and pure, often in the sense that little or no processing or manipulation has occurred after the primary production of the food and its ingredients.

"*Sensory properties*" constituted the fourth dimension. Taste was an important dimension for TFPs, with a distinct taste emerging as one of the strongest characteristics of TFPs. The importance of sensory characteristics as a quality dimension in determining consumers' acceptance or rejection has been pointed out in a large amount of previous studies and is widely accepted. Sensory characteristics were mentioned as one of the simplest and easiest ways to recognize and identify the authenticity and traditional character of a food product.

1.3.2   Words Associated with Tradition in the Food Context

Family, *Old*, *Habit*, *Christmas*, and *Grandmother* were among the most frequently elicited words in the Truefood word association study (Guerrero et al. 2010). Frequency of elicitation has been related with the strength or importance of a concept in the consumers' minds (Guerrero et al. 2000). Accordingly, consumers related the concept of traditional in a food context to their family and familial situation (*Family* and *Grandmother*) and with repeated practice (*Habit*), although it was also linked to special occasions such as *Christmas*. Family and traditions are closely related. Family is the most natural and common way of transmitting norms and values from one generation to another and to build a cultural identity (Abad and Sheldon 2008). According to Nelms (2005), habit is also an important element of traditions because it may create new ones such as special foods, different activities, bedtime, or mealtime routines. The meaning of the word *Old* is not so obvious. *Old* may have a neutral, a positive, or a negative connotation, positive as something authentic, well established and proven to be wholesome, and that has to be preserved versus negative as something outdated, old-fashioned, and not very useful or attractive anymore nowadays.

Comparing the results from the different countries indicated some similarities and differences. Gastronomic associations (*Restaurant*, *Cooking*, *Meal*, *Recipe*, and *Dish*) were positioned closer to France in a simple correspondence analysis. For Polish consumers, traditional seemed to be mostly linked to sensory properties (*Tasty*), *Family*, and *Dinner*. Further, *Old-fashioned*, *Quality*, *Restaurant*, and *Culture* were closer to Belgium; *Home*, *Good*, and *Habit* closer to Spain; *Christmas*, *Rural*, *Country*, and *Good* closer to Norway; and *Homemade*, *Natural*, and *Old* closer to Italy. Notwithstanding these country specificities, in general and from a qualitative point of view, the perception of the word traditional was quite similar in all countries. Words such as *Family*, *Good*, *Grandmother*, *Healthy*, *Natural*, *Regional*, *Restaurant*, or *Simple* were frequently elicited in all countries.

The elicited words constituted ten dimensions. These dimensions were *Sensory* (includes words like *Tasty*, *Taste*, or *Flavor*), *Health* (includes words like *Healthy*, *Unhealthy*, *Heavy*, or *Nutritious*), *Elaboration* (includes words like *Handmade*, *Homemade*, *Elaboration*, or *Laborious*), *Heritage* (includes words like *Ancestors*, *Old*, *Family*, *Culture*, or *Everlasting*), *Variety* (includes words like *Variety*, *Boring*, or *Choice*), *Habit* (includes words like *Habitual/typical*), *Origin* (includes words like *Country/origin*), *Basic/simple*, *Special occasions* (included words like *Celebration*, *Holidays*, or *Christmas*), and *Marketing* (includes words like *Expensive*, *Store/shop*, or *Distribution*). The *Sensory* and *Health* dimensions were closer to Poland and Italy; *Heritage* was closer to Spain; *Special occasions*, *Basic/simple*, and *Origin* were closer to Norway; *Elaboration* was closer to France; and *Habit*, *Marketing*, and *Variety* were closer to Belgium. These differences reflect differences in food cultures in relation to traditional food across European countries. Importantly, these dimensions corroborate well with the definition for the concept of TFPs obtained in the previous section using focus group discussions in the same six countries.

1.3.3 Quantitative Consumer-Driven Definition of TFPs

Based on the insights from the qualitative exploratory research, 13 statements reflecting different elements of the consumers' conception of TFP were included in the quantitative consumer study (Vanhonacker et al. 2010a). The list of statements is presented in Table 1.2. Survey participants were asked to indicate their agreement with each statement on a seven-point Likert scale, where a score of "1" corresponded with *totally disagree*, "4" corresponded with *neither agree nor disagree*, and "7" corresponded to *totally agree*. Each of the 13 statements received a mean score significantly higher than the scale's midpoint. This indicates that, on average, all elements were relevant for a consumer-driven definition of TFPs. The highest mean score was obtained for "grandparents already ate it" and the lowest score for "natural, low processed."

Based on the scores given by the study participants, the following consumer-driven definition for the concept of TFPs was set forth: "*A traditional food product is a product frequently consumed or associated to specific celebrations and/or seasons, transmitted from one generation to another, made in a specific way according to gastronomic heritage, naturally processed, and distinguished and known because of its sensory properties and associated to a certain local area, region or country*" (Vanhonacker et al. 2010a).

Cross-country differences were analyzed through performing simple correspondence analysis (Fig. 1.1). This analysis takes into account the frequency of occurrence of the answers on the positive side of the seven-point scale (response categories

Table 1.2 Consumers' agreement with statements reflecting the concept of TFPs

Statement	Mean value	SD
When I think about traditional food, I think about food products that my parents and *grandparents already ate*, i.e., food that has been available for a long time	5.93	1.17
I consider traditional food as *well-known* food	5.66	1.25
Traditional food has an *authentic recipe*, i.e., an original, since long known recipe	5.54	1.33
To me, a traditional food product is associated with *specific sensory properties*	5.51	1.31
The availability of traditional food is strongly *dependent on the season*	5.50	1.42
A traditional food product is typically produced *in grandmothers' way*	5.34	1.42
Traditional food has an *authentic origin of raw material*, i.e., use of the same kind of raw material as originally used when the product was developed	5.30	1.37
According to me, traditional food is typically something one *can eat very often*	5.22	1.45
Traditional food has an *authentic production process*, i.e., following the original production process, established when the product was developed	5.11	1.40
A traditional food product must *contain a story*	5.07	1.60
The key steps of the production of traditional food must be *local*	4.85	1.57
When I think about traditional food, I think about *special occasions* and/or celebrations	4.70	1.68
When it comes to food products, for me traditional food means *natural, low processed*	4.66	1.68

Mean values and standard deviations (SD) on seven-point Likert scales
Statements are ranked according to their mean value ($n = 4828$)

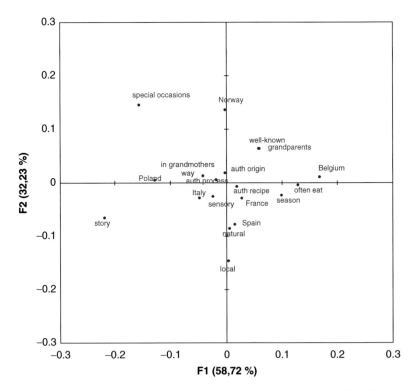

Fig. 1.1 Simple correspondence analysis plot for the frequency of occurrence of positive associations with TFPs (agreement or scores 5, 6, or 7 on seven-point scale). *Source*: Vanhonacker et al. (2010a)

"5," "6," and "7"). Elements of the definition that are located close to a country in the correspondence analysis plot indicate a strong association of the element with TFPs in that particular country; a large distance indicates a weak(er) association. A country that is located close to the plot's center indicates a broader conceptualization of the concept of TFPs, i.e., an association of TFPs with multiple elements, without particular elements being dominant. A central location of an element in the map suggests less between-country variation in the association scores. A remote position of a country in the map points to dominance of a particular element, whereas for an element a remote position suggests a higher degree of between-country variation.

In Belgium—a country with a rather remote location in the correspondence plot—TFPs were strongly associated with food that exists for a very long time, that is consumed very regularly, and that depends on the season. France, Italy, and Spain were more centrally located in the correspondence plot, which indicates that TFPs are perceived as a very broad concept in these countries, without a strong emphasis on specific elements. Differentiation between these countries can be found in the way that French and Spanish consumers emphasized the long existence and the daily character of TFPs, whereas Italian consumers tended to value authenticity and specialty relatively higher.

Regarding the Norwegian results, a very strong focus was placed on the long existence and knowledge of TFPs, which seems to decrease the relative importance of other elements. For Poland, gastronomic heritage received much emphasis. Further, the high association with the long existence as opposed to the low association with frequent consumption was striking in Poland. This suggests an association with products that used to be daily food products in the past, but that have evolved to products with a special character and that are consumed mainly at special occasions in more recent times. These results indicate that the use of the term "traditional food" should be handled carefully across countries or in an international context, given the observed differences in the meaning of TFPs and its associations across countries.

1.3.4   Profile of Typical Traditional Food Consumers

The Truefood consumer study participants were also provided with a list of 18 so-called character profiles to allow gauging the image that consumers have of a typical consumer of TFPs as well as of a typical nonconsumer of TFPs (Vanhonacker et al. 2010b). For each character profile, participants were asked to indicate on a seven-point scale, the type of food they thought these people would be likely to consume. The response category "1" corresponded with "*a person who almost exclusively uses nontraditional food*, while a score of "7" indicated *a person who almost exclusively uses traditional food*." The midpoint of the scale represented a neutral point.

Overall, the image of a typical traditional food consumer was most strongly associated with "people living in the countryside" (mean=5.95) and "people loving national or regional cuisine" (mean=5.89) (Fig. 1.2). Additional image associations with a traditional food consumption pattern were found for "old-fashioned people," "people who enjoy cooking," and "housewives." There was only a small difference

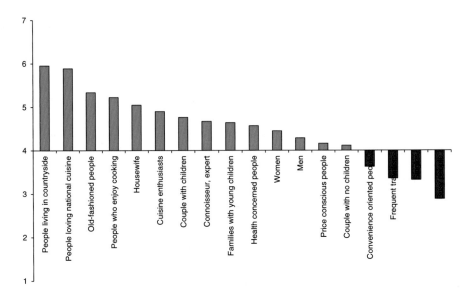

Fig. 1.2 Image of a traditional food product consumer in Europe. Mean score on seven-point scale: "1"="a person who almost exclusively uses nontraditional food," "7"="a person who almost exclusively uses traditional food." *Based on*: Vanhonacker et al. (2010b)

in the mean value for women (mean=4.45) versus men (mean=4.29). Families with children were thought more likely to be TFP consumers than families without children, whereas the children's age had only a minor impact. Similar results were obtained for "families with young children" and "couples with children at home."

Only four character profiles from the list were perceived as more likely to be consumers of nontraditional food: "busy people" (mean=2.89), "singles" (mean=3.33), "frequent travelers" (mean=3.36), and "convenience-oriented people" (mean=3.63). Thus, the typical projected image that European consumers have of consumers of TFPs appeared to be strongly determined by notions of locality, a positive attitude toward food preparation and consumption, and a traditional way of living in which stability is preferred over change and excitement. This type of profile closely fits with the role model of a traditional family. By contrast, consumers with a typical nontraditional food consumption pattern were seen as younger people, singles, and those with less time available, who live busy lives, travel quite often, and do not prioritize time for food shopping and for cooking.

1.3.5 Consumer Perception and Image of Traditional Food

The general image of TFPs was assessed in the Truefood consumer study through asking participants to indicate their level of agreement with the statement "When you think about the image you have of traditional food in general, how would you describe your personal opinion/feelings about it?" A seven-point measurement scale was used, anchored with "1"=*very negative*, "4"=*neither positive nor negative* at the midpoint, and "7"=*very positive*. A second question is related to the consumer perception of a series of intrinsic and extrinsic product attributes of TFPs. This question was probed as: "Please indicate to what extent traditional food has the following characteristics according to you?" A set of 15 items with seven-point semantic differential scales was presented to the study participants, with a negative anchor to the left (e.g., *low in quality*) and a positive anchor to the right (e.g., *high in quality*). Attributes were quality, quality consistency, taste, ordinary versus special taste, appearance, health, ease of preparation, availability, time of preparation, safety, nutritional value, price, width of assortment, environmental friendliness, and supportiveness of local economy.

The general image of TFPs was clearly positive across the six European countries, scoring above 5.5 on average on the seven-point scale (Almli et al. 2011). The two highest mean scores were obtained in Spain (6.04) and Poland (6.01), while the lowest mean scores were observed for Belgium (5.51) and France (5.62). The mean score for the total pooled pan-European sample was 5.80, with a standard deviation of 1.06. Only 81 participants (1.7 %) out of the 4765 pan-European valid answers collected for this question utilized the negative side of the response scale, i.e., scores from "1" to "3" on the seven-point scale, when evaluating their personal image of traditional foods. An analysis of score distributions by gender and age groups showed no effect of gender, but a tendency of differences across age groups: the older the consumers were, the stronger their positivity toward TFPs was.

Cross-national similarities and differences in the attribute perceptions of TFPs were observed. On average, Spanish and Italian consumers gave similar attribute perception scores. To them, TFPs are characterized mainly by a good and special taste, a high and consistent quality, a good appearance, a high nutritional value, and healthiness. Belgian consumers perceived TFPs as having a good taste, a high quality, and good availability. French consumers found TFPs to be of high quality but rather expensive. Polish consumers characterized TFPs by a good and special taste, a high (though yet not highly consistent) quality, a high environmental friendliness, a good support for the local economy, a high preparation time, and rather high prices. Finally, Norwegian consumers characterized TFPs with a good taste, a high quality, a high safety, but a relatively low healthiness and a long preparation time.

A clear distinction between countries was observed for six of the attributes. First, a special taste was attributed to TFPs mostly in Italy, Spain, and Poland, but not particularly so in Norway, France, and Belgium. Polish and Belgian consumers differed substantially with regard to their perceptions of availability, ease of preparation, and time of preparation related to TFPs. This may be explained by the divergent conceptions of traditional food in the two countries and the distinct examples of traditional foods that consumers had in mind. Whereas Polish consumers defined traditional food mainly as specialty dishes consumed on festive occasions, Belgian consumers considered traditional food as familiar food with a daily character.

Furthermore, Norway demarked itself with a relatively low score on *healthy* and a relatively high score on *safety*. The low score on healthiness corroborates findings reported by Pieniak et al. (2009), who found a negative association of weight control with the general attitude to TFPs in Norway. Also from the Truefood focus group discussions, it was concluded that traditional foods in Norway were recognized as rather fatty. As regards the attribute *safety*, the high perception of this attribute is consistent with earlier studies where it was shown that Norwegian consumers feel particularly confident that the governmental food controls secure safe food in Norway (Kjaernes et al. 2007).

Although consumers in all countries reported that traditional food is time-consuming to prepare, scoring below the midpoint in Norway and Poland, this trend was not observed in Belgium. This relates probably to the fact that Belgians defined TFPs rather as familiar food with a daily character, which are often more rapid to prepare than festive dishes.

Last but not least, TFPs were perceived as rather expensive in France, Poland, and Norway, with scores of 3.9 on the seven-point scale on average, but not in the other countries. This may reflect the wide presence of specialty products like "Produits du terroir" in France and the definition of TFPs as festive foods in Poland and Norway.

1.4   Conclusions

TFPs constitute an integral and growing part of European consumers' diets. This chapter presented insights obtained from the Truefood (EU FP6) exploratory and conclusive descriptive consumer research performed in 2006–2007. These studies aimed at providing a consumer-driven definition of TFPs. The presented definition is multifaceted, i.e., it consists of several dimensions in which elements relating to habit, naturalness, heritage, taste, and locality occupy a primary position. TFPs have been shown to benefit from an overall favorable image and positive attribute perceptions among European consumers. Last but not least, traditional food consumers have been profiled as people with a strong interest in locality, food, and tradition. The definitions, conceptualizations, perceptions, and consumer profiles obtained in these studies entail a large potential for further product development, segmentation, targeting, market positioning, and marketing communication in the European traditional food sector.

Cross-cultural differences have been identified, which associate with differences in the market presence of traditional foods, as well as differences in gastronomic traditions and eating habits. Despite a positive image of TFPs, there is still a potential for further image improvement and subsequent sales growth. Based on the insights from the Truefood consumer studies, the traditional food sector is encouraged to safeguard and capitalize on the current high-quality, high-value image of TFPs which satisfies quality- and specialty-seeking food consumers. An enlarged assortment and attention to convenience aspects in product development may also allow increasing sales volumes within the existing consumer segments, though notions of exclusivity are important for some market segments. Another recommendation pertains to a focus on developing healthier TFPs, such as low-fat and/or salt-reduced products (e.g., meat or dairy products), which may be appealing to particular consumer segments and national markets, e.g., the Norwegian one. Finally, the strong environment-friendly, supportive of local economy model, as valued highly by Polish traditional food consumers, may be transferred and "exported" to other countries.

References

Abad NS, Sheldon KM (2008) Parental autonomy support and ethnic culture identification among second-generation immigrants. J Fam Psychol 22:652–657

Almli VL, Verbeke W, Vanhonacker F, Næs T, Hersleth M (2011) General image and attribute perception of traditional foods in Europe. Food Qual Prefer 22:129–138

Askegaard S, Madsen TK (1998) The local and the global: exploring traits of homogeneity and heterogeneity in European food cultures. Int Bus Rev 7.549–568

Avermaete T, Viaene J, Morgan EJ, Pitts E, Crawford N, Mahon D (2004) Determinants of product and process innovation in small food manufacturing firms. Trends Food Sci Technol 15:474–783

Becker T (2009) European food quality policy: the importance of geographical indications, organic certification and food quality assurance schemes in European countries. Estey Centre J Int Law Trade Policy 10:111–130

Bertozzi L (1998) Tipicidad alimentaria y dieta mediterranea. In: Medina A, Medina F, Colesantie G (eds) El color de la alimentacion mediterranea: Elementos sensoriales y culturales de la nutricion. Icaria, Barcelona, pp 15–41

Caporale G, Policastro S, Carlucci A, Monteleone E (2006) Consumer expectations for sensory properties in virgin olive oils. Food Qual Prefer 17:116–125

Cayot N (2007) Sensory quality of traditional foods. Food Chem 101:154–162

Cembalo L, Cicia G, Del Giudice T, Scarpa R, Tagliafierro T (2008) Beyond agropiracy: the case of Italian pasta in the United States retail market. Agribusiness 24:403–413

Chambers S, Lobb A, Butler L, Harvey K, Traill B (2007) Local, national and imported foods: a qualitative study. Appetite 49:208–213

EU (2006) Council Regulation (EC) No 509/2006 of 20 March 2006 on agricultural products and foodstuffs as traditional specialities guaranteed. Off J Eur Union L 93/1

EU DOOR (2011) EU agricultural product quality policy. DOOR (Database of Origin and Registration). http://ec.europa.eu/agriculture/quality/door/list.html. Accessed May 2011

EU DOOR (2015) EU agricultural product quality policy. DOOR (Database of Origin and Registration). http://ec.europa.eu/agriculture/quality/door/list.html. Accessed May 2015 (Retrieved May 2015 from EU DOOR. 2011. "EU agricultural product quality policy—DOOR Database of Origin and Registration")

EuroFIR (2007) FOOD-CT-2005-513944. EU 6th Framework Food Quality and Safety Programme. http://www.eurofir.net

European Commission (2007) European research on traditional foods. http://ftp.cordis.europa.eu/pub/fp7/kbbe/docs/traditional-foods.pdf

Guerrero L, Colomer Y, Guardia MD, Xicola J, Clotet R (2000) Consumer attitude towards store brands. Food Qual Prefer 11:387–395

Guerrero L, Guàrdia MD, Xicola J, Verbeke W, Vanhonacker F, Zakowska S, Sajdakowska M, Sulmont-Rossé C, Issanchou S, Contel M, Scalvedi LM, Granli BS, Hersleth M (2009) Consumer-driven definition of traditional food products and innovation in traditional foods. A qualitative cross-cultural study. Appetite 52:345–354

Guerrero L, Claret A, Verbeke W, Enderli G, Zadowska-Biemans S, Vanhonacker F, Issanchou S, Sajdakowska M, Granli BS, Scalvedi L, Contel M, Hersleth M (2010) Perception of traditional food products in six European countries using free word association. Food Qual Prefer 21:225–233

Guinard JX, Yip D, Cubero E, Mazzucchelli R (1999) Quality ratings by experts, and relation with descriptive analysis ratings: a case study with beer. Food Qual Prefer 10:59–67

Iaccarino T, Di Monaco R, Mincione A, Cavella S, Masi P (2006) Influence of information on origin and technology on the consumer response: the case of soppressata salami. Food Qual Prefer 17:76–84

Jordana J (2000) Traditional foods: challenges facing the European food industry. Food Res Int 33:147–152

Kjaernes U, Roe E, Bock B (2007) Societal concerns on farm animal welfare. PowerPoint presentation on the Second Stakeholder Conference 'Assuring animal welfare: from societal concerns to implementation', Welfare Quality®, Berlin, 3–4 May. www.welfarequality.net

Krueger RA (1988) Focus groups: a practical guide for applied research. Sage, Newbury Park, 215 p

Kuznesof S, Tregear A, Moxey A (1997) Regional foods: a consumer perspective. Br Food J 99:199–206

Lobb A, Mazzocchi M (2007) Domestically produced food: consumer perceptions of origin, safety and the issue of trust. Acta Agric Scand C Food Econ 4:3–12

Ministero Agricoltura (1999) Decreto Legislativo 30 Aprile 1998 n. 173. Decreto Ministero Agricoltura 8 settembre 1999 n. 350, Italy

Montanari M (1994) The culture of food. Wiley-Blackwell, Hoboken, 232 p

Nelms BC (2005) Giving children a great gift: family traditions. J Pediatr Health Care 19:345–346

Nosi C, Zanni L (2004) Moving from "typical products" to "food-related services". Br Food J 106:779–792

Pieniak Z, Verbeke W, Vanhonacker F, Guerrero L, Hersleth M (2009) Association between traditional food consumption and motives for food choice in six European countries. Appetite 53:101–108

Platania M, Privitera D (2006) Typical products and consumer preferences: the "soppressata" case. Br Food J 108:385–395

Resurreccion AVA (1998) Consumer sensory testing for product development. Aspen, Gaitherburg, 254 p

Roininen K, Arvola A, Lahteenmaki L (2006) Exploring consumers' perceptions different qualitative of local food with two techniques: laddering and word association. Food Qual Prefer 17:20–30

Sanzo MJ, Belen del Rio A, Iglesias V, Vazquez R (2003) Attitude and satisfaction in a traditional food product. Br Food J 105:771–790

Schamel G (2007) Auction markets for specialty food products with geographical indications. Agric Econ 37:257–264

Stefani G, Romano D, Cavicchi A (2006) Consumer expectations, liking and willingness to pay for specialty foods: do sensory characteristics tell the whole story? Food Qual Prefer 17:53–62

Vanhonacker F, Lengard V, Hersleth M, Verbeke W (2010a) Profiling European traditional food consumers. Br Food J 112:871–886

Vanhonacker F, Verbeke W, Guerrero L, Claret A, Contel M, Scalvedi L, Zakowska-Biemans S, Gutkowska K, Sulmot-Rossé C, Raude J, Gransli BS, Hersleth M (2010b) How European consumers define the concept of traditional food: evidence from a survey in six countries. Agribusiness 26:453–476

Verbeke W, Pieniak Z, Guerrero L, Hersleth M (2012) Consumers' awareness and attitudinal determinants of European Union quality label use on traditional foods. Bio-Based Appl Econ 1:89–105

Weatherell C, Tregear A, Allinson J (2003) In search of the concerned consumer: UK public perceptions of food, farming and buying local. J Rural Stud 19:233–244

Wycherley A, McCarthy M, Cowan C (2008) Speciality food orientation of food related lifestyle (FRL) segments in Great Britain. Food Qual Prefer 19:498–510

Chapter 2
Consumer's Valuation and Quality Perception of Kid's Meat from Traditional "Cabrito da Gralheira": Protected Geographical Indication

António Lopes Ribeiro, Ana Pinto de Moura, and Luís Miguel Cunha

2.1 Introduction

During the last years, there has been an increased interest in traditional food products (TFP), which are linked to a place or region of origin (Verbeke and Roosen 2009). Two main drivers may explain this trend: the increasing policy support, particularly within the European Union (EU), and the consumer demand for TFP (Pieniak et al. 2009).

2.1.1 European Policy Toward Traditional Food Products

The EU's Common Agricultural Policy (CAP), in 1992, led to a policy change orientation, from price supports and increasing food quantity policies promoting rural development, in part, through increasing food quality (Becker 2009). In 1992, the EU introduced a system to protect and promote traditional and regional food products which are linked to the territory or to a production method, the EU system of

A.L. Ribeiro
Department of Sciences and Technology (DCeT), Universidade Aberta, Porto, Portugal

A.P. de Moura
Department of Sciences and Technology (DCeT), Universidade Aberta, Porto, Portugal

LAQV, Rede de Química e Tecnologia (REQUIMTE), University of Porto, Porto, Portugal

L.M. Cunha (✉)
Department of Geosciences, Environment and Spatial Plannings (DGAOT),
Faculty of Sciences, University of Porto, Campus Agrário de Vairão, Rua Padre Armando
Quintas, 7, Vila do Conde, 4485-661, Vairão, Portugal

LAQV, Rede de Química e Tecnologia (REQUIMTE), University of Porto, Porto, Portugal
e-mail: lmcunha@fc.up.pt

© Springer Science+Business Media New York 2016
K. Kristbergsson, J. Oliveira (eds.), *Traditional Foods*, Integrating Food
Science and Engineering Knowledge Into the Food Chain 10,
DOI 10.1007/978-1-4899-7648-2_2

geographical indications that allows three different forms of protection: Protected Designation of Origin (PDO), Protected Geographical Indication (PGI), and Traditional Specialties Guaranteed (TSG). The first two categories of protection are established by Regulation (EEC) No. 2081/92 (EU 2002a), which was later replaced by Council Regulation (EC) No. 510/2006 (EU 2006b), while TSGs are protected by Regulation (EEC) No. 2082/92 (EU 2002b), later replaced by Council Regulation (EC) No. 509/2006 (EU 2006a).

The aim of Regulation (EEC) No. 2081/92 was the protection of geographical indications as names for food products, other than wine and spirits. The fundamental difference between PDO and PGI depends how strongly the product is linked to a specific geographical place. For the PDO, the quality or characteristics of the product must be essentially or exclusively due to the particular geographical environment of the place of origin, where the geographical environment is understood to include natural and human factors, such as climate, soil quality, particular skills, social patterns, practices, and perceptions (Bérard and Marchenay 2007). Additionally, production, processing, and preparation of the raw materials, up to the stage of the finished product, must take place in the defined geographical area. Regarding PGI, the link between the product and the geographical area the product is named upon is not as restricted as in the case of PDO. The requirement for PGI is that the product possesses a specific quality, reputation, or other characteristic attributable to the geographic region. Moreover, it is enough that one of the stages of production, processing, or preparation has taken place in the defined area. Furthermore, under the rules for PGIs, it is enough that a specific quality, reputation, or other characteristic is attributable to the geographical origin. In sum, the production of a PDO is fully realized in a territory, whereas a PGI can be more or less delocalized while retaining certain geographic meaningful features (Larson 2007).

Products protected by these EU quality schemes receive legal protection, EU financial support, and member state financial aid for their promotion is a possibility. If a product is registered as either PDO or PGI, the legal protection of the name is much more comprehensive than the protection for a brand name, because not only is the name protected against unfair competition, but also the mere use of the name in any commercial context is prohibited. This high level of legal protection to the names of registered products enables to encourage the rural development (Avermaete et al. 2004; Williams and Penker 2009), particularly by communities engaged in traditional agricultural practices, improving the income of farmers and retaining the rural population in these areas (EU 2006b). Additionally, due to their protected tradition and to their increased level of recognition in the global market, producers differentiate TFP based on attractive consumer criteria, promoting their higher market prices (Ittersum et al. 2007; Dagne 2010).

Consumers look for products that are authentic, with a solid tradition behind them. This reflects concerns toward food products expected to be safe, healthy, with a good quality and respecting the environment, while promoting local communities (Cunha and Moura 2004). Moreover, the excessive homogenization reinforced by the globalization movement promotes that consumers' attempt to differentiate themselves through cultural identity (Jordana 2000). In fact, according to literature, there exist different motivations to purchase and consume TFPs (Platania and

Privitera 2006; Wycherley et al. 2008; Pieniak et al. 2009; Vanhonacker et al. 2010b). Europe cannot be regarded as a homogeneous food country (Rozin 1990; Guerrero et al. 2010), particularly regarding TFPs and traditional cuisine that are mainly based on natural resources locally available (Vanhonacker et al. 2010b).

According to Becker (2009), Southern European countries have a high number of collective quality marks, which can be regarded as possible candidates for registration as a PDO or a PGI. These countries have a more traditional food character due to a greater market share of small companies and a better climate, which supports a widespread availability of TFPs (Jordana 2000). As a result, Southern European consumers are more likely to be confronted and to be familiar with traditional foods and use them in their delectable and elaborate diets (Jordana 2000; Trichopoulou et al. 2007; Pieniak et al. 2009).

In Portugal, Spain, and Greece, the number of products registered is lower than in Italy and France, but still higher than in the other EU countries. Portugal holds the fourth position considering the number of registered products, after Italy, France, and Spain (EU Database of Origin and Registration 2011): the number of registered products in 2011 already totaled 116 (58 PDOs and 58 PGIs). In this context, Portugal is a PDO-/PGI-oriented country, considering the presence of a high number and a medium growth rate of PDO/PGI products, namely, with a high number of meat products registered as PDO or PGI (Becker 2009).

In Spain and Portugal, there is an important demand for meat from young goats (milk-fed kids), often slaughtered near 60 days of age (Jiménez-Badillo et al. 2009). In Portugal, in 2011, there was one PDO kid meat, *Cabrito Transmontano*, and there were four PGI kid meats, named *Cabrito das Terras Altas do Minho*, *Cabrito do Barroso*, *Cabrito da Beira*, and *Cabrito da Gralheira* (EU Database of Origin and Registration 2011). In 2008, Portugal had the second position in number of registered fresh products, after France and before Spain (52, 27, and 13 registered products, respectively); nevertheless the production was estimated at 2257 tons (120,785 for France and 37,311 for Spain), and the turnover was estimated at 12.3 € million (517.1 € million for France and 190.4 € million for Spain), reinforcing weakness in production and commercialization (EU Database of Origin and Registration 2011).

2.1.2 Drivers of Traditional Food Product Consumption

From European consumers' point of view, a TFP is a "product frequently consumed or associated to specific celebrations and/or seasons, transmitted from one generation to another, made in a specific way according to the gastronomic heritage, naturally processed, distinguished and known because of their sensory properties and associated to a certain local area region or country" (Guerrero et al. 2009, p. 348). This definition reflects broadness and subjective opinions and beliefs about TFP (Vanhonacker et al. 2010b), that are generally associated with a positive general image and consumption (Pieniak et al. 2009; Vanhonacker et al. 2010b; Almli et al. 2011). Further, when investigating the TFP choice motives, at least four interrelating factors linked to TFP attributes emerged (Platania and Privitera 2006; Wycherley et al. 2008; Pieniak et al. 2009; Vanhonacker et al. 2010a; Almli et al. 2011): (a) sensory attributes (quality, taste, authenticity-uniqueness), (b) health attributes (nutritional value, safety, natural content), (c) purchase-consumption attributes (price, availability-convenience, familiarity consumption/festive occasions), and (d) ethical attributes (environmental friendliness, support from the local economy).

Taste is recognized as an important factor influencing food choice in general (Steptoe et al. 1995; Glanz et al. 1998; Alves et al. 2005; Eertmans et al. 2006; Cardello et al. 2007) and in TFP choice context (Iaccarino et al. 2006; Platania and Privitera 2006; Vanhonacker et al. 2010b). TFPs taste good, in the sense that they have a unique taste. In fact, distinct taste appeared as one of the strongest characteristics that consumers associated with TFPs (Guerrero et al. 2009, 2010; Vanhonacker et al. 2010b; Almli et al. 2011). This unique taste is strongly identified with a specific region because TFP have, by their very nature, a land-based geographical origin sourcing of indigenous raw materials and with cultural and gastronomic heritage (Tregear et al. 1998; Chambers et al. 2007; Chrysochoidis et al. 2007; Guerrero et al. 2009, 2010).

Moreover, TFPs are perceived to have a higher quality (Fandos and Flavián 2006; Chambers et al. 2007; Ittersum et al. 2007; Vanhonacker et al. 2010b; Almli et al. 2011), as they taste good and are considered as safe. Consequently, TFPs could be perceived either as good for health, reinforcing their perceived natural content and authenticity (no chemical modification, no additives), or as bad for health, due to their potential high-fat content and energy density and risk of microbial contaminations, resulting from their minimal preservation, processing, or packaging (Kuznesof et al. 1997; Cayot 2007; Trichopoulou et al. 2007; Guerrero et al. 2009; Pieniak et al. 2009; Almli et al. 2011).

In fact, TFPs are often associated with special dishes consumed on festive occasions (Christmas, Easter), reinforcing their hedonic attributes rather than health benefits (Pieniak et al. 2009; Almli et al. 2011). Nevertheless, TFPs are also linked to familiar situations and perceived as food products that are eaten quite frequently and linked to family eating habits (Platania and Privitera 2006; Conter et al. 2008; Guerrero et al. 2009, 2010; Pieniak et al. 2009; Vanhonacker et al. 2010b).

Additionally, consumers associated TFPs with ethical concerns (Almli et al. 2011). Consumers may prefer local products to foreign ones as they help support the vitality of rural areas (Platania and Privitera 2006; Roininen et al. 2006; Chambers et al. 2007; Ittersum et al. 2007) and due to their environmental friendliness production (Åsebø et al. 2007; Risku-Norja et al. 2008; Almli et al. 2011). However, according to Pieniak et al. (2009), these ethical concerns do not have a significant relation with general attitudes toward TFP and TFP consumption.

Furthermore, TFPs are associated with higher prices and fail to appeal to consumers in terms of perceived convenience (Chambers et al. 2007; Almli et al. 2011). Curiously, those attributes are also perceived by consumers having a positive image toward TFPs (Vanhonacker et al. 2010a; Almli et al. 2011). For these consumers, more time and effort spent on preparing TFP-based meals for their family may engage into positive feelings for these products, as they consider cooking as taking care of their family. Additionally, European consumers are aware of the price premiums associated with traditional foods. However, the literature is not consensual at this point, and in other studies, price has not emerged as an important product attribute of TFPs (Platania and Privitera 2006; Pieniak et al. 2009).

In sum, European consumers may trade off the inconvenience and the TFPs' higher prices in order to enjoy their unique taste and safety, reasons that linked TFP to a higher food product quality.

The aim of this study was to investigate Portuguese consumers' quality perceptions toward kid's meat from traditional *Cabrito da Gralheira* (PGI), considering the main factors valorized when buying this TFP. This knowledge is essential for the implementation of successful smaller-scale marketing strategies, namely, the communication of transparent messages, considering that the major part of the kid's meat from traditional *Cabrito da Gralheira* (PGI) is produced by small food businesses.

2.2 The Local Context and *Cabrito da Gralheira*: PGI

The Commission Regulation (EC) No. 1107/96 of 12 June 1996 recognized the PGI *Cabrito da Gralheira*. This appellation comprises chilled carcasses of kids obtained from Serrana breed goats (*Capra hircus*), which populate the northern area of Beira Litoral, Portugal. This Serrana breed is perfectly suited to the specific conditions of the area delimited by the Serrana massifs of Caramulo, Montemuro, Nave, and Lapa, over the quota of 700 m (EU Database of Origin and Registration 2011). 4373 goat breeders produce 230 tons of kid meat per year (Qualigeo GIs in the World 2011).

The area of development is confined to the municipalities of Arouca, Vale de Cambra, S. Pedro do Sul, Oliveira de Frades, Vila Nova de Paiva, and Castro Daire, in the Aveiro and Viseu districts (center interior of Portugal). As in other areas of Portugal, where the goat breeding is an important activity, these animals suit the marginal land areas very well, where they can graze on wild grasses and shrubs. This gives the meat its much appreciated and characteristic flavor (Qualigeo GIs in the World 2011).

The slaughter of animals (males and females) is made up to 1 year of age with a weight less than or equal to 10 kg. Refrigerated carcasses shall have the following characteristics: (a) weight by 6 kg, including head and pluck; (b) dark brownish red color; (c) meat texture, firm, hard, and rigid, very tough to cut; (d) grain, coarse and shallow; (e) smell, *sui generis*; (f) fat, yellowish, sparse distribution subcutaneous and perirenal abundant, velvety texture (EU Database of Origin and Registration 2011).

Cabrito da Gralheira (PGI) is a very tender and lean meat. It is pink in color and has a characteristic flavor which can be attributed to its natural diet. This kid meat is only sold as *Cabrito da Gralheira* (PGI). It is sold whole, with the offal prepacked separately and without the tail, the lungs, or the liver (EU Database of Origin and Registration 2011). It plays a leading part in the gastronomy of the delightful

Serrana of Beira Litoral because of its unique high quality. It also symbolizes many ancient traditions and local customs and is eaten at many popular and religious festivals such as at Christmas time and Easter time (Qualigeo GIs in the World 2011). The labeling shall meet the requirements of the legislation, which must include the words *Cabrito da Gralheira* (PGI), beyond the certification mark for their private inspection body and certification (EU Database of Origin and Registration 2011).

2.3 Material and Methods

2.3.1 Subjects and Questionnaire

This investigation used a survey methodology. 238 questionnaires were distributed from November 2009 to March 2010 at different professional meetings by the first author. The questionnaire consisted of 21 questions, organized into four groups including:

(a) Sociodemographic consumer characterization
(b) Consumption habits and purchase behavior of kid meat
(c) Consumer knowledge of *Cabrito da Gralheira* (PGI)
(d) Consumer attitudes toward *Cabrito da Gralheira*, evaluated through a set of eleven items concerning product valorization, using a 5-point scale, with 1, "not at all important," and 5, "very important"

2.3.2 Statistical Analysis

Data on consumer attitudes toward *Cabrito da Gralheira* was analyzed using the principal components method to reduce the original items into different factors, with the process being optimized by means of a *varimax* rotation. Suitability of the data to fit under such procedure is taken through the Kaiser-Meyer-Olkin (KMO) measure of sample adequacy. High values (between 0.5 and 1.0) indicate factor analysis is adequate (Malhotra 2007) and through the amount of total variance that is explained by the factors. Moreover, the internal consistency of each of the resulting factors was inspected using the Cronbach's alpha coefficient (Cronbach 1951). This allows measuring how well a set of items reflects a single one-dimensional latent construct.

Scores for the resulting factors were computed by averaging (unweighted) item ratings per factor, yielding values from 1 = "not at all important" to 5 = "very important." Nonparametric tests of comparison were employed where necessary because of the skewed nature of the data, and correlations between the factors were investigated using Spearman's correlation (r_S). All statistical procedures were performed using IBM SPSS for Windows v.20 (IBM 2011).

2.4   Results

From the 238 distributed questionnaires, 173 were collected, with only 114 answering to all the questions related to consumer attitudes. Demographic details of the achieved sample are summarized in Table 2.1.

Table 2.1 Demographic data and behavior and knowledge toward kid's meat

Variables	n	%
Demographics		
Sex ($n=171$)		
Male	112	65.5
Female	59	34.5
Age group ($n=168$)		
18–29 years	36	21.4
30–39 years	22	13.1
40–54 years	60	35.7
55+ years	50	29.8
Level of education ($n=172$)		
Basic level (up to 9 years of school)	48	27.9
Secondary level or technical course (up to 12 years of school)	66	38.4
Higher education	58	33.7
Behavior		
Consumption of kid's meat ($n=173$)		
Yes	148	85.5
No	25	14.5
Place of consumption ($n=167$)		
Home	135	80.8
Restaurant	32	19.2
Place of purchase[a] ($n=169$)		
Butcher shop	98	58.0
Hyper/supermarket	42	24.9
Other	80	47.3
Perceived preferred purchase format ($n=155$)		
Whole	58	37.4
Half parts	34	21.9
Quarters, trays, etc.	63	40.7
Acceptable price premium for *Cabrito da Gralheira* ($n=146$)		
Up to 5 %	55	37.7
Up to 10 %	66	45.2
More than 10 %	25	17.1
Consumption of *Cabrito da Gralheira* ($n=173$)		
Yes	113	65.3
No	60	34.7

(continued)

Table 2.1 (continued)

Variables	n	%
Consuming *Cabrito da Gralheira* for ($n=83$)		
Less than 2 years	26	31.3
2–5 years	17	20.5
More than 5 years	40	48.2
Knowledge		
Do you know what a PGI is? ($n=173$)		
Yes	124	71.7
No	49	28.3
Have you heard about *Cabrito da Gralheira*? ($n=173$)		
Yes	119	68.8
No	54	14.5

[a]Multiple response question

Table 2.2 Mean values (and standard deviation) of individual items and factors regarding consumer attitudes toward consumption of *Cabrito da Gralheira*

Factor/item	Loadings	Mean (± std. dev.)
Factor 1, perceived quality (var. = 38.7 %; $\alpha = 0.78$)		*4.3 (±0.6)*
Quality	0.78	4.3[a] (±0.8)
Juiciness	0.74	4.2[a,b] (±0.8)
Taste	0.69	4.5[a] (±0.6)
Warrant	0.66	4.3[a,b] (±0.8)
Tenderness	0.62	4.3[a] (±0.8)
Factor 2, quality assurance (var. = 15.1 %; $\alpha = 0.71$)		*4.2 (±0.8)*
Quality assurance	0.79	4.3[a,b] (±1.0)
Place of origin	0.76	4.0[b,c] (±1.0)
Food safety	0.58	4.3[a] (±0.9)
Factor 3, tradition (var. = 9.40 %; $\alpha = 0.65$)		*3.9 (±0.8)*
Knowledge of the breed	0.84	3.8[c] (±1.1)
Knowledge of the production system	0.67	3.9[c] (±0.9)
Animal health		*4.4[a] (±0.9)*

Explained variance (*var.*) and loadings resulting from principal component analysis with *varimax* rotation

[a,b,c]Homogenous groups according to the nonparametric Wilcoxon test, at 95 % confidence level

Results showed that only a small fraction of the respondents is willing to pay a price premium above 10 %, which may be explained by the reduced number of consumers expressing a longer experience of consumption of *Cabrito da Gralheira* (Table 2.1).

Ranking of the different attitudinal items toward valorization of *Cabrito da Gralheira* consumption yielded taste as the most valued characteristic, together with animal health, meat tenderness, and meat quality (see Table 2.2).

Application of the exploratory factor analysis to consumer attitude data leads to the exclusion of one of the original 11 items and yielded three factors (Table 2.2), which accounted for 63.2 % of the total variance and presented a KMO value of 0.789. The factors were named perceived quality, quality assurance, and tradition, all yielding a considerably high internal consistency, with Cronbach's alpha (α) values ranging from 0.67 to 0.78. Item on animal health was considered as a one-item factor due to its high mean value (Table 2.2).

Significant associations ($p < 0.001$) were observed between all four factors, with Spearman's correlation values ranging from 0.313 to 0.479, with the most prominent values being observed between perceived quality and quality certification ($r_S = 0.479$) and animal health and quality certification ($r_S = 0.459$).

In general, there is no significant effect of the sample characteristics on the evaluated attitudinal factors. Women, in general, give higher ratings to all factors, and respondents with a high education level gave higher scores to quality assurance (Table 2.3). Those expressing the willingness to pay a higher price premium also evaluate perceived quality at a higher level. Contrarily to the expectations, respondents that have not consumed or/and not heard about *Cabrito da Gralheira* are the ones giving higher values for perceived quality.

Table 2.3 Mean (and standard deviation) for perceived attitudes toward valorization of consumption of Cabrito da Gralheira, according to demographical, behavioral, and knowledge characterization of the sample

Variables (group size)	Perceived quality	Quality assurance	Tradition	Animal health
Demographics				
Sex				
Male (85)	4.3[b] (±0.6)	4.0[b] (±0.8)	3.8 (±0.9)	4.3[b] (±1.0)
Female (42)	4.5[a] (±0.5)	4.5[a] (±0.5)	4.0 (±0.8)	4.8[a] (±0.4)
Age group				
18–29 years (21)	4.4 (±0.6)	4.4 (±0.6)	4.0 (±0.7)	4.6 (±0.7)
30–39 years (14)	4.2 (±0.7)	4.2 (±1.0)	3.7 (±0.9)	4.3 (±1.0)
40–54 years (45)	4.3 (±0.7)	4.2 (±0.6)	4.0 (±0.8)	4.5 (±0.8)
55+ years (44)	4.4 (±0.4)	4.1 (±0.8)	3.8 (±0.9)	4.4 (±0.9)
Level of education				
Basic level (41)	4.3 (±0.5)	4.0[b] (±0.8)	3.9 (±0.7)	4.5 (±0.7)
Secondary level or technical course (42)	4.4 (±0.6)	4.4[a] (±0.6)	4.0 (±0.8)	4.4 (±0.9)
Higher education (45)	4.4 (±0.6)	4.2[a,b] (±0.8)	3.7 (±1.0)	4.5 (±0.9)
Behavior				
Consumption of kid's meat				
Yes (120)	4.3 (±0.6)	4.2 (±0.7)	3.9 (±0.8)	4.5 (±0.8)
No (9)	4.6 (±0.5)	4.0 (±1.3)	3.6 (±1.0)	4.4 (±1.1)
Place of consumption				
Home (96)	4.4 (±0.5)	4.2 (±0.7)	3.9 (±0.9)	4.5 (±0.8)
Restaurant (31)	4.1 (±0.8)	4.1 (±0.8)	3.7 (±0.8)	4.2 (±1.0)

(continued)

Table 2.3 (continued)

Variables (group size)	*Perceived quality*	*Quality assurance*	*Tradition*	*Animal health*
Perceived preferred purchase format				
Whole (44)	4.3 (±0.5)	4.1 (±0.7)	3.9 (±0.9)	4.4 (±1.0)
Half parts (29)	4.4 (±0.7)	4.3 (±0.6)	3.9 (±0.9)	4.6 (±0.6)
Quarters, trays, etc. (49)	4.3 (±0.6)	4.3 (±0.8)	3.8 (±0.8)	4.4 (±0.9)
Acceptable price premium for *Cabrito da Gralheira*				
Up to 5 % (41)	4.2[b] (±0.5)	4.0[b] (±0.7)	4.0 (±0.7)	4.5 (±0.8)
Up to 10 % (62)	4.5[a] (±0.6)	4.3[a] (±0.7)	3.8 (±0.9)	4.4 (±0.9)
More than 10 % (24)	4.3[a,b] (±0.6)	4.3[a] (±0.8)	3.6 (±0.9)	4.5 (±0.7)
Consumption of *Cabrito da Gralheira*				
Yes (56)	4.2[b] (±0.5)	4.2 (±0.6)	3.9 (±0.7)	4.5 (±0.8)
No (73)	4.4[a] (±0.6)	4.2 (±0.8)	3.8 (±1.0)	4.5 (±0.9)
Consuming *Cabrito da Gralheira* for				
Less than 2 years (25)	4.2 (±0.5)	4.3 (±0.6)	3.7 (±0.7)	4.4 (±0.8)
2–5 years (17)	4.4 (±0.5)	4.2 (±0.7)	4.1 (±0.6)	4.5 (±0.5)
More than 5 years (39)	4.1 (±0.6)	4.0 (±0.8)	3.7 (±0.9)	4.5 (±1.0)
Knowledge				
Do you know what a PGI is?				
Yes (106)	4.4 (±0.5)	4.2 (±0.7)	3.8 (±0.8)	4.5 (±0.8)
No (23)	4.3 (±0.7)	4.1 (±0.7)	4.0 (±0.9)	4.4 (±0.9)
Have you heard about *Cabrito da Gralheira*?				
Yes (99)	4.3[b] (±0.6)	4.2 (±0.7)	3.9 (±0.8)	4.4 (±0.9)
No (30)	4.6[a] (±0.5)	4.3 (±0.8)	3.9 (±0.9)	4.6 (±0.7)

[a,b]Homogenous groups according to the nonparametric Mann–Whitney test, at 95 % confidence level

2.5 Discussion

The purpose of this research was to evaluate Portuguese consumer quality perceptions toward kid's meat from traditional *Cabrito da Gralheira* (PGI), considering the main factors that they valorized when buying this TFP.

It emerges that a vast majority of the interviewed consumers have already consumed kid's meat. They bought this meat product from specialized retailers (butcher shop), and they prepare and consume this kind of meat at home rather than consuming at restaurants. One may suspect that considering kid's meat, these consumers use strategies to minimize their risk perception in order to make a safe purchase and consumption, respectively. This approach is in accordance with the psychometric paradigm (Slovic 1993), in the sense that consumers associate greater risk with circumstances and practices which they perceive are controlled by others, such as eating in restaurants, compared with situations in which they have perceived control, such as preparing and eating food at home (Yeung and Morris 2001), namely, kid's meat.

Additionally, consumers of this study were aware of the PGI denomination for typical *Cabrito da Gralheira* and knew its distinctive PGI label. This is consistent with the fact that consumers from Southern countries are more familiar with the EU system of geographical indications, as these countries present the highest number of registered products with a geographical indication (EU Database of Origin and Registration 2011) and where culinary traditions and TFP are prominent in society (Becker 2009; Pieniak et al. 2009; Guerrero et al. 2010; Vanhonacker et al. 2010b).

Nevertheless, only 65.3 % of the interviewees have already bought *Cabrito da Gralheira* (PGI), probably due to its high price and restricted distribution. In fact, less than half of respondents were able to pay a price premium of 5–10 % for this product, sustaining that price could be a barrier for buying of more TFP (Chambers et al. 2007).

Furthermore, preferred format for purchase of *Cabrito da Gralheira* (PGI) usually was as a whole piece (refrigerated carcasses present a maximum of 6 kg, head included), followed by half parts, revealing that *Cabrito da Gralheira* (PGI) is associated with collective consumption. In fact, particularly in Portugal, kid's meat is associated with a strong traditional and festive consumption, namely, at Easter and Christmas (Rodrigues and Teixeira 2009). In fact, the general image of TFP may be typically described as special occasion foods rather than everyday foods (Guerrero et al. 2009, 2010), even if they can be consumed in ordinary and everyday meals (Vanhonacker et al. 2010b; Almli et al. 2011).

Next, the results indicated that regarding *Cabrito da Gralheira* (PGI), taste, animal health, the meat tenderness, and the quality of the product were the most valorized factors by interviewees. This corroborates Cayot (2007), who proposes that consumers demand for safe and tasteful TFP. In fact, TFP are usually bought due to their special taste and quality (Platania and Privitera 2006; Ittersum et al. 2007; Vanhonacker et al. 2010b), and for European consumers, these products may be distinguished and known because of its sensory proprieties and high quality (Sanzo et al. 2003; Iaccarino et al. 2006; Chambers et al. 2007; Guerrero et al. 2009, 2010; Almli et al. 2011).

Through principal component analysis, three main components were retained and identified as "perceived quality," "quality assurance," and "tradition." This means that *Cabrito da Gralheira* (PGI) is perceived as a product with an assured quality and a product linked to knowledge of production systems and animal breed. As referred by Ittersum et al. (2007), consumers have a favorable image of regional certification labels, which significantly influences their willingness to buy TFP, through consumers' quality perceptions. The quality assurance enhances the perceived quality of TFP, while tradition strengthens their attitudes connected with cultural knowledge, framed globally on agriculture tradition approach (Platania and Privitera 2006). These findings are in accordance with Guerrero et al. (2009) in the sense that consumers perceived TFP in opposition to processed food products, from whom little or no processing or manipulation has occurred after a primary production, thus preserving their natural proprieties.

Evaluation of the *Cabrito da Gralheira* (PGI) factors at the individual level, in socioeconomic terms, showed that women had the highest level of the three factors.

This emphasizes the role of women, in Southern countries, as the gatekeepers of the household food domain (Moura and Cunha 2005). Additionally, "quality assurance" is less valorized by interviewees with low level of education, probably reflecting their lack of knowledge to evaluate a food quality assurance scheme. Unexpectedly, consumers who have never eaten *Cabrito da Gralheira* (PGI) gave more importance to "perceived quality" than experienced consumers. This could be explained by the fact that they have higher expectations regarding the product quality with a PGI designation. According to Hofstede (2001), the effects of regional certification labels may be larger for consumers who are unaware of TFP certification. Finally, as expected, respondents who were not able to pay more for *Cabrito da Gralheira* (PGI) did not valorize "perceived quality" and "quality assurance" factors. In making their buying decision, these consumers probably give less attention to quality criteria and more to other factors, such as price, for instance. In fact, from other studies, price emerged as a barrier to TFP purchase (see Sect. 2.1.2).

To sum up, consumers' valorization of *Cabrito da Gralheira* (PGI) is related to high perceived quality and safety and traditional production process. These dimensions are those taken on EU system to protect and promote TFP (see Sect. 2.1) and confirms previous European studies, revealing a positive TFP consumer image (Guerrero et al. 2009, 2010; Vanhonacker et al. 2010b; Almli et al. 2011). Thus, this industry is encouraged to maintain high-quality standards in order to develop favorable product attitudes, as proposed by Ittersum et al. (2007) and Almli et al. (2011) in a TFP general context. Nevertheless, more primary producers should adopt an efficient commercialization system through the producer group activity in order to ensure a sustainable *Cabrito da Gralheira* (PGI) production and increase sales volumes.

The authors stress that the findings obtained in this research are not generalizable to a larger population, considering the convenience nature of the sample.

Acknowledgments Authors Moura, A. P. and Cunha, L. M. acknowledge financial support by Fundação para a Ciência e a Tecnologia (FCT), Ministry of Education and Science, through program PEst-C/EQB/LA0006/2011.

References

Almli VL, Verbeke W, Vanhonacker F, Næs T, Hersleth M (2011) General image and attribute perceptions of traditional food in six European countries. Food Qual Prefer 22(1):129–138
Alves H, Cunha LM, Lopes Z, Santos MC, Costa Lima R, Moura AP (2005) Motives underlying food choice: a study of individual factors used by the Portuguese population. In: Pangborn Sensory Science Symposium, 6, Abstract Book, Elsevier, Harrogate North Yorkshire, p 19
Åsebø K, Jervell AM, Lieblein G, Svennerud M, Francis C (2007) Farmer and consumer attitudes at farmers markets in Norway. J Sustain Agric 30(4):67–93
Avermaete T, Viaene J, Morgan EJ, Pitts E, Crawford N, Mahon D (2004) Determinants of product and process innovation in small food manufacturing firms. Trends Food Sci Technol 15:474–483
Becker T (2009) European food quality policy: the importance of geographical indications, organic certification and food quality assurance schemes in European countries. Estey Centre J Int Law Trade Policy 10(1):111–130
Bérard L, Marchenay P (2007) Localized products in France: definition, protection and value-adding. Anthropol Food S5. http://aof.revues.org/index415.html. Accessed 24 Jan 2011

Cardello AV, Schutz HG, Lesher LL (2007) Consumer perceptions of foods processed by innovative and emerging technologies: a conjoint analytic study. Innov Food Sci Emerg Technol 8:73–83

Cayot N (2007) Sensory quality of traditional foods. Food Chem 102:445–453

Chambers S, Lobb L, Butler L, Harvey K, Trailla WB (2007) Local, national and imported foods: a qualitative study. Appetite 49:208–213

Chrysochoidis G, Krystallis A, Perreas P (2007) Ethnocentric beliefs and country-of-origin (COO) effect: impact of country, product and product attributes on Greek consumers' evaluation of food products. Eur J Mark 41(11/12):1518–1544

Conter M, Zanardi E, Ghidinia S, Pennisi L, Vergara A, Campanini G, Ianieri A (2008) Consumers' behaviour toward typical Italian dry sausages. Food Control 19(6):609–615

Cronbach LJ (1951) Coefficient alpha and the internal structure of test. Psychometrics 16(3):297–334

Cunha LM, Moura AP (2004) Conflicting demands of agricultural production and environment protection: consumers' perception on quality and safety of food. In: Filho WL (ed) Ecological agriculture and food production in central and eastern Europe-risks associated with industrial agriculture, vol 44, Nato scientific series. IOS, Amsterdam, pp 137–157

Dagne T (2010) Law and policy on intellectual property, traditional knowledge and development: protecting creativity and collective rights in traditional knowledge based agricultural products through geographical indications. Estey J Int Law Trade Policy 11(1):78–127

Eertmans A, Victoir A, Notelaers G, Vansant G, Van Den Berg O (2006) The food choice questionnaire: factorial invariant over western urban populations? Food Qual Prefer 17:344–352

EU (2002a) Council Regulation (EEC) No. 2081/92 of 14 July 1992 on the protection of geographical indications and designations of origin for agricultural products and foodstuffs. Off J L 208 of 24 July 1992, 1–8

EU (2002b) Council Regulation (EEC) No. 2082/92 of 14 July 1992 on certificates of specific character for agricultural products and foodstuffs. Offi J L 208 of 24 July 1992, 9–14

EU (2006a) Council Regulation (EC) No. 509/2006 of 20 March 2006 on agricultural products and foodstuffs as traditional specialties guaranteed. Off J Eur Union, L 93 of 31 March 2006, 1–11

EU (2006b) Council Regulation (EC) No. 510/2006 of 20 March 2006 on the protection of geographical indications and designations of origin for agricultural goods and foodstuffs. Off J Eur Union, L 93 of 31 March 2006, 12–25

EU Database of Origin and Registration (2011) EU agricultural product quality policy—DOOR Database of origin and Registration. http://ec.europa.eu/agriculture/quality/door/list.html. Accessed 24 Jan 2011

Fandos C, Flavián C (2006) Intrinsic and extrinsic quality attributes, loyalty and buying intention: an analysis for a PDO product. Br Food J 108:646–662

Glanz K, Basil M, Maibach E, Goldberg J, Snyder D (1998) Why Americans eat what they do: taste, nutrition, cost, convenience, and weight control concerns as influences on food consumption. J Am Diet Assoc 98(10):1118–1126

Guerrero L, Guardia MD, Xicola J, Verbeke W, Vanhonacker F, Zakowska-Biemans S, Sajdakowska M, Sulmont-Rossé C, Issanchou S, Contel M, Scalvedi ML, Granli BS, Hersleth M (2009) Consumer-driven definition of traditional food products and innovation in traditional foods. A qualitative cross-cultural study. Appetite 52:345–354

Guerrero L, Claret A, Verbeke W, Enderli G, Issanchou S, Zakowska-Biemans S, Vanhonacker F, Sajdakowska M, Granli BS, Scalvedi L, Contel M, Hersleth M (2010) Perception of traditional food products in six European countries using free word association. Food Qual Prefer 21(2):225–233

Hofstede G (2001) Culture's consequences, 2nd edn. Sage, London

Iaccarino T, Di Monaco R, Mincione A, Cavella S, Masi P (2006) Influence of information on origin and technology on the consumer response: the case of soppressata salami. Food Qual Prefer 17(1–2):76–84

IBM (2011) IBM SPSS statistics base 20. IBM, New York

Ittersum KV, Meulenberg MTG, Trijp HCM, Candel MJJM (2007) Consumers' appreciation of regional certification labels: a pan-European study. J Agric Econ 58(1):1–23

Jiménez-Badillo MR, Rodrigues S, Sañudo C, Teixeira A (2009) Non-genetic factors affecting live weight and daily gain weight in Serrana Transmontano kids. Small Rumin Res 84(1–3):125–128

Jordana J (2000) Traditional foods: challenges facing the European food industry. Food Res Int 33(3–4):147–152

Kuznesof S, Tregear A, Moxey A (1997) Regional foods: a consumer perspective. Br Food J 99(6):199–206

Larson J (2007) Relevance of geographical indications and designations of origin for the sustainable use of genetic resources. Global Facilitation Unit for Underutilized Species, Rome. http://www.underutilized-species.org/Documents/PUBLICATIONS/gi_larson_lr.pdf. Accessed 24 Jan 2011

Malhotra NK (2007) Marketing research, 5th International Edition. Pearson, Upper Saddle River

Moura AP, Cunha LM (2005) Why consumers eat what they do: an approach to improve nutrition education and promote healthy eating. In: Doyle D (ed) Consumer citizenship: promoting new responses, taking responsibility, vol 1. Forfatterne, Norway, pp 144–156

Pieniak Z, Verbeke W, Vanhonacker F, Guerrero L, Hersleth M (2009) Association between traditional food consumption and motives for food choice in six European countries. Appetite 53:101–108

Platania M, Privitera D (2006) Typical products and consumer preferences: the "soppressata" case. Br Food J 108(5):385–395

Qualigeo GIs in the World (2011) http://qualigeo.eu/HomeSearch.aspx. Accessed 24 Jan 2011

Risku-Norja H, Hietala R, Virtanen H, Ketomaki H, Helenius J (2008) Localisation of primary food production in Finland: production potential and environmental impacts of food consumption patterns. Agric Food Sci 17(2):127–145

Rodrigues S, Teixeira A (2009) Effect of sex and carcass weight on sensory quality of goat meat. J Anim Sci 87:711–715

Roininen K, Arvola A, Lahteenmaki L (2006) Exploring consumers perceptions of local food with two different qualitative techniques: laddering and word association. Food Qual Prefer 17:20–30

Rozin P (1990) The importance of social factors in understanding the acquisition of food habits. In: Capaldi ED, Powley TL (eds) Taste, experience, and feeding habits. American Psychology Association, Washington DC, pp 255–269

Sanzo MJ, del Río AB, Iglesias V, Vázquez R (2003) Attitude and satisfaction in a traditional food product. Br Food J 105(11):771–790

Slovic P (1993) Perceived risk, trust and democracy. Risk Anal 13(6):675–682

Steptoe A, Pollard TM, Wardle J (1995) Development of a measure of the motives underlying the selection of food: the food choice questionnaire. Appetite 25:267–284

Tregear A, Kuznesof S, Moxey A (1998) Policy initiatives for regional foods: some insights from consumer research. Food Policy 23(5):383–394

Trichopoulou A, Soukara S, Vasilopoulou E (2007) Traditional foods: a science and society perspective. Trends Food Sci Technol 18:420–427

Vanhonacker F, Lengard V, Hersleth M, Verbeke W (2010a) Profiling European traditional food consumers. Br Food J 112(8):871–886

Vanhonacker F, Verbeke W, Guerrero L, Claret A, Contel M, Scalvedi L, Żakowska-Biemans S, Gutkowska K, Sulmont-Rossé C, Raude J, Granli BS, Hersleth M (2010b) How European consumers define the concept of traditional food: evidence from a survey in six countries. Agribusiness 26(4):453–476

Verbeke W, Roosen J (2009) Market differentiation potential of country-of-origin, quality and traceability labeling. Estey Centre J Int Law Trade Policy 10(1):20–35

Williams R, Penker M (2009) Do geographical indications promote sustainable rural development? Jahrbuch der Österreichischen Gesellschaft für Agrarökonomie 18(3):147–156, http://oega.boku.ac.at/fileadmin/user_upload/Tagung/2008/Band_18/18_3__Williams_Penker.pdf. Accessed 24 Jan 2011

Wycherley A, McCarthy M, Cowan C (2008) Speciality food orientation of food related lifestyle segments in Great Britain. Food Qual Prefer 19(5):498–510

Yeung RMW, Morris J (2001) Food safety risk: consumer perception and purchase behavior. Br Food J 103(3):170–187

Chapter 3
Traditional Fermented Foods in Thailand

Busaba Yongsmith and Wanna Malaphan

3.1 Introduction

Most traditional food products are produced through the process of drying, salting, pickling, and fermenting or a combination of these methods. Thailand, like other countries with seasonal tropical climate, is quite warm and highly humid except for a few months of the rainy season. Food storage and preservation are important in maintaining food supplies especially in rural areas. Numerous Thai fermented products are made from fish, meat, cereals, legumes, fruits, and vegetables. In Thai daily life, people usually keep more than one fermented product in their households. They consume these products every day in one form or another, either as food or seasoning throughout the country. The main ones are Ka-Nom-Jeen (Thai fermented rice noodles), Nam-Pla (fish sauce), Ka-Pi (shrimp paste), Pla-Raa (fermented fish with high sea salt), and so forth. These unique Thai products are normally and traditionally produced at home through wisdom handed down from generation to generation. The qualities of these products vary greatly from community to community and vary also from the simple to complicated processes. As time goes by, the production of some fermented products such as Naem or Nham (fermented minced pork) and Nam-Pla have been changed to small-scale and large-scale industries, respectively. We review here the Thai salted and non-salted fermented food products. Some Thai fermented drinks are also presented. Production method and key microorganisms involved will be focused on some unique Thai fermented products.

B. Yongsmith (✉) • W. Malaphan
Department of Microbiology, Faculty of Science, Kasetsart University,
Bangkok 10900, Thailand
e-mail: fscibus@ku.ac.th

© Springer Science+Business Media New York 2016
K. Kristbergsson, J. Oliveira (eds.), *Traditional Foods*, Integrating Food
Science and Engineering Knowledge Into the Food Chain 10,
DOI 10.1007/978-1-4899-7648-2_3

31

3.2   Thai Traditional Fermented Foods

Generally, Thai traditional fermented products can be grouped into two main categories: salted and non-salted.

3.2.1   Thai Salted Fermented Foods

This category includes various fermented products of fishery, animals, fruits, vegetables, and soybean with various levels of salt concentration ranging from 0.5 to over than 20 % as shown in Table 3.1.

3.2.1.1   Fishery Products

The fermentation processes of salted fermented products vary from very simple to very complicated. For some products, such as fish sauce, raw materials, processing methods, and composition are very similar to products found in other Asian countries.

Fishery products can be categorized according to main processing techniques and ingredients (Phithakpol et al. 1995). Some products have very similar processing techniques, but vary with the minor ingredients added to cater to local preferences or availability of materials. Other products are similar but have different local names. Fish fermentation in Thailand is mainly associated with either high or low salt content. A further category can be identified by the addition of carbohydrate sources and a small number by the addition of fruits. The classification is shown in Fig. 3.1.

Fish with a Large Proportion of Salt

There are many products in this category but only a few are well known throughout Thailand, e.g., Nam-Pla, Ka-Pi, and Bu-Du. Most of these products are made from marine fish in the coastal provinces, especially the east coast. Some description involved and production methods of these products are as follows.

Nam-Pla (น้ำปลา)

General name	Fish sauce
Nature of product	Clear liquid, brown color, salty taste
Main utilization	Condiment, flavoring agent
Raw materials	Fish, salt

Table 3.1 Type of Salted fermented foods in Thailand

Product name	Thai name	Typical characteristics	% Salt
Fishery products			
Fish			
Bu-Du	บูดู	Muslim sauce, fish sauce	19.4–20.6
Jing-Jang	จิงจัง	Fermented fish	13.2–24.8
Ka-Pi-Pla	กะปิปลา	Fermented fish paste	10.9–19.8
Khem-Mak-Nat	เค็มมักนัต	Fermented fish with pineapple	10.9–13.9
Nam-Pla	น้ำปลา	Fish sauce	22.8–26.2
Pla-Chao	ปลาเจ่า	Thai sweetened fermented fish	5.0–6.4
Pla-Jorm	ปลาจ่อม	Fermented fish	2.6–10.2
Pla-Mam	ปลามัม	Fermented fish with pineapple	6.2–7.3
Pla-Paeng-Daeng	ปลาแป้งแดง	Red fermented fish	2.3–4.0
Pla-Raa	ปลาร้า	Fermented fish	11.5–23.9
Pla-Som	ปลาส้ม	Fermented fish	4.0–10.7
Pla-Too-Khem	ปลาทูเค็ม	Salted mackerel	8.5–13.5
Pla-Uan	ปลาอวน	Fermented fish	4.0
Som-Fak	ส้มฟัก	Thai fermented fish cake	3.6–3.9
Som-Khai-Pla	ส้มไข่ปลา	Fermented fish egg	0.9–4.1
Tai-Pla	ไตปลา	Fermented fish viscera	13.5–25.3
Shrimp			
Ka-Pi	กะปิ	Fermented shrimp paste	14.0–40.1
Koei-Cha-Loo	เคยชะลู	Fermented shrimp	6.5–8.0
Koei-Nam	เคยน้ำ	Fermented shrimp	3.1–4.5
Koong-Chao	กุ้งเจ่า	Thai sweetened fermented shrimp	5.0–6.4
Koong-Jorm	กุ้งจ่อม	Fermented shrimp	3.2–9.4
Koong-Som	กุ้งส้ม	Fermented shrimp	7.5–10
Others			
Hoi-Kraeng-Dorng	หอยแครงดอง	Fermented clam	5.0–7.2
Hoi-Ma-Laeng-poo-Dorng	หอยแมลงภู่ดอง	Fermented sea mussel	11.4–12.5
Hoi-Siap-Dorng	หอยเสียบดอง	Fermented shellfish	14.3–23.2
Hoi-Som	หอยส้ม	Fermented shell	3.5–4.4
Maeng-Ka-Proon-Dorng	แมงกะพรุนดอง	Dried fermented jelly fish	0.5
Nam-Poo	น้ำปู	Fermented crab paste	10.3
Poo-Khem	ปูเค็ม	Fermented crab	7.8–17.3
Animal products			
Khai-Khem	ไข่เค็ม	Salted eggs	
Mam	มัม	Fermented beef or pork sausage	2.3–7.1
Naang	หนาง	Fermented pork or beef	1.2–3.4
Naem	แหนม	Fermented pork/beef	2.5–2.8
Nang-Khem	หนังเค็ม	Fermented buffalo skin	1.2–2.0
Sai-Krok-Prieo	ไส้กรอกเปรี้ยว	Fermented sausage	1.1–1.9
Som-Dteen-Wooa	ส้มตีนวัว	Fermented ox hoof and hock	little

(continued)

Table 3.1 (continued)

Product name	Thai name	Typical characteristics	% Salt
Som-Neua	ส้มเนื้อ	Fermented beef	3.1
Fruit products			
Buay-Dorng	บ๊วยดอง	Fermented Japanese apricots	18.3
Kra-Thorn-Dorng	กระท้อนดอง	Pickled santol	0.5–1.7
Loog-Jan-Ted-Dorng	ลูกจันทร์เทศดอง	Pickled nutmeg fruits	0.2–0.6
Loog-Pling-Dorng	ลูกปลิงดอง	Pickled bilimbi	4.1
Loog-Tor-Dorng	ลูกท้อดอง	Fermented peach	23
Lum-Pee-Dorng	หลุมพีดอง	Fermented Lumpee	1.0–1.8
Ma-Dan-Dorng	มะดันดอง	Pickled garcinia	0.4–2.7
Ma-Kaam-Dorng	มะขามดอง	Fermented tamarind	2.4–2.5
Ma-Kaam-Pom-Dorng	มะขามป้อมดอง	Fermented Indian gooseberry	1.2–1.6
Ma-Kork-Nam-Dorng	มะกอกน้ำดอง	Pickled Spanish plums	3.7–4.2
Ma-Muang-Dorng	มะม่วงดอง	Fermented green mango	1.7–11.3
Ma-Nao-Dorng	มะนาวดอง	Pickled lime	12.4–18.6
Ma-Pring-Chae-Im	มะปริงแช่อิ่ม	Sweetened plum mango	1.3
Ma-Yom-Dorng	มะยมดอง	Pickled star gooseberry	0.9–9.7
Poot-Sa-Dorng	พุทราดอง	Pickled jujube	0.7–2.5
Sa-Mor-Dorng	สมอดอง	Fermented olive	1.2
Too-Rian-Prieo	ทุเรียนเปรี้ยว	Fermented durian	7.1
Vegetable products			
Dton-Horm-Dorng	ต้นหอมดอง	Fermented spring shallots	1.5–2.3
Horm-Dorng	หอมดอง	Pickled shallots	2.0
Hooa-Pak-Kaat-Dorng	หัวผักกาดดอง	Dried salted Chinese radish	11.9–13.2
Ka-Lam-Dorng	กระหล่ำดอง	Pickled cabbage	1.7
Khing-Dorng	ขิงดอง	Pickled ginger	4.0–5.3
Kra-Tiam-Dorng	กระเทียมดอง	Pickled garlic bulbs	3.2–5.0
Lam-Peuak-Dorng	ลำเผือกดอง	Fermented taro stalk	3.0
Loog-Kra-Dorng	ลูกกระดอง	Fermented perah seeds	0.8
Loog-Niang-Dorng	ลูกเนียงดอง	Fermented djenkol bean	2.7
Loog-Riang-Dorng	ลูกเรียงดอง	Fermented parkia seeds	1.7–3.4
Ma-Keua-Dorng	มะเขือดอง	Pickled eggplant	2.6–3.8
Ma-Keua-Proh-Dorng	มะเขือเปราะดอง	Fermented garden eggplants	
Miang	เมี่ยง	Fermented tea leaf, pickled tea leaf	0.1–1.5
Nor-Mai-Dorng	หน่อไม้ดอง	Fermented bamboo shoot	0.5–6.4
Pak-Kaat-Dorng	ผักกาดดอง	Pickled mustard greens	1.6–7.3
Pak-Kaat-Dorng-Haeng	ผักกาดดองแห้ง	Dried fermented mustard greens	39.8
Pak-Koom-Dorng	ผักกุ่มดอง	Pickled crateava	1.5–2.6
Pak-Naam-Dorng	ผักหนามดอง	Pickled lasia	0.7–2.3
Pak-Sian-Dorng	ผักเสี้ยนดอง	Fermented wild spider flower or leaf tips	0.8–2.1

(continued)

Table 3.1 (continued)

Product name	Thai name	Typical characteristics	% Salt
Sa-Tor-Dorng	สะตอดอง	Fermented sator seed	1.3–3.7
Tang-Chai	ตั้งฉ่าย	Dried fermented cabbage	15.8
Tooa-Li-Song-Dorng	ถั่วลิสงดอง	Fermented peanut sprouts	2.1
Tooa-Ngork-Dorng	ถั่วงอกดอง	Pickled mungbean sprouts	1.1
Soybean products			
See-Iu	ซีอิ๊ว	Soy sauce	19.2–21.7
Tao-Hoo-Yee	เต้าหู้ยี้	Fermented soybean cheese	12.6–19.6
Tao-Jieo	เต้าเจี้ยว	Soybean paste	15.7–19.0

Note: High salted products, 15.5–20 % or above; medium salted products, 4.5–15 %; and low salted products, 0.5–4 %

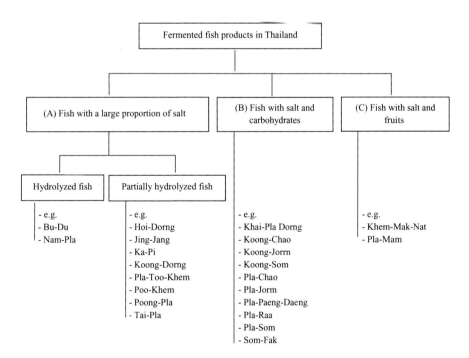

Fig. 3.1 Classification of fermented fish products in Thailand

This product is produced in most Asian countries but is different in raw materials, salt concentration, and the fermentation process and in name as shown in Table 3.2. For Thai fish sauce, the process is illustrated in Fig. 3.2, and the chemical composition is presented in Table 3.3.

Table 3.2 Similar or related products of fish sauce in Asian countries

Countries	Product names
Thailand	Nam-Pla, Nam-Pla-Dee, Nam Pla-Sod
Burma	Ngapi
Cambodia	Tuuk-prahoe, nuoc-mam-gau-ca
Indonesia	Ketjab ikan, kecap ikan
Japan	Uwo-shoyu or ounage, shottsura, kaomi, and ounago
Laos	Nam-Pla
Malaysia	Bu-Du
Philippines	Patis
Vietnam	Nuoc-mam

Small fresh fish (mostly *Clupeoids* sp.)
↓
Washing
↓
Place in wooden or cement tank or earthen jar
↓
Mix with salt, 3:1 (w/w)
cover the top layer with salt
↓
Remove fluid from the top of the tank after 3-4 days
↓
Press the fish layers firmly
↓
Ferment for at least 6 months
↓
Expose to sunlight, 2-4 weeks
↓
Filter

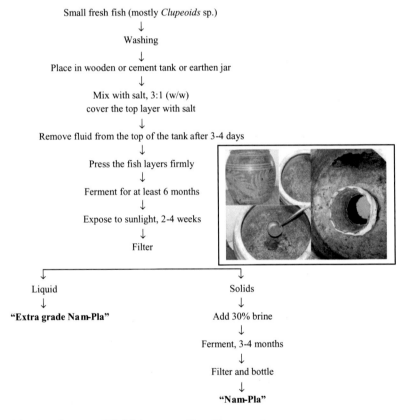

Liquid   Solids
↓   ↓
"Extra grade Nam-Pla"   Add 30% brine
↓
Ferment, 3-4 months
↓
Filter and bottle
↓
"Nam-Pla"

Fig. 3.2 Flow diagram of Thai fish sauce or Nam-Pla processing

Bu-Du (บูดู)

This product is another kind of hydrolyzed fish with high percentage of salt and usually produced in the southern part of Thailand. The fermentation of Bu-Du was similar to fish sauce but it has and extra ingredients such as palm sugar or coconut

Table 3.3 Chemical composition of Thai fish sauce (Nam-Pla)

Compositions	Rattagool et al. (1983)	Saisithi (1984)	Liptasiri (1975) High grade	Low grade
NaCl (%)	26.058	26–31	28.15	28.9
Total-N (TN) (%)	2.299	0.06–4.41	1.92	0.92
Formaldehyde N (FN) (%)	1.74		1.13	0.83
FN/TN	74.4			
Organic-N (%)			1.64	0.62
Ammonia-N (%)		10–30[a]	0.28	0.28
Amino-N (%)		40–60[a]	0.85	0.55
Trimethylamine-N (mg %)			20	20
pH	5.3	5.4–6.6		
Specific gravity	1.24	1.19–1.22		

[a]% of Total-N

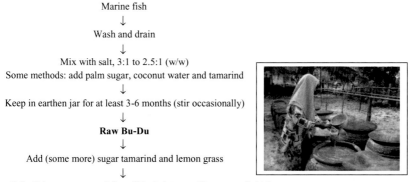

Marine fish
↓
Wash and drain
↓
Mix with salt, 3:1 to 2.5:1 (w/w)
Some methods: add palm sugar, coconut water and tamarind
↓
Keep in earthen jar for at least 3-6 months (stir occasionally)
↓
Raw Bu-Du
↓
Add (some more) sugar tamarind and lemon grass
↓
Boil until the fish meat separated from fish skeleton and become colloidal
↓
Filter and bottle
↓
"Bu-Du"

Fig. 3.3 Flow diagram of Thai Muslim sauce (Bu-Du) processing

water and tamarind which give a unique taste to the products. The flow diagram of fermentation process is shown in Fig. 3.3.

Kapi (**)

This product is an example of partially hydrolyzed fish with high percentage of salt but it is usually mixed with small shrimp. The process of fermentation is rather shorter and drier than fish sauce and Bu-Du because the water was drained before fermentation took place. The flow diagram of fermentation process is shown in Fig. 3.4. These fermented

Small shrimp or fish
(*Acetes* sp., *Myses* sp., *Copepoda* sp., *Lacifer* sp.)
↓
Mix with salt, 4:1 or 5:1 ratio (w/w)
↓
Place in bamboo basket overnight to drain
↓
Spread on bamboo mat to dry in the sunlight
↓
Grind to fine paste
↓
Pack tightly in earthen jar or wooden tub
↓
Ferment for at least 4 months
↓
"Ka-Pi"

Fig. 3.4 Flow diagram of Thai fermented small shrimp or fish paste processing of Ka-Pi

products are commonly used to prepare a shrimp paste sauce and served with fried mackerel and boiled vegetables which are quite healthy foods for low-calorie dishes.

Fish with Salt and Carbohydrates

There are also many kinds of products in this category. The fish used are mainly fresh water and also freshwater shrimp. Some products mix small fish and shrimp together. Small fish are normally used as the large ones are sold in the local markets or salted and sun dried for local family consumption. Carbohydrate is provided by rice, cooked, roasted, or fermented, according to individual products. The rice is a source of carbohydrate for the microorganisms involved in the fermentation, mainly lactic acid bacteria, and also gives a characteristic flavor to the products. The roasted rice also gives a brown color. There are two types of fermented rice used, Khao-Maak, which is mold and yeast fermented, and Ang-Kak rice, which is fermented using the mold *Monascus purpureus*, which gives a red color as well as a character-istic flavor. Four kinds of this product category, Pla-Raa, Pla-Som, Pla-Paeng-Daeng, and Som-Fak, with their production methods are described as follows.

Pla-Raa (ปลาร้า)

This fermented fish product is quite common for people in the northeastern part of Thailand but not for others. The smell is quite strong and usually added to papaya salads which are a popular dish for local people. It is different from fermented fish in Group A which is mentioned earlier by supplementation with carbohydrates to

Fresh water fishes; the most common varieties used.

> Pla-ka-dee (*Trichogastor trichopterus*),
> Pla-cha-lard(*Notopterus notopterus*),
> Pla-chon (*Ophicephalus striatus*),
> Pla-dug(*Clarias batrachus*)

> Scale and remove the entrails
> (head, tail and fins will be removed in case of big fishes)

Mix with salt in the ratio of 3:1 (w/w) and leave over night

> then packed tightly in a small neck earthen jar,
> press and lock at the neck of the jar with bamboo sticks
> flood with saturated brine. Keep for 15 days to three months
> (incubation period is depended on a requirement to continue the next step)

Take out the fish from the jar
(some method wash and drain first)
and mix with ground roasted rice in the ratio of 1:10 (w/w)

> repack in the jar as described above and flood with brine,
> keep for more than 6 months
> (longer incubated product will be a better-tasted 'Pla-Raa")

"Pla-Raa"

Fig. 3.5   Flow diagram of Thai fermented fish processing of Pla-Raa

give a better taste due to the organic acids produced by lactic acid bacteria. The process of fermentation is shown in Fig. 3.5.

Pla-Som (ปลาส้ม)

Compared to Pla-Raa, Pla-Som is mostly popular for city people, and it is widely sold in the supermarket. Most people usually fried Pla-Som and eat with hot cooked rice due to its salty and sour taste. The minced garlic was also added to improve taste. Therefore, the smell of herbs was superb. The process of fermentation is shown in Fig. 3.6.

Pla-Paeng-Daeng (ปลาแป้งแดง)

General name	Red fermented fish
Nature of product	Whole fish, semisolid, sour and salty, lightly alcoholic, reddish color
Main utilization	Main dish
Raw materials	Fish, salt, cooked rice, ang-kak (red rice)

This kind of fermented fish is different from Pla-Raa and Pla-Som in color due to the addition of red rice, which gives it a unique characteristic. This kind of product is also found in other Asian countries as shown in Table 3.4. The process of fermentation is shown in Fig. 3.7. It is a nutritious food as shown by chemical composition in Table 3.5

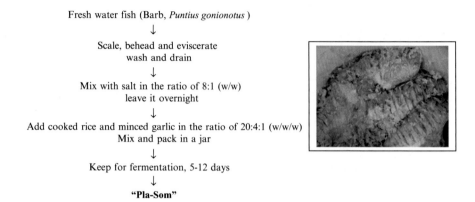

Fresh water fish (Barb, *Puntius gonionotus*)
↓
Scale, behead and eviscerate
wash and drain
↓
Mix with salt in the ratio of 8:1 (w/w)
leave it overnight
↓
Add cooked rice and minced garlic in the ratio of 20:4:1 (w/w/w)
Mix and pack in a jar
↓
Keep for fermentation, 5-12 days
↓
"Pla-Som"

Fig. 3.6 Flow diagram of Thai fermented fish processing of Pla-Som

Table 3.4 Similar or related products of fermented fish (Pla-Paeng-Daeng)

Countries	Product names
Thailand	Pla-Paeng-Daeng
Cambodia	Paak or mam-chao
Indonesia	Makassar fish or red fish
Philippines	Burong isda (red variety), burong dalag, buro

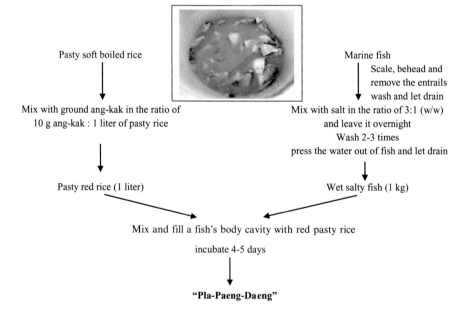

Pasty soft boiled rice

Marine fish
  Scale, behead and
  remove the entrails
  wash and let drain

Mix with ground ang-kak in the ratio of
10 g ang-kak : 1 liter of pasty rice

Mix with salt in the ratio of 3:1 (w/w)
and leave it overnight
Wash 2-3 times
press the water out of fish and let drain

Pasty red rice (1 liter)

Wet salty fish (1 kg)

Mix and fill a fish's body cavity with red pasty rice

incubate 4-5 days

"Pla-Paeng-Daeng"

Fig. 3.7 Flow diagram of Thai fermented fish processing of Pla-Paeng-Daeng

Table 3.5 Chemical composition of Thai fermented fish (Pla-Paeng-Daeng)

Compositions	Dhamaharee et al. (1984)	Nutrition Division, Department of Health, Thailand (1978)	NRCT (1982)
Moisture (%)	62.49–78.82		70.7
Protein (%)	9.48–15.66	4.35–8.48	4.9
Fat (%)	0.68–4.37	4.37–12.80	2.4
Fiber (%)	nil–0.39		0.1
Ash (%)	4.74–9.24		3.9
Total invert sugar (%)	2.28–19.57		
NaCl (%)	2.51–7.18	4.49–9.20	
Lactic acid (%)	2.04–4.06	1.42–2.10	
pH (%)	4.0–5.3	3.9–5.2	
Carbohydrate (%)			18.0

Som-Fak (ส้มฟัก)

General name	Fermented fish cake
Synonyms	Som-Fuk, Som-Doc (Northeastern Thailand)
	Pla-Fug (Northern Thailand)
	Pla-Mug (Eastern Thailand)
	Fug-Som (some part of Northeastern Thailand)
Nature of product	Minced fish flesh, solid, sour, and salty
Main utilization	Main dish
Raw materials	Fish flesh, salt, cooked rice, garlic

This kind of fermented fish has a similar taste to Pla-Paeng-Daeng but is different in form of raw materials. Som-Fak used minced fish, while whole fish was used for Pla-Paeng-Daeng. Moreover, Som-Fak is quite similar to fermented pork sausage (Naem) for the way they are packed. After mixing, it can be packed in plastic bags or wrapped in banana leaves. The process of fermentation is shown in Fig. 3.8, and the chemical composition is presented in Table 3.6.

Fish with Salt and Fruits

There are only two fermented fish products made with fruits, Pla-Mam and Khem-Mak-Nat, which have pineapples added.

Khem-Mak-Nat (เค็มหมักนัด)

Fermented fish with salt and fruit is a typical fermented product for people in the Northeast of Thailand. The process of Khem-Mak-Nat fermentation is shown in Fig. 3.9.

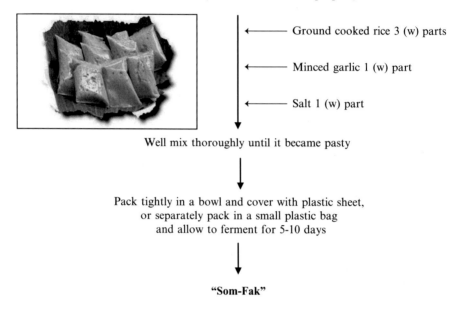

Minced or ground fish meat
("Pla-cha-do", *Ophicephalus lucius*)
is the best variety of fish used for this purpose)

← Ground cooked rice 3 (w) parts

← Minced garlic 1 (w) part

← Salt 1 (w) part

Well mix thoroughly until it became pasty

Pack tightly in a bowl and cover with plastic sheet,
or separately pack in a small plastic bag
and allow to ferment for 5-10 days

"Som-Fak"

Fig. 3.8 Flow diagram of Thai fermented fish cake processing of Som-Fak

Table 3.6 Chemical composition of Thai fermented fish cake (Som-Fak)

Compositions	Dhamaharee et al. (1984)[a]	Adams et al. (1985)
Moisture (%)	65.64–76.21	67.68
Protein (%)	17.01–18.43	2.18[b]
Fat (%)	trace-1.80	6.56
Fiber (%)	0.15–0.92	
Ash (%)	4.45–4.67	
Total invert sugar (%)	Nil	
NaCl (%)	3.64–3.83	5.58
Lactic acid (%)	1.27–1.46	1.12
pH	4.15–5.0	5.0

[a]Range of two commercial samples
[b]% Total- N

3.2.1.2 Animal Products

There are few animal products, and the classification system is similar to that for fishery products. Cooked rice or roasted rice is used in most products as a source of carbohydrate, and some spices, e.g., fresh garlic and pepper, are added to the meat for flavor. Some products are almost unknown to the young generation, e.g., Nang-Khem. Some vegetables are added to some products to reduce cost as well as for consumer preference.

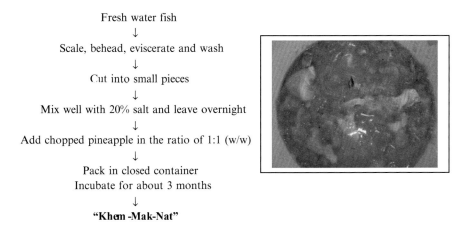

Fresh water fish
↓
Scale, behead, eviscerate and wash
↓
Cut into small pieces
↓
Mix well with 20% salt and leave overnight
↓
Add chopped pineapple in the ratio of 1:1 (w/w)
↓
Pack in closed container
Incubate for about 3 months
↓
"Khem -Mak-Nat"

Fig. 3.9 Flow diagram of Thai fermented fish with pineapple processing of Khem-Mak-Nat

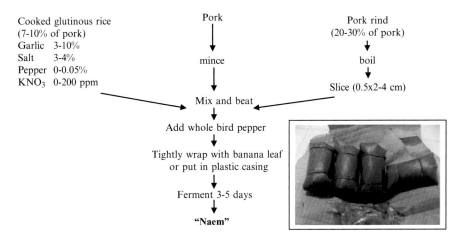

Cooked glutinous rice
(7-10% of pork)
Garlic   3-10%
Salt   3-4%
Pepper   0-0.05%
KNO$_3$ 0-200 ppm

Pork
↓
mince
↓

Pork rind
(20-30% of pork)
↓
boil
↓
Slice (0.5x2-4 cm)

Mix and beat
↓
Add whole bird pepper
↓
Tightly wrap with banana leaf
or put in plastic casing
↓
Ferment 3-5 days
↓
"Naem"

Fig. 3.10 Flow diagram of Thai fermented pork processing of Naem

Naem or Nham (แหนม)

Naem is a fermented meat product which is usually consumed in all parts of Thailand (Khieokhachee et al. 1997). The ingredients of Naem produced in different area are quite similar but local wisdom are much different. This product is classified as a low-salt product, but it has a unique characteristic similar to fermented dry sausage (salami) in Western countries. It is always consumed raw or uncooked as appetizer dishes. The fermentation process of Naem is shown in Fig. 3.10.

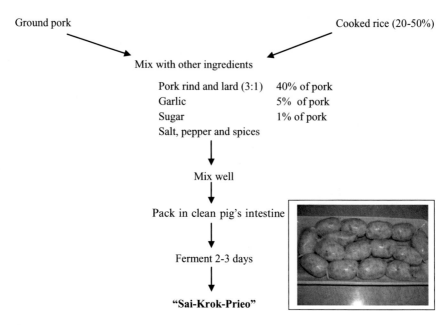

Fig. 3.11 Flow diagram of Thai fermented sausage processing of Sai-Krok-Prieo

Sai-Krok-Prieo (ไส้กรอกเปรี้ยว) or Sai-Krok-Isan (ไส้กรอกอีสาน)

This product is quite similar to Naem but no pork rind and cooked rice added is much greater than in Naem. After mixing, it is packed in pig's intestine and tied as a small meatball. This product is not eaten raw but usually grilled before serving. The fermentation process of Sai-Krok-Prieo is shown in Fig. 3.11.

3.2.1.3 Fruit and Vegetable Products

Salt-fermented fruit products are especially popular among Thai girls who eat them as a snack rather than cake or biscuits. The fermented fruit is normally dipped in a mixture of sugar, salt, and ground chili to give a better taste. Some fermented fruits need further preparation before selling or eating, for example, fermented green mango, which is peeled, cut into pieces, and soaked in syrup overnight to reduce the salty taste. Any kinds of fruits can be fermented before eating, even unexpected fruits such as grape. This is because of girls' preferences rather than oversupply of the fruits.

Various kinds of vegetables are also fermented, mainly by mixing with salt for a few hours and then adding rice-wash water. They can be eaten the next day, usually with Nam-Prik, a Thai traditional dish of chili paste. The storage life is short, but some products with high salt content can be kept for many months.

Tea leaves
↓
Steamed
↓
Wrapped tightly in individual bundle
↓
Packed into container
(A small basket for young tea leaves,
large underground cement wells for nature tea leaves)

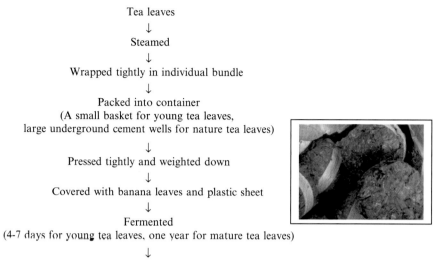

↓
Pressed tightly and weighted down
↓
Covered with banana leaves and plastic sheet
↓
Fermented
(4-7 days for young tea leaves, one year for mature tea leaves)
↓
"Miang"

Fig. 3.12 Flow diagram of Thai fermented tea leaves (Miang) processing

Miang (เมี่ยง)

Thai fermented tea leaves called Miang, is similar to *lak hpak* in Myanmar. It is popular in the north of Thailand. It is usually consumed together with peanuts, ginger, garlic, or roasted coconut sliced as a side dish. The fermentation process is shown in Fig. 3.12.

3.2.2 Thai Non-Salted Fermented Foods

Thai Non-salted fermented foods includes fermented products made from all agricultural raw materials high in carbohydrate content such as rice, fruits, and cassava or high in protein such as soybean without the addition of salt as shown in Table 3.7.

3.2.2.1 Non-alcoholic Products

Ka-Nom-Jeen (ขนมจีน)

General name	Fermented rice noodle
Nature of product	Solid, white noodle, bland taste
Main utilization	Main dish (with curry)
Raw materials	Rice

Thai non-salted fermented rice product is called Ka-Nom-Jeen. It is usually consumed as a main dish served with various kinds of curry. Sometimes it is served

Table 3.7 Type of non-salted fermented foods in Thailand

Product types	Thai names	Product names
Alcoholics		
From rice	ลูกแป้ง	Loog-Paeng
	ข้าวหมาก	Khao-Maak
	น้ำขาว	Nam-Khao
	อุ	Ou
From fruits	ไวน์	Wine
	น้ำตาลเมา	Nam-Dtaan-Mao
From cassava	ตา-แป	Taa-Pae
Non-alcoholics		
From rice	อั้งคัก	Ang-Kak
	ขนมจีน	Ka-Nom-Jeen
	ขนมผักบัว	Ka-Nom-Fak-Booa
	ขนมถ้วยฟู	Ka-Nom-Thuai-Fu
	ขนมตาล	Ka-Nom-Dtaan
From beans	ถั่วเน่า	Thua-Nao
From coconut water	วุ้นมะพร้าว	Woon-Ma-Prao
From pineapple	วุ้นสับปะรด	Woon-Sap-Paroat

together with papaya salads or can be used as a raw material for fusion foods such as Vietnamese foods. The fermentation process of Ka-Nom-Jeen is shown in Fig. 3.13, and the chemical composition is shown in Table 3.8.

Ka-nom-Thuai-Fu (ขนมถ้วยฟู)

This product is a typical kind of Thai dessert. The word "kanom" in Thai language means dessert. Most of Thai desserts are usually made with rice flour and coconut cream and then steamed until cooked. But Ka-Nom-Thuai-Fu is special because look-pang (yeast starter culture) is also added for fermentation before being steamed. So it gives a fluffy character and good aroma. The fermentation process of Ka-Nom-Thuai-Fu is shown in Fig. 3.14.

Khao-Daeng (ข้าวแดง) or Ang-Kak

This product is similar to red yeast rice or red fermented rice in Chinese which fermented by *Monascus* sp. The fermentation process of Khao-Daeng is shown in Fig. 3.15.

Thua-Nao (ถั่วเน่า)

Thai fermented soybean called Thua-Nao is similar to Japanese natto but slightly different in texture. After fermentation, it is usually grinded and made as a paste and then sun dried or roasted. It is used to prepare a seasoning sauce. This product is quite popular in the north of Thailand. The fermentation process of Thua-Nao is shown in Fig. 3.16.

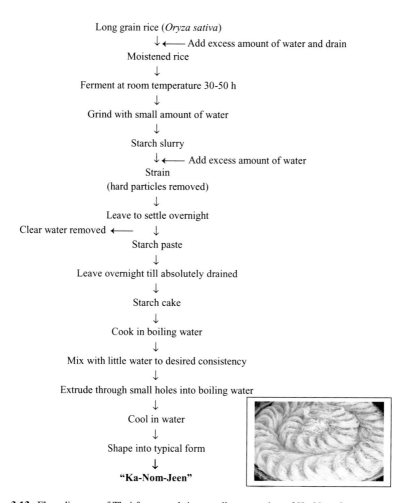

Long grain rice (*Oryza sativa*)
↓ ←—— Add excess amount of water and drain
Moistened rice
↓
Ferment at room temperature 30-50 h
↓
Grind with small amount of water
↓
Starch slurry
↓ ←—— Add excess amount of water
Strain
(hard particles removed)
↓
Leave to settle overnight
Clear water removed ←—— ↓
Starch paste
↓
Leave overnight till absolutely drained
↓
Starch cake
↓
Cook in boiling water
↓
Mix with little water to desired consistency
↓
Extrude through small holes into boiling water
↓
Cool in water
↓
Shape into typical form
↓
"Ka-Nom-Jeen"

Fig. 3.13 Flow diagram of Thai fermented rice noodle processing of Ka-Nom-Jeen

Table 3.8 Chemical composition of Thai fermented rice noodle (Ka-Nom-Jeen)

Compositions	Kraidej et al. (1977)	Nutrition Division, Department of Health, Thailand (1984)	Nutrition Division, Department of Health, Thailand (1978)
Moisture (%)	79.19	80.7	77.4
Protein (%)	6.1	0.9	1.5
Fat (%)	0.42	0.1	
Invert sugar (%)	0.005	0.1	
Ash (%)	0.466	0.1	
Carbohydrate (%)		18.2	20.9

Rice flour (360g)
↓
Add 130 g sugar and about 240 ml water
↓
Mix Thoroughly
↓
Add 10 g Look-pang
(powdered)
↓
Ferment overnight
↓
Pour batter into small cups
(10-15 ml)
↓
Steam 15 minutes
↓
"Kanom-Thuai-Fu"

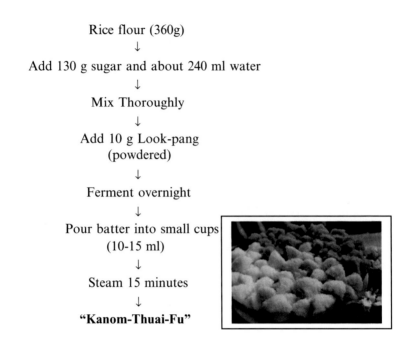

Fig. 3.14 Flow diagram of Thai fermented rice cake processing or Ka-nom-Thuai-Fu

Rice
↓
Soak in water, drain
↓
Steam
↓
Cool
↓
Inoculate with suspension of *M. purpureus* in water
↓
Stir
↓
Incubate at room temperature (10-20 days)
Add occasionally sterile water as fermentation progresses
↓
Red rice
↓
Dry at 40°C
↓
"Khao-Daeng"

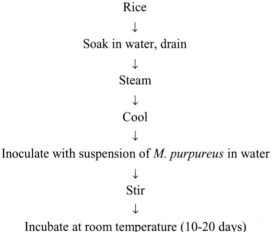

Fig. 3.15 Flow diagram of Khao-Daeng processing

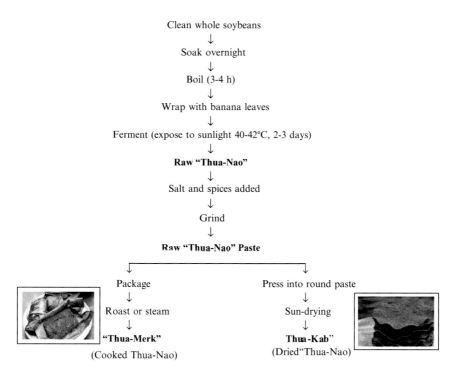

Clean whole soybeans
↓
Soak overnight
↓
Boil (3-4 h)
↓
Wrap with banana leaves
↓
Ferment (expose to sunlight 40-42°C, 2-3 days)
↓
Raw "Thua-Nao"
↓
Salt and spices added
↓
Grind
↓
Raw "Thua-Nao" Paste

Package | Press into round paste
↓ | ↓
Roast or steam | Sun-drying
↓ | ↓
"Thua-Merk" | **Thua-Kab"**
(Cooked Thua-Nao) | (Dried"Thua-Nao)

Fig. 3.16 Flow diagram of Thai fermented soybean processing of Thua-Nao

3.2.2.2 Alcoholic Products

Traditional fermented alcoholic products in Thailand are classified into three kinds which are (1) rice wine (Sa-To or Ou), (2) palm wine (Nam-Dtaan-Mao), and (3) distilled spirit from rice (Nam-Khao or Lao-Khao).

Basically, the common method for making rice wine is using powdered Loog-Paeng (starter) spread over steamed glutinous rice and fermented in a wide-mounted jar for the saccharifying process and transferred to a narrow-mouthed jar for the alcoholic fermentation. In some places, rice husk is applied during the fermentation process. The product is usually consumed by drinking through long fine bamboo straws. Production methods of Loog-Paeng, Sa-To/Ou/Nam-Khao, and Nam-Dtaan-Mao are as follows.

Khao-Maak (ข้าวหมาก)

This product is a Thai traditional alcoholic fermented product which is in semi-solid form. It is made from fermented glutinous rice with Loog-Paeng (yeast starter culture) and then wrapped in banana leaves or lotus leaves. After fermentation

Glutinous rice (*Oryza glutinou s*)
↓
Soak overnight
↓
Steam
↓
Wash off sticky material
↓
Drain
↓
Mix

Loog-paeng 0.2-0.3%
(on dry weight of rice)
↓
grind ⟶ Mix

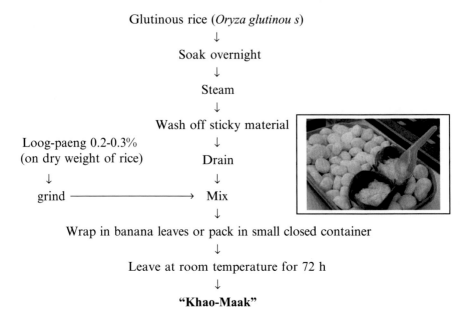

Wrap in banana leaves or pack in small closed container
↓
Leave at room temperature for 72 h
↓
"Khao-Maak"

Fig. 3.17 Flow diagram of Thai sweetened rice processing of Khao-Maak

complete, it is kept cool in a refrigerator before serving. The cold product gave a refreshment dessert with alcoholic taste and ester aroma. The fermentation process of Khao-Maak is shown in Fig. 3.17.

Loog-Paeng (ลูกแป้ง)

Loog-Paeng is prepared from ground rice mixed with a lot of Thai spices and shaped as a small flattened ball and then sprinkled with old Loog-Paeng (fungal starter culture). When the fungi grow all over the surface of the small ball, let it dry under the sun and keep in refrigerator until used as a starter culture for Khao-Maak fermentation. The making process of Loog-Paeng is shown in Fig. 3.18.

Sa-To (สาโท) and Nam-Khao (น้ำขาว) or Lao-Khao (เหล้าขาว)

Sa-To and Nam-Khao are Thai traditional alcoholic beverage (Chaownsungket 1978). The fermentation process of Sa-To and Nam-Khao is quite similar to Khao-Maak except water is added for a second fermentation. It has two distinct colors which are white or red depending on the type of glutinous rice (white or black). If the fermented product is not filtered, it is quite turbid and is called Sa-To. It is called Nam-Khao when the product is filtered. The fermentation process of Sa-To is shown in Fig. 3.19.

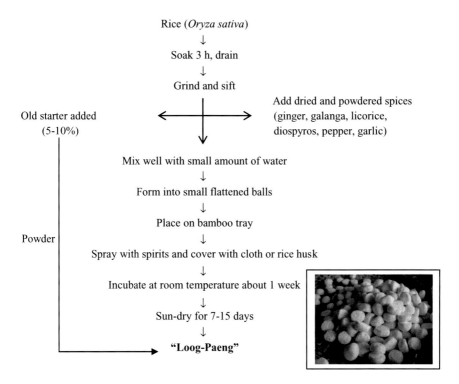

Rice (*Oryza sativa*)
↓
Soak 3 h, drain
↓
Grind and sift

Old starter added   ←——————————→   Add dried and powdered spices
(5-10%)   (ginger, galanga, licorice,
                                                diospyros, pepper, garlic)
↓

Mix well with small amount of water
↓
Form into small flattened balls
↓
Place on bamboo tray
Powder   ↓
Spray with spirits and cover with cloth or rice husk
↓
Incubate at room temperature about 1 week
↓
Sun-dry for 7-15 days
↓
"Loog-Paeng"

Fig. 3.18 Flow diagram of Thai mold rice processing of Loog-Paeng

Ou (อุ)

Ou is fermented from polished rice mixed with Loog-Paeng and rice husk and then packed in a clay jar. The fermented product is served by putting a straw into the jar instead of pouring to the glass. The fermentation process of Ou is presented in Fig. 3.20.

Nam-Dtaan-Mao (น้ำตาลเมา)

Nam-Dtaan-Mao is fermented from coconut palm juice with a piece of bark which gives a special aroma from bark. The fermentation process of Nam-Dtaan-Mao is presented in Fig. 3.21.

3.3   Microorganisms Associated with the Fermentation

Professor Dr. Davi Yanasugondha, the former Dean of Faculty of Science, Kasetsart University, initiated the study of applied (non-medical) microbiology in Thailand in 1965. He had surveyed microorganisms associated with a variety of fermented

Glutinous rice
(*Oryza glutinos a*)
↓
Washing
↓
Steeping
(6-12 h)
Excess water drained
↓
Steaming
(30-60 min.)
↓
Cooling to room temp.
↓
Washing ———→ Excess water drained
↓
Mix with 0.1-0.2% of Loog-paeng
↓
Incubate at room temp.
(1-3 days)
↓
Add water and let it ferment for 4-14 days
↓
Sato
↓
Solid removed ←——— Filter
↓
"Nam-Khao"
or
(Lao-Khao)

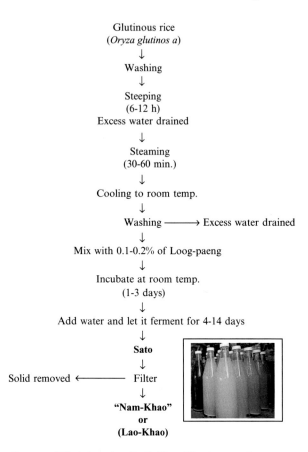

Fig. 3.19 Flow diagram of Thai rice wine (Sa-To/Nam-Khao) processing

foods all around the country. Differences in typical fermented foods in each part of Thailand had been noted, as shown in Fig. 3.22. Naem, Thua-Nao, Miang, and Nam-Poo are typical fermented foods of Northern Thailand; Pla-Raa, Pla-Som, Sai-Krok-Prieo are typical fermented foods of the northeastern part; Pla-Jorm, Hoi-Dorng, Pak-Sian-Dorng, and Ma-Muang-Dorng are typical fermented foods of the eastern part; while Bu-Du, Ka-Pi, Khem-Mak-Nat and Pla-Plang-Daeng are typical fermented foods of Southern Thailand. Nam-Pla, Som-Fak, and Ka-Nom-Jeen are the general traditional fermented foods in Central and in all parts of Thailand. Later, this study had extended to others like the Faculty of Fishery, Agro-industry, and so on, and results collected in Appendix of Proceedings of the 1st International Symposium and Workshop on Insight into the World of Indigenous Fermented Foods for Technology Development and Food Safety (IWIFF), 2003, and are presented in Table 3.9.

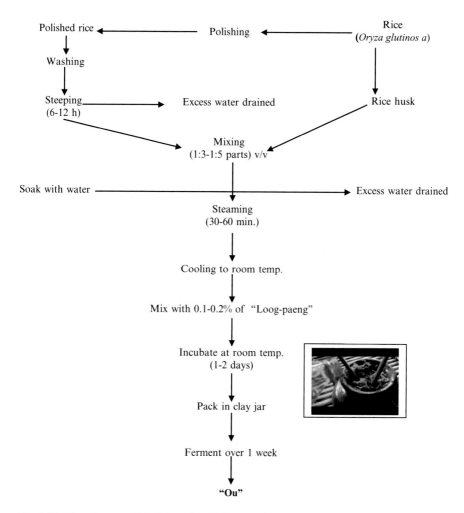

Fig. 3.20 Flow diagram of Thai rice wine (Ou) processing

3.4 Development and Future Trends

Desirable improvements of traditional fermented foods are to shortening aging time with better flavor, prolonged shelf life, etc. It requires the development of starter preparation as well as improved equipment.

3.4.1 Development of Pure Starter Microorganisms

Food microorganisms deliver a range of unique products which cannot be mimicked by chemicals. Some Thai traditional fermented foods become industrialized due to increasing market such as soy sauce, Naem, and Nam-Pla.

Coconut palm
↓
Select suitable flower apathes
↓
Slit with sharp knife, 10-15 cm from the pointed apex
↓
Spathe end sliced off, 4-5 mm
(once a day, 3-5 days)
↓
Sap flow
↓
Collect in bamboo tube*
(remove a thin-4 mm-slice from the cut end twice,
daily to ensure regular flow)
↓
Fermented sap
↓
Filter
↓
"Nam-Dtaan-Mao"

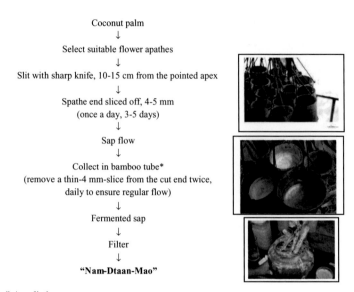

* Contain small piece of bark
from mai-payom (*Shorea floribunda*), mai-kiam (*Cotybium lanceolatum*), mai-maglur (*Diospyros mollis*) or mai-takean
(*Hopea adorata*) to prevent spoilage from microorganisms.

Fig. 3.21 Flow diagram of palm rice wine processing of Nam-Dtaan-Mao

North
Naem
Thua-Nao
Miang
Nam-Poo

NORTH

Northeast
Pla-Raa
Pla-Som
Sai-Krok-Prieo
Khem-Mak-Nat

Central
Nam-Pla
Som-Fak
Ka-Nom-Jeen

CENTRAL

NORTHEAST

East
Pla-Jorm,
Hoi-Dorng
Pak-Sian-Dorng
Ma-Muang-Dorng

EAST

South
Bu-Du, Ka-Pi
Pla-Paeng-Daeng

SOUTH

Fig. 3.22 Typical traditional fermented foods in each part of Thailand

Table 3.9 List of microorganisms found in Thai fermented foods

Microorganisms	Thai fermented foods
Actinomucor elegans	Tao-Hoo-Yee
Amylomyces rouxii	Nam-Khao, Khao-Maak
Aspergillus oryzae	See-Iu, Tao-Jieo
Aspergillus sojae	See-Iu
Aspergillus sp.	Khao-Maak
Bacillus amyloliquefaciens	Thua-Nao
Bacillus cereus	Nor-Mai-Dorng
Bacillus laterosporus	Bu-Du
Bacillus licheniformis	Ka-Pi, Pla-Raa
Bacillus pantothenticus	Nam-Pla
Bacillus polymyxa	Loog-Niang-Dorng, Pla-Paeng-Daeng, Tai-Pla
Bacillus sp.	Ka-Pi, Koong-Som, Loog-Riang-Dorng, Naang-Neua, Nam-Pla, Nor Mai Dorng, Pak-Sian-Dorng, Pla-Chao, Pla-Som, Poot-Sa-Dorng, See-Iu, Som-Fak
Bacillus subtilis	Bu-Du, Pla-Paeng-Deaeng, Pla-Raa, Tai-Pla, Ka-Nom-Jeen
Bacillus sphaericus	Ka-Pi
Candida sp.	Dton-Horm-Dorng, Koong-Som, Look-Riang-Dorng, Pla-Som, Som-Fak, Sa-Mor-Dorng, Too-Rian-Prieo, Thua-Nao
Chlamydornucor sp.	Khao-Maak
Corynebacterium sp.	Ka-Pi
Coryneform bacteria	Bu-Du, Nam-Pla
Issatchenkia orientalis	Loog-Paeng
Debaryomyces sp.	Pak-Kaat-Dorng, Sai-Krok-Prieo
Klebsiella pneumoniae	Thua-Nao
Endomycopsis sp.	Khao-Maak, Pla-Chao
Endomycopsis fibuligera	Loog-Paeng, Khao-Maak
Halobacterium salinarum	Nam-Pla
Hansenula anomala	Khao Maak
Lactobacillus acidophilus	Naem
Lactobacillus brevis	Naem, Pak-Kaat-Dorng, Som-Fak, Pla-Som
Lactobacillus casei var. *casei*	Koong-Som, Ka-Nom-Jeen
Lactobacillus casei var. *alactosus*	Koong-Som
Lactobacillus cellobiose	Naem
Lactobacillus delbrueckii	Tao-Jieo, Naem
Lactobacillus salivarius	Sai-Krok-Prieo
Lactobacillus fermentum	Som-Fak
Lactobacillus paracasei	Sai-Krok-Prieo
Lactobacillus plantarum	Dton-Horm-Dorng, Koong-Som, Loog-Kra-Dorng, Loog-Riang-Dorng, Naang-Moo, Naem, Nor-Mai-Dorng, Pak-Kaat-Dorng, Pak-Naam-Dorng, Pak-Sian-Dorng, Pla-Som, Poot-Sa-Dorng, Sa-Tor-Dorng, Som-Fak, Too-Rian-Prieo, Pla-Som, Ka-Nom-Jeen, Sai-Krok-Prieo

(continued)

Table 3.9　(continued)

Microorganisms	Thai fermented foods
Leuconostoc pentosus	Naem
Lactobacillus sp.	Dton-Horm-Dorng, Ka-Pi, Miang, Naang-Moo, Lum-Pee-Dorng, Loog-Riang-Dorng, Naang-Neua, Nor-Mai-Dorng, Pak-Sian-Dorng, Ra-Kam-Dorng, Sa-Tor-Dorng, Sai-Krok-Prieo, Som-Kai-Pla, Too-Rian-Prieo, Nam-Pla
Lactobacillus mesenteroides	Naem
Micrococcus sp.	Bu-Du, Ka-Pi, Koong-Jorm, Koong-Som, Mam, Naang-Neua, Nam-Pla, Pla-Chao, Pla-Jorm, Pla-Raa, Pla-Som, Poot-Sa-Dorng, Tai-Pla
Micrococcus morrhuae	Ka-Pi
Micrococcus varians	Naem, Nam-Pla
Mucor sp.	Khao-Maak, Nam-Khao
Pediococcus acidilactici	Sai-Krok-Prieo
Pediococcus cerevisiae	Naem, Pak-Kaat-Dorng, Pla-Chao, Pla-Som, Som-Fak
Pediococcus halophilus	Bu-Du, Hoi-Ma-Laeng-Poo-Dorng, Hoi-Som, Jing-Jang, Ka-Pi, Koong-Jorm, Koong-Som, Pla-Chao, Pla-Jorm, Pla-Paeng-Daeng, Pla-Raa, Pla-Som, See-Iu, Tao-Jieo, Tai-Pla, Nam-Pla
Pediococcus pentosaceus	Dton-Horm-Dorng, Naang-Neua, Pak-Naam-Dorng, Pak-Sian-Dorng, Som-Fak, Naem, Sai-Krok-Prieo
Pediococcus sp.	Nam-Pla, Pla-Paeng-Daeng, Pla-Raa, Tai-Pla, Koong-Som
Pichia anomala	Loog-Paeng
Pichia burtonii	Loog-Paeng
Pichia farinosa	Ka-Nom-Jeen
Pichia terriocola	Ka-Nom-Jeen
Pichia sp.	Koong-Som, Lum-Pee-Dorng, Pak-Kaat-Dorng, Pak-Sian-Dorng, Pla-Som, Poot-Sa-Dorng, Ra-Kam-Dorng, Sa-Tor-Dorng
Proteus sp.	Bu-Du
Rhizopus oryzae	Nam-Khao
Saccharomyces cerevisiae	Koong-Jorm, Nam-Dtaan-Mao, Nam-Khao, Pla-Chao, Pla-Jorm, Pla-Paeng-Daeng
Saccharomyces hansenula	Tao-Jieo
Saccharomyces rouxii	Tao-Jieo
Saccharomyces sp.	Khao-Maak, Koong-Som, Pla-Som, See-Iu
Saccharomycopsis fibuligera	Loog-Paeng
Sarcina sp.	Bu-Du, Nam-Pla, Pla-Chao
Staphylococcus aureus	Bu-Du, Ka-Pi, Loog-Riang-Dorng, Tai-Pla, Koong-Jorm, Hoi-Ma-Lang-Poo-Dorng
Staphylococcus epidermidis	Bu-Du, Ka-Pi, Koong-Jorm, Koong-Som, Mam, Pla-Jorm, Pla-Paeng-Daeng, Pla-Raa, Pla-Som, Poot-Sa-Dorng, Sa-Mor-Dorng, Tai-Pla, Hoi-Ma-Lang-Poo-Dorng
Staphylococcus sp.	Ka-Pi, Nam-Pla, Pla-Chao, Pla-Raa, Pla-Som, Som-Fak

(continued)

Table 3.9 (continued)

Microorganisms	Thai fermented foods
Staphylococcus saprophyticus	Nam-Pla
Streptococcus faecalis	Koong-Som, Pla-Som, Sai-Krork-Prieo, Som-Fak, Tao-Jieo
Streptococcus avium	Ka-Nom-Jeen
Streptococcus lactis	Ka-Nom-Jeen
Torulaspora globosa	Loog-Paeng
Trichosporon sp.	Hoi-Ma-Lang-Poo-Dorng
Vibrio fischeri	Pla-Paeng-Daeng, Tai-Pla

In case of soy sauce industries, the screening of the pure koji mold had been the collaboration of Kasetsart University and Mahidol University under ASEAN Project since 1975. They succeeded in strain selection and spore inoculum production in plastic bag for contribution to the local industries. Later, Mahidol University with German Grant (200 MB) established a Quality Central and Training Center (QCTC). This unit facilitates the introduction of improved *Aspergillus oryzae* starter for not only soy sauce fermentation but also to Thai rice wine in the near future.

BIOTEC (National Center for Genetic Engineering and Biotechnology) has developed the pure lactic acid bacteria starter in medium-scale technologies for Naem production in the aspect of food safety from salmonellosis or staphylococcus enterotoxin outbreaks (Valyasevi and Rolle 2002).

It is believed that traditional fermented foods are the best sources for potential and safe microorganisms for other applications. The microorganisms (yeast and molds) in Loog-Paeng cause changes of Thai sticky rice into many good volatile compounds. Kasetsart University has developed strain improvement of *Monascus* sp. for rice products such as, red *Monascus* rice (RMR), golden brown *Monascus* rice (GBMR), and white *Monascus* rice (WMR) which exhibited strong antimutagenic, anticholesterol, and antioxidant activities (Kruawan et al. 2005; Chayawat et al. 2009; Yongsmith et al. 2000, 2013). Many lactic acid bacteria are probiotic and have anticholesterol and immune activator producing capabilities. Large-scale production of these strains is needed for either fermented food industries as well as other applications. They will stimulate the development of national economy as a whole.

3.4.2 Improved Technologies or Equipments

Improvement of inoculum chamber with fiberglass tanks for koji inoculum and moromi stage resulted in soy sauce product quality and consistency (Mongkolwai, et al. 1998). Another good example of modified technology is the utilization of the iodized salting in the Pla-Raa fermentation. This process improved the product nutrition and has been accepted warmly in the market. Such technologies should be modified for other traditional fermented foods.

3.5   Safety Aspects of Thai Fermented Foods

Although the benefits of fermented foods are enhanced organoleptic properties and prolonged shelf life, insufficient control may lead to unsafe products. Unsanitary environment and unhygienic of food handler can affected the food risky especially for spontaneous fermentation. Therefore, starter culture technology may help to initiate and control the fermentation which will result in more safety of products. The Thai Food and Drug Administration and Ministry of Health try to persuade the small-scale producers to start with the primary GMP (good manufacturing practices) while medium- and large-scale producers must follow the GMP Law. Thai national policy has also encouraged the academic institutions to cooperate with industries to do research which helps to improve the quality as well as food safety. Risk analysis should be criticized which enable the risk assessment, risk management, and risk communication to convey properly to all stakeholders for sustainable development.

3.6   Summary

Study on traditional fermented foods in Thailand has been initiated by Kasetsart University since 1970.

Formerly, Thai fermented foods were consumed only within the country. Recently, the Thai government has declared her policy for Thai foods as "Kitchen of the World." Some fermented food products are now quite popular in international markets. Food safety as well as quality issues in processing must be emphasized. Potential microbial strains can be isolated and used as practical starters. Since starter culture technology is the driving force for the development of many SME's or SML's fermented food industries, scaling up as well as technological improvement should be aimed for. Integration of technical know-how of the cottage (local) producers and technical know-why of the scientists are necessary, which leads to increased economic development as widely as possible from the smallest to the largest scales.

References

Adams MR, Cooke RD, Rattagool P (1985) Fermented fish products of South East Asia. Trop Sci 25:61–73
Chaownsungket M (1978) Selection of the yeast and mold strains for rice wine production. M.S. Thesis, Kasetsart University, Bangkok
Chayawat J, Jareonkitmongkol S, Songsason A, Yongsmith B (2009) Pigments and anti-cholesterol agent production by *Monascus kaoliang* KB9 and its color mutants in rice solid cultures. Kasetsart J (Nat Sci) 43:696–702

Dhamaharee B, Pithakpol B, Reungmaneepaitoon S, Varanyanond W (1984) Chemical composition of fermented foods in Thailand. In: 22nd Annual conference, Kasetsart University

Khieokhachee T, Praphailong W, Chowvalimititum C, Kunawasen S, Kumphati S, Chavasith V, Bhumiratana S, Valyasevi R (1997) Microbial interaction in the fermentation of Thai pork sausage. In: Proc. 6th ASEAN Food Conference, Singapore, 24–27 Nov, pp 312–318

Kraidej L, Wongkhaluang C, Muangnoi M, Watana P, Patarakulpong P, Chimanag P, Phoopat S, Yongmanitchai W (1977) "Kanom-jeen" symposium on indigenous fermented foods, GIAM V, Bangkok, 21–26 Nov

Kruawan K, Kangsadalampai K, Yongsmith B, Sriyapai T (2005) Effects of Monascus colorants on the mutagenecity of nitrite treated 1-aminopyrene using Ames test. Thai J Pharm Sci 29(1–2):29–41

Liptasiri S (1975) Studies on some properties of certain bacteria isolated from Thai fish sauce. M.S. Thesis, Kasetsart University, Bangkok

Mongkolwai T, Assavanig A, Amnajsongsiri C, Flegel T, Bhumiratana A (1998) Technology transfer for small and medium soy sauce fermentation factories in Thailand a consortium approach. Food Res Int 30:555–563

Nutrition Division, Department of Health (1978) Nutrient composition tables of Thai foods. Ministry of Public Health, Bangkok, Thailand. 48 pp. (In Thai)

Nutrition Division, Department of Health (1984) Nutrient composition tables of Thai foods. Ministry of Public Health, Bangkok. 48 pp. (In Thai)

Phithakpol B, Varanyanond W, Reungmaneepaioon S, Wood H (1995) The traditional fermented foods of Thailand. ASEAN Food Handling Bureau, Kuala Lumpur, pp 6–9

Rattagool P, Merhahip P, Wongchinda N (1983) Fermented fishery products in Thailand. In: Joint FAO/FTDD/IFRPD meeting on fermented fishery products, Bangkok

Saisithi P (1984) Fisheries industry process. Institute of Food Research and Products Development, Kasetsart University, Bangkok

NRCT (1982) Thai traditional fermented food research project, phase I. National Research Council of Thailand, Bangkok

Valyasevi R, Rolle RS (2002) An overview of small-scale, food fermentation technologies in developing countries with special reference to Thailand: scope for their improvement. Int J Food Microbiol 75:231–239

Yongsmith B, Kitprechavanich V, Chitradon L, Chaisrisook C, Budda N (2000) Color mutants of Monascus sp. KB9 and their comparative glucoamylases on rice solid culture. J Mol Catal B Enzym 10:263–272

Yongsmith B, Thongpradis P, Klinsupa W, Chantrapornchai W, Haruthaithanasan V (2013) Fermentation and quality of yellow pigments from golden brown rice solid culture by a selected Monascus mutant. Appl Microbiol Biotechnol 97(20):8895–8920

Chapter 4
Traditional Food in Romania Integrated in a Protected Geographical Designations System

Gabriela Nedita, Nastasia Belc, Lucia Romanescu, and Roxana Cristina Gradinariu

4.1 Introduction

This paper presents the situation of traditional food in Romania and the infrastructure set up to protect product designations.

Named traditional products are part of a national inventory made by the Ministry of Agriculture and Rural Development (MADR), through its territorial infrastructure. Each product is assessed based on the declared specifications and is granted with an attestation. To qualify for protection, such as protected designation of origin (PDO), protected geographical indication (PGI) and traditional speciality guaranteed (TSG), the food product should be registered at community level, as a result of the application made by a group. Today, the problem of 'group' which means any association, irrespective of its legal form or composition, working with the same agricultural product or foodstuff, is the most sensitive aspect. Possibly due to recent historical reasons, small farmers and producers may be reluctant to engage in

G. Nedita • N. Belc (✉)
Institute of Food Bioresources, 6 Dinu Vintila Street, 2nd District, Bucharest 021102, Romania
e-mail: neditag@yahoo.com; nastasia.belc@bioresurse.ro

L. Romanescu
National Office of the Romanian Traditional and Ecological Products,
22 Michael Weiss Street, Brasov, Romania
e-mail: lucia.romanescu@gmail.com

R.C. Gradinariu
Ministry of Agriculture, Forests and Rural Development,
24 Carol I Bvd., 3rd District, OP 37, Bucharest 020921, Romania
e-mail: roxana.gradinariu@madr.ro

© Springer Science+Business Media New York 2016
K. Kristbergsson, J. Oliveira (eds.), *Traditional Foods*, Integrating Food
Science and Engineering Knowledge Into the Food Chain 10,
DOI 10.1007/978-1-4899-7648-2_4

associative entities. Notwithstanding, the strong dynamics in this domain are an indicator that opportunities and advantages for local communities to keep their traditions in a sustainable way are being used.

4.2 Key Aspects of Systems of Traditional Products and Geographical Designations

4.2.1 Consistence of GIs Approach with Other Policies

Geographical indications (GIs) of food products, defined by the Word Trade Organization (WTO) on the basis of quality, reputation or other characteristics due, essentially, to their geographical origin (RO 2006/IB/AG-04 TL), is at the same time an intellectual property right, with full autonomy, as with patents and trademarks, mentioned in the WTO agreements.

As GIs generate economic interest, due to expected potential benefits for producers, they are increasingly used in many countries around the world, such as China (Long Jin tea, Shaoxing yellow rice wine, Xuan Weï ham, Maotaï), Vietnam (Nuoc Mam fish sauce from Phu Quoc), and India (Thaï jasmine Hom Mali Rice, Darjeeling tea), as well as for specific products, such as coffee (Indonesia, Kenya, Central America, etc.) (RO 2006/IB/AG-04 TL). It is important to note that GIs should not be confused with a mere indication of the source (which provides information only of the location, without guarantee of a particular quality associated with place of origin or production models), but GI is a collective right if the use is reserved exclusively for products meeting the specifications (book of requirements) defined and approved by the competent authorities (INAO 2005).

In Europe today, the GI concept refers to processed and unprocessed agricultural products and has led to the development of two categories: protected designation of origin (PDO) and protected geographical indication (PGI). There is also the concept of traditional speciality guaranteed (TSG).

It is extremely important to bear in mind the distinction between the traditional products recognition system and the designations protection system. The system of the traditional products attestation provides identification at national level and supports farmers to join and apply to register the same product.

4.2.2 Difference between National System of Traditional Products and Geographical Designations

The concept of 'traditional' has a general positive image and is linked with habitual consumption; special occasions; specific origin; made in a specific way; transmitted from one generation to another; known for sensory properties; related with health; expensive, good quality and availability; providing diversity (Verbeke et al. 2008).

In March 2006 the new EC regulations (Regulation no. 510/2006/EC and Regulation no. 509/2006/EC) in the field of geographical designations were adopted. The EU Protected Food Name Scheme identifies regional and traditional foods whose authenticity and origin can be guaranteed. Under this system, a named food or drink (separate arrangements exist for wines and spirits) registered at a European level will be given legal protection against imitation throughout the EU.

Products with protected name status fall into three categories:

Protected Designation of Origin: Effective from 1 May 2010

Open to products produced, processed and prepared within a specific geographical area and with features and characteristics attributable to that area.

Protected Geographical Indication

Open to products produced or processed or prepared within a specific geographical area and with features or qualities attributable to that area.

Traditional Speciality Guaranteed

Open to products that are traditional or have customary names and have features that distinguish them from other similar products. These features need not be attributable to the geographical area the product is produced in nor entirely based on technical advances in the method of production

Consequently Romania has implemented Government Decision H.G. nr. 828/2007, establishing the National System for protection of the names of origin and geographical indications of food and agriculture products (published in MO no. 556/2007).

In addition to European legislation, since January 2005 Ordin no. 690/2005/MADR has been adopted as a national regulation with the aim to support the identification of traditional products in Romania.

The named traditional products represent a national inventory made by the Ministry of Agriculture and Rural Development (MADR), through its territorial infrastructure. First of all the products have to pass two stages of approval to obtain national recognition as being a traditional product:

1. Approval from the MADR, based on a file evaluation with elements describing, among other requirements, the traditional character of the methods of production, ingredients (raw materials or/and primary products) and history of the products.
2. The above is without prejudice to all applicable EU food law requirements, including the general objectives laid down in Article 5(1) of Regulation (EC) no. 178/2002. Notwithstanding, some derogations from Regulation 852/2004/EC have been approved by the National Authority for Food Safety and Veterinary Medicine (ANSVSA) for companies manufacturing traditional food.

Fig. 4.1 National logo for traditional products

Table 4.1 Situation of traditional product registration in different countries of the EU

	RO	IT	AT	HU	PL	ES	UK	DE	FR
Year of legislative framework set-up	2004	2000	–	1997	–	–	–	–	–
Year beginning attestation	2005	2000	2005	1998	–	–	–	–	–
Total number of attested products	*2185*	*4366*	*72*	*376*	*550*	*526*	*429*	*297*	*890*

Source: http://www.onpter.wfg.ro, accessed on 2.11.2009

Each product is assessed based on the declared specifications and is granted with an attestation. The recognised traditional products can use the national logo (Fig. 4.1) on their labels.

By setting up a national system of protection, it is necessary to have a central office within the Ministry of Agriculture and Rural Development for the management of the national protection of designations of origin and indications for agricultural products. In 2008, through the Ordin no. 42/2008/MADR, the National Office of Ecological and Traditional Products from Romania (ONPTER) was founded with the remit of strictly monitoring the registration, licensing and promotion of traditional and ecological Romanian products that will be traded domestically while ensuring a high quality of products, by origin, local composition and production method.

1. Substantiv

   (a) Advance
   (b) Pass
   (c) Preferment
   (d) Remove

National protection is not mandatory but is widespread in the EU to support consumer credibility. Countries such as Italy, Spain, Portugal, France, Germany and Denmark give particular attention to this area because of the high relevance of the food industry to the economy. Table 4.1 provides details of active national protection systems in Europe.

4.3 Traditional Foods in Romania

4.3.1 Dynamics of the Traditional Food Sector

Since its inception in 2008, ONPTER has recognised 2738 products as traditional foods, belonging to the nine categories (Nedita et al. 2009) shown in Fig. 4.2. The predominant ones are dairy products (40 %), meat products (27 %) and bakery products (20 %), followed by drinks (9 %, excepting wines) and fruits and vegetables (2 %). Less represented are fish products (1 %) and miscellaneous. The number

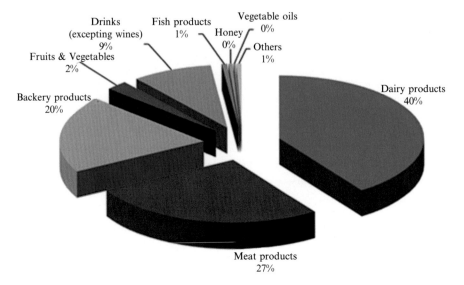

Fig. 4.2 Distribution of registered Romanian traditional products by category

of registration of honey and the number of registration of vegetable oils are both below 1 %.

A suggestive picture of the distribution of the different types of traditional food products at national level shows a predominant number in the south-central area, followed by the north region. Lowest interest is manifested in the Galati and Timis areas, as can be seen in the Fig. 4.3.

4.3.2 Examples of Romanian Traditional Products

Examples of registered traditional products are given in Table 4.2. Some are further elaborated next.

4.3.2.1 Traditional Dairy Products

1. Fermented cheese from sheep milk, kept in a crust tree—fir tree [Fig. 4.4a, b]. This product is common in mountain areas and it is sold in local markets. It is prepared especially in micro units or in sheep farms (Nedita et al. 2008).
2. Fermented cheese from cow milk, sheep milk, goat milk or a mixture, preserved in natural membranes (animal stomach) [Fig. 4.5]. It is made in small- or

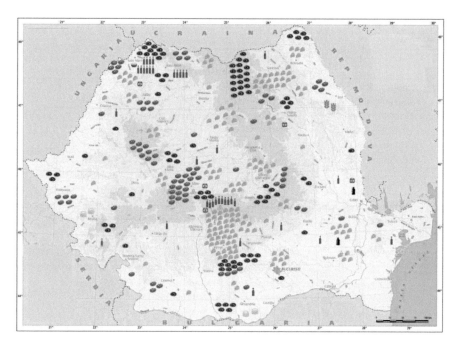

LEGEND (RO / EN)

PANIFICAŢIE/BAKERY
MEAT
DAIRY
SPIRITS
MHONEY
CANNED FRUITS-VEGETABLES
CONFECTIONERY
SOFT DRINKS
MINIMALLY PROCESSED PRODUCTS
MILLING PRODUCTS
BAKERY

Fig. 4.3 Map of traditional food products in Romania (*Source*: *www.onpter.ro*, *accessed on November 2010*)

Table 4.2 Examples of Romanian traditional products

No.	Categories	Products
1.	Dairy products	Cas, cascaval, braza telemea, branza de burduf, branza topita, kefir, urda, unt
2.	Meat products	Sangerete, caltabosi, slanina, Vasli, pasta de mici, pastrama, sloi, sunca, muschi, toba, birsoaita, drob jumari, soric, piftie, pate, coaste afumate, pasatic, pecie, rulada, moietic, untura, chisca, cotlet, parizer, kaiser, rasoale, costita, piftie, babic

(continued)

Table 4.2 (continued)

No.	Categories	Products
3.	Bakery products	Paine cu cartofi, cozonac, placinte, colaci, lipie, covrigi, pita de malai oseneasca, pasca, turta, mamaliga, hencles, placinta cu varza
4.	Fruits and vegetables	Magiun, fructe deshidratate, dulceata, zacusca, sirop, compot, gem, bullion, muraturi
5.	Drinks (except wines)	Tuica, palinca, horinca, braga, bere, afinata
6.	Fish products	Salata icre, pastrav afumat
7.	Miscellaneous	Bors, otet, sarmale

Fig. 4.4 (**a** and **b**) Fermented cheese in a crust tree (*Romanian name*: *Brânză de burduf in coaja de brad*)

Fig. 4.5 Fermented cheese in a natural membrane (*Romanian name*: *Brânză de burduf in membrana naturala*)

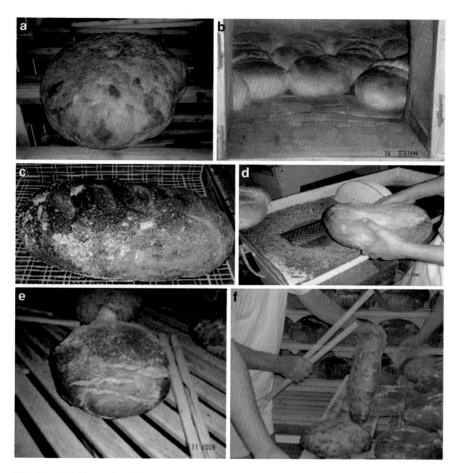

Fig. 4.6 (a–f) Beaten bread crust (*Romanian name*: *Pâine bătută de coajă*)

medium-sized food enterprises, and it is sold in local markets, in big shops and even in supermarkets (Nedita et al. 2008).

4.3.2.2   Traditional Bakery Products

The most traditional bakery product found to match all criteria of a traditional product is the *beaten bread crust* [Fig. 4.6a]. This can be in different forms: with or without potatoes. It is specific of the west part of Romania, being produced in all small bakeries in Transylvania. This kind of bread is usually made in large sizes (1.5, 2 or 3 kg) [Fig. 4.6a, b, c]. The specific operation is the removal of the crust after approximately 10 min when the bread is taken off from the oven [Fig. 4.6b)]. The operation is made in different ways: with wood sticks [Fig. 4.6e, f)] or with an electric machine [Fig. 4.6d)]. The shelf life of this bread is longer than common bread (5 days instead of 1–2 days) and has a special flavour, being preferred to industrial products by consumers (Nedita et al. 2008).

4.4 Conclusions

It is important to have an operational control system enhanced so that any operator who meets the EC requirements can take advantage of such a system of quality valorisation. The functionality of the control system and proper assessment of compliance with the GIs is expressed in the consumer's trust for origin; manufacturing method; specific product characteristics as are approved and guaranteed by the authorities; identification, presented as a clear and informative label, approved by the public authorities; and traceability, guaranteed by third-party control.

In Romania, the system has a history of not more than 6 years, and although still requiring some completion of all mechanisms and procedures envisaged, it has already been used extensively by small- and medium-sized producers to obtain a competitive advantage to their products by complying with the requirements for quality, tradition and origin.

References

INAO (2005) Applicants' Guide. INAO, France

Nedita G et al (2008) Tradiţie şi siguranţă alimentară. DO & DO Prepress Printing, Bucuresti, 150 p. ISBN 978-973-88-314-1-4

Nedita G, et al. (2009) Contribution of traditional food to local and regional sustainability. In: Book of abstracts EFFoST conference new challenges in food preservation. Processing, safety, sustainability, section sustainability—sustainable products [P390], 11–13 Nov, Budapest, p 336

Regulation (EC) no. 509/2006 on agricultural products and foodstuffs as traditional specialties guaranteed

Regulation (EC) no. 510/2006 on the protection of geographical indications and designations of origin for agricultural products and foodstuffs

RO 2006/IB/AG-04 TL, Proiect twinning RO-FR (2009) Dezvoltarea sistemului de calitate pentru protecţia produselor agricole şi alimentare (DOP-denumire de origine protejată, IGP—indicaţie geografică protejată, STG—specialitate tradiţionala garantată). MADR, februarie-iunie

Verbeke et al. (2008) Defining traditional foods and innovation in traditional foods. Presented in third European conference on sensory and consumer research

Chapter 5
Traditional Foods in Slovakia

Jan Brindza, Dezider Toth, Radovan Ostrovsky, and Lucia Kucelova

5.1 Popular Meals Based on Natural Plants and Fungal Species

Slovak ancestors utilized products of over 300 different species growing freely—among them were approximately 68 herbs, 80 species of fungi, 20 forest fruits, 95 medical plants applied in popular medicine and other 35 species as products for preparation of miscellaneous foodstuffs in fresh, dried or other processed form.

5.1.1 Herbs

In several regions many parts of wildly growing plants were used to produce traditional foods as well. Some of these types were collected only in emergency cases—like famine or crop failure—and therefore such species were marked as poverty plants. In the spring months, the nettle (*Urtica dioica*) leaves were collected to be used for soups, salads, sauces and teas. Similarly wood sorrel (*Rumex acetosa*) and orache (*Atriplex hortensis*) were applied. These species were also suitable for several other meals. It is important to mention that they had anti-scorbut effects. Leaves of black mustard (*Brassica nigra*); horseradish (*Armoracia rusticana*); burdock (*Arctium lappa*); dandelion (*Taraxacum officinale*); flowers of coltsfoot (*Tussilago farfara*); flowers, leaves and root of succory (*Cichorium intybus*); leaves of bittercress (*Cardamine amara*); leaves of lesser celandine (*Ranunculus ficaria*); leaves of valerian (*Valerianella olitoria*); common mallow (*Malva neglecta*); leaves of thistle

J. Brindza (✉) • D. Toth • R. Ostrovsky • L. Kucelova
Slovak University of Agriculture, Tr. A. Hlinku 2, Nitra 949 01, Slovak Republic
e-mail: jan.brindza@uniag.sk; brindza.jan@gmail.com

© Springer Science+Business Media New York 2016
K. Kristbergsson, J. Oliveira (eds.), *Traditional Foods*, Integrating Food
Science and Engineering Knowledge Into the Food Chain 10,
DOI 10.1007/978-1-4899-7648-2_5

(*Sonchus arvensis*); leaves of purple clover (*Trifolium pratense*); Jerusalem sage (*Pulmonaria officinalis*); wool mullein (*Verbascum thapsiforme*); hazel (*Corylus avellana*); and many other so-called poverty plants were used by people to make some archaic meals. Even today some of these resources are used as additives or in cooking special game dishes.

5.1.2 Forest Fruits

Forest plants represented an important feeding source. There were collected fruits and leaves of dog rose (*Rosa canina* L.), whortleberry (*Vaccinium myrtillus* L.), cranberry (*Vaccinium vitis-idaea* L.), alpine strawberry species (*Fragaria vesca, F. moschata, F. viridis*), raspberry (*Rubus idaeus* L.), black elder (*Sambucus nigra*), Cornelian cherry (*Cornus mas*), sea buckthorn (*Hippophae rhamnoides*), wildly growing pears (*Pyrus* spp.), apples (*Malus* spp.), morello (*Prunus cerasifera*), cramp bark (*Viburnum opulus*), hazel (*Corylus avellana*), chestnut (*Castanea sativa*), juniper (*Juniperus communis*), blackthorn (*Prunus spinosa*), barberry (*Berberis vulgaris*), hawthorn (*Crataegus monogyna*) and common beech (*Fagus sylvatica*). The total value of forest fruits collected in Slovakia represented in 2006 55.19 millions € and 2 years later 70.90 millions €, but in the past these amounts were significantly higher.

5.1.3 Spices and Seasonings

The majority of the ancient population in Slovakia believed that some spices and seasoning additives can give the consumer magic abilities. The most often applied spices of the Slovak kitchen included:

(a) Ground ivy (*Glechoma hederacea*)—added to soup, browned flour, goulash, stewed dishes or simply for decoration of meals
(b) Sweet marjoram (*Majorana hortensis*)—goulash, some soups, to fried, stewed and roasted meats and to sausages or liver sausages
(c) Caraway (*Fructus carvi*)—soups, stewed or roasted meat, sausages and bread and used as flavouring in homemade alcoholic beverages
(d) Wild thyme (*Thymus serpyllum*)—meat dishes, sauces or added as subsidiary agent supporting good sleeping
(e) Common dill (*Anethum graveolens*)—sheep cheese and sauce flavouring and/ or as disinfectant of containers for cabbage fermentation
(f) Garden ginger (*Zingiber*)—meat dishes, cakes and flavouring of homemade spirits
(g) Juniper (*Juniperus communis*)—meat (especially ram and game) marinade component and spirit additive

(h) Roots of yellow gentian (*Gentiana lutea*)—additive to alcoholic apéritifs in mountain resorts
(i) Yellow seed (*Lepidium campestre*)—sauces, spreads and alcoholic beverage flavouring
(j) Seeds of opium poppy (*Papaver somniferum*)—as a long-year favourite of Slovak kitchen, it is used in stuffings, powderings, cake ingredients, additives to ripening doughs, noodle seasoning and many other dishes
(k) Fir needles—in several mountain villages, it was added to kettle goulash and game dishes
(l) Stinging nettle (*Urtica dioica*)—sauces, spreads and syrups
(m) Fresh or dried fruit of cranberry (*Vaccinium vitis-idaea*)—sour sauces, stewed meat, game and side dishes
(n) Garlic (*Allium sativum*)—soups, stewed and fried meat, important additive to products prepared for hog feast and many other meals
(o) Onion (*Allium cepa*)—majority of soups, sauces, goulash, smoked meats, spreads and several other purposes
(p) Chive (*Allium schoenoprasum*)—soups, fillings, spreads flavouring, side dish and/or consumable decoration
(q) Sweet basil (*Ocimum basilicum*)—spreads and ram meat heat-treating process
(r) Pot marjoram (*Origanum vulgare*)—stewed fatty meat and spaghetti
(s) Wormwood (*Artemisia abrotanum*)—fatty meal preparation and/or healing teas
(t) Summer savoury (*Satureja hortensis*)—substitute for black pepper

5.1.4 Fungi

Fungi represented one of the most important components of available natural products suitable to nourish the population. To keep them for later consumption, fungi were preserved in dried or fermented forms. It is valid primarily for the following species: flat-bulb mushroom (*Agaricus abruptibulbus*), field mushroom (*Agaricus campestris*), blushing wood mushroom (*Agaricus silvaticus*), tawny grisette (*Amanita fulva*), blusher (*Amanita rubescens*), grey spotted amanita (*Amanita spissa*), honey mushroom (*Armillariella mellea*), penny-bun boletus (*Boletus edulis*), scarletina bolete (*Boletus erythropus*), St. George's mushroom (*Calocybe gambosa*), yellow chanterelle (*Cantharellus cibarius*), trooping funnel (*Clitocybe geotropa*), shaggy mane (*Coprinus comatus*), velvet shank (*Flammulina velutipes*), slimy spike (*Gomphidius glutinosus*), Judas' ear (*Hirneola auricula-judae*), hedgehog (*Hydnum repandum*), conifer tuft (*Hypholoma capnoides*), copper spike (*Chroogomphus rutilus*), sheathed woodtuft (*Kuehneromyces mutabilis*), amethyst deceiver (*Laccaria amethystina*), deceiver (*Laccaria laccata*), false saffron milkcap (*Lactarius deterrimus*), orange-red lactarius (*Lactarius volemus*), orange birch bolete (*Leccinum aurantiacum*), birch bolete (*Leccinum scabrum*), wood blewit

(*Lepista nuda*), common puffball (*Lycoperdon perlatum*), parasol (*Macrolepiota procera*), shaggy parasol (*Macrolepiota rhacodes*), fairy ring mushroom (*Marasmius oreades*), garlic mushroom (*Marasmius scorodonius*), black morel (*Morchella conica*), classic yellow morels (*Morchella esculenta*), golden bootleg (*Phaeolepiota aurea*), oyster mushroom (*Pleurotus ostreatus*), deer mushroom (Pluteus cervinus), dryad's saddle (*Polyporus squamosus*), gypsy (*Rozites caperata*), charcoal burner (*Russula cyanoxantha*), bare-toothed russula (*Russula vesca*), crab brittlegill (*Russula xerampelina*), cauliflower mushroom (*Sparassis crispa*), bovine bolete (*Suillus bovinus*), weeping bolete (*Suillus granulatus*), larch bolete (*Suillus grevillei*), slippery Jack (*Suillus luteus*), sand boletus (*Suillus variegatus*), bay bolete (*Xerocomus badius*), red-cracked boletus (*Xerocomus chrysenteron*) and suede bolete (*Xerocomus subtomentosus*).

In Slovakia fungi collection, activities reached such a level that it could be marketed as fungi tourism. In the top season, maybe half of the Slovakia population moves to the forests and meadows eagerly gathering edible fungi. This trend is illustratively documented by the total financial value of forest fungi collected—in 2006 it represented 83.75 million € and in 2008 52.08 million € (Tutka et al. 2009).

5.1.5 Medicinal Plants

Traditional food and medicine were always closely interconnected. Many individuals possessing traditional knowledge of plants with healing effects would exist in ancient populations. These plants were continually collected and applied in popular medicine enabling the healing of several diseases. Housewives in many households stored dried parts (roots, leaves, flowers, fruits, seeds, bark) or whole plants and prepared mainly some healing or wound-cleansing teas, extracts in alcohol and concoctions for bathing.

Among the collected plants, the following species should be mentioned—balm (*Mellissa L.*), thyme (*Thymus vulgaris*), common doder (*Sambucus nigra* L.), camomile (*Matricaria recutita*), pimpinella (*Pimpinella* L.), mullein (*Verbascum* L.), hawthorn (*Crataegus* L.), gentian (*Gentiana* L.), comfrey (*Symphytum* L.), strawberry (*Fragaria* L.), common lime flower (*Tilia petiolaris*), perforated Saint John's wort (*Hypericum perforatum* L.), red raspberry (*Rubus idaeus* L.), crimson bramble (*Rubus arcticus*), wild chicory (*Cichorium intybus*), woundwort (*Stachys* L.), common lungwort (*Pulmonaria officinalis* L.), coltsfoot (*Tussilago farfara* L.), dandelion (*Taraxacum officinale*), primrose (*Primula* L.), garden sage (*Salvia officinalis* L.), goldenrod (*Solidago*), horsetail (*Equisetum* L.), narrow-leaved plantain (*Plantago lanceolata* L.), deadnettle (*Lamium* L.) and agrimony (*Agrimonia* L.). Interestingly, even now many people are engaged in healing plants collection as documented for the year of 2008, the total financial value of medicinal plants collected individually for own needs represented 8.82 million € (Tutka et al. 2009).

5.2 Traditional Food Resources Based on Cultivated Species and Animal Husbandry

In the traditional agrosystems, Slovak inhabitants exploited around 200 cultivated plant species, classified as cereals (10), legumes (7), oil-bearing plant (7), hoed crops (9), vegetables (52), fruit plants (25) and 65 other crops.

Among the most used cereals were wheat (*Triticum aestivum* L.), einkorn wheat (*T. monococcum* L.), hard wheat (*T. durum* L.), rye (*Secale cereale* L.), barley (*Hordeum vulgare* L.), oat (*Avena sativa* L.), corn (*Zea mays* L.), sorghum (*Sorghum bicolor*), buckwheat (*Fagopyrum esculentum* L.) and millet (*Panicum miliaceum* L.).

Fava bean (*Faba vulgaris*), chickpea (*Cicer arietinum*), bean (*Phaseolus vulgaris*), grey pea (*Pisum sativum* L. subsp. *sativum* convar. *speciosum*), field pea (*Pisum sativum*), tuberous pea (*Lathyrus sativus*) and lentil (*Lens esculenta*) were the mostly utilized leguminous plants by the former inhabitants of Slovakia.

The leading oil-bearing plants were formed by white mustard (*Sinapis alba*), mustard greens (*Sinapis juncea* L.), fibre flax (*Linum usitatissimum*), opium poppy (*Papaver somniferum*), rape (*Brassica napus* L. spp. *oleifera*), sunflower (*Helianthus annuus*) and soya (*Glycine max*).

Widely cultivated hoed-crop species are wild chicory (*Cichorium intybus*), turnip cabbage (*Brassica oleracea* convar. *gongylodes*), borecole (*Brassica oleracea* var. *acephala*), carrot (*Daucus carota*), beetroot (*Beta vulgaris* convar. *crassa*), topinambur (*Helianthus tuberosus*), sugar beet (*Beta vulgaris* spp. *vulgaris* var. *altissima*), rutabaga (*Brassica napus* convar. *napobrassica*) and potato (*Solanum tuberosum*).

Vegetable species were generally utilized, but prevailing are the eggplant (*Solanum melongena*), broccoli (*Brassica oleracea* convar. *Botrytis italica*), fava bean (*Vicia faba* subsp. *major*), garlic (*Allium sativum*), onion (*Allium cepa*), scalvion (*Allium ascalonicum Strand*), beetroot (*Beta vulgaris* provar. *conditiva*), wild chicory (*Cichorium intybus*), endive (*Cichorium endivia*), watermelon (*Citrullus lanatus*), fennel (*Foeniculum vulgare*), bean (*Phaseolus vulgaris*), scorzonera (*Scorzonera hispanica*), pea (*Pisum sativum*), horse radish (*Armoracia rusticana*), kohlrabi/turnip cabbage (*Brassica oleracea* convar. *gongylodes*), cabbage (*Brassica oleracea* var. *capitata*), Peking cabbage (*Brassica pekinensis*), cauliflower (*Brassica oleracea* var. *botrytis*), broccoli (*Brassica oleracea* L. var. *italica*), Savoy cabbage (*Brassica oleracea* var. *sabauda*), common dill (*Anethum graveolens*), coriander (*Coriandrum sativum*), corn (*Zea mays* convar. *saccharata*), sweet marjoram (*Majorana hortensis*), mangold (*Beta vulgaris* var. *vulgaris*), melon (*Melo sativus*), carrot (*Daucus carota*), turnip (*Brassica rapa* var. *esculenta*), red pepper (*Capsicum annuum*), parsnip (*Pastinaca sativa* subsp. *sativa*), custard squash (*Cucurbita pepo* var. *patisson Duch.*), chive (*Allium schoenoprasum* subsp. *schoenoprasum*), parsley (*Petroselinum crispum*), leek (*Allium porrum*), tomato (*Lycopersicon esculentum*), caraway (*Carum carvi*), rhubarb (*Rheum rhabarbarum* L.), garden rhubarb (*Rheum undulatum*), radish (*Raphanus sativus* var. *major*), oriental radish (*Raphanus sativus* subsp. *niger*), head lettuce (*Lactusa sativa* var. *capitata*), asparagus (*Asparagus officinalis*), spinach (*Spinacia oleracea*), gourd (*Cucurbita*), Malabar gourd (*Cucurbita ficifolia*), cucumber (*Cucumis sativus*), corn salad (*Valerianella locusta*), wild celery (*Apium graveolens*) and pepperwort (*Lepidium sativum*).

Many cultural fruit species and berry plants were traditionally cultivated on Slovakia—among them should be listed, primarily the peach (*Prunus persica*), peach-almond crossbreed (*Prunus amygdalus* × *Prunus persica*), cherry (*Cerasus avium*), sweet cherry (*Prunus avium*), blackberry (*Rubus fruticosus*), quince (*Cydonia oblonga*), gooseberry (*Ribes uva-crispa*), chestnut tree (*Castanea sativa*), pear tree (*Pyrus communis*), apple tree (*Malus domestica*), strawberry (*Fragaria* L.), rowan tree (*Sorbus aucuparia* L.), common mountain ash × Siberian hawthorn (*Sorbus aucuparia* × *Crataegus sanguinea*), red raspberry (*Rubus idaeus*), almond (*Amygdalus communis*), apricot (*Prunus armeniaca*), medlar (*Mespilus germanica*), walnut tree (*Juglans regia*), sea buckthorn (*Hippophae rhamnoides*), currant (*Ribes* L.), apple rose (*Rosa villosa*), cherry plum (*Prunus cerasifera*), garden plum (*Prunus domestica*), vine grape (*Vitis vinifera*) and sour cherry (*Prunus cerasus*).

Different types of tea, extracts and curative remedies resulted from the following species: angelica (*Archangelica officinalis*), sweet basil (*Ocimum basilicum*), blessed thistle (*Cnicus benedictus*), wool mullein (*Verbascum densiflorum*), thorn apple (*Datura stramonium*), common thyme (*Thymus vulgaris*), yellow gentian (*Gentiana lutea*), Indian hemp (*Hibiscus cannabinus*), hollyhock (*Althaea rosea*), marshmallow (*Althaea officinalis*), goatroot ononis (*Ononis arvensis*), horehound (*Marrubium vulgare*), goat's rue (*Galega officinalis*), downy hemp nettle (*Galeopsis segetum*), lime flower (*Tilia cordata, Tiliae flos*), English lavender (*Lavandula angustifolia*), lovage (*Levisticum officinale*), great burdock (*Arctium lappa*), St. John's wort (*Hypericum perforatum*), belladonna (*Atropa belladonna*), peppermint (*Mentha* × *piperita*), bee balm (*Melissa officinalis*), common bearberry (*Arctostaphylos uva-ursi*), wormwood (*Chenopodium ambrosioides*), woolly foxglove (*Digitalis lanata*), pot marigold (*Calendula officinalis*), horse-heal (*Inula helenium*), dragon sagewort (*Artemisia dracunculus*), Roman wormwood (*Artemisia pontica*), chamomile (*Matricaria chamomilla*), milk thistle (*Silybum marianum*), common agrimony (*Agrimonia eupatoria*), roman chamomile (*Anthemis nobilis* L.), Chamomile (*Matricaria recutita*), apple rose (*Rosa villosa*), summer savoury (*Satureja hortensis*), fenugreek (*Trigonella foenum-graecum*), plantain (*Plantago lanceolata*), common licorice (*Glycyrrhiza glabra*), mallow flower (*Malva mauritiana*), garden sage (*Salvia officinalis*), valerian (*Valeriana officinalis*) and hyssop (*Hyssopus officinalis*).

Animal products were gained from the:

- Husbandry of cows, pigs, sheep and goats; in the group of domestic fowl, geese, ducks, hens, turkey hens, pigeons and rabbits were included
- Hunting of game—roebucks, deers, fallow deers, wild boars, hares, pheasants, quails and wild ducks

5.3 Traditional Foods in Popular Nutrition

To describe, characterize and illustrate the traditional Slovak kitchen, practice would require an extensive publication. It could be said that any region, even any village, had some particular uniqueness, own modification and specific approaches, methods of preparation, use of natural resources or cultivated plants with different independent characters, and there are so many specificities that is quite impossible to generalize. In many kitchens, the housewives applied an original way of how to do it or technologies allowing to reach aesthetical and gustatory qualities which are simply unique. Therefore, only some generalized knowledge on traditional kitchen products is presented here.

5.3.1 Soups

Historically, soups are ranked among the oldest foods; they were served for daily, holiday or festive occasions. Soups were generally subdivided into thin (caraway or noodle soups) and thick (densified). In poor families dense soups prevailed, usually covering the only meal for breakfast, lunch and often for supper as well. Around 350 miscellaneous soups are recorded. The most known and widespread type is the caraway soup, closely followed by potato and leguminous types (green bean, lentil, field bean, chick pea, peavine) and then cereals (wheat, millet, maize, barley grits—*lohaza*), vegetable (made of one vegetable species or a combination of several ones), fungal, fruit and meat soups (giblets of geese, ducks, turkeys, hens, quails, cows, pigs, rams, goats and other kinds of meat).

The following is an example of a method to prepare a special simple soup called *demikat*:

(a) Slice an old bread and several onions on a plate; they should be piled in several layers.
(b) Add on the top of every layer crumbled sheep cheese, some pig lard and red pepper.
(c) The layers should be flushed by boiling water and then let them a few minutes to scald. Poor families used water from boiled potatoes or noodles.

Another traditional soup often used in the past is *kisel*, which consisted of a coarse-grained cereal mixture with acid milk and boiled with addition of flour.

5.3.2   Mash

The second oldest group of meals are represented by different mashes. Originally they were made of whole, crushed or milled cereal grains, which were heated in water or milk, which by gradual boiling came to thickening. Mashes were made of barley, wheat, rye, maize, oat and/or buckwheat flours. Sometimes other additives were combined with the flour, such as fungi, vegetable, fruit or even meat. The dish was prepared at home using stony mills and then stored in a wooden case, where it was kept covered by bags filled with grains.

The maize mash was served in most families during the Christmas feast. The oldest type of mash was prepared from the millet grains or grits. This food had different names—mostly *psheno* or *gold mash*—but in the central Slovakia region *kulasha* and in western region *jahli*. Similarly different names were given even in neighbouring villages to mash made of barley groats—*krupi, gershli, lohadza* and/ or *pencaki*. Many mashes made of different flours, milled maize or mixture of several cereal species are also found. Favourite mashes include those made of legumes (lentil, green bean, field bean, chick pea, peavine, garden pea) combined with further vegetable species and cereals (Babilon 1870, Stolicna 1997, Vansova 1998, Stolicna 2000).

5.3.3   Pasta Dishes

Pasta dishes prepared from flour made of different cereals traditionally formed an important component of nutrition. Pasta was used to thicken some soups or as side dish combined with meat and often as the main dish. The oldest type of pasta is called *trhance*—it could be prepared from the dough by hand on small pieces or throwing small amounts of thin dough directly in the boiling water (using common or wooden spoon, eventually a knife).

For another traditional pasta called *mrvenica*—crumbled dough—the dish was prepared by crumbling the dough with the fingers getting small particles of a maize grain size. *Mrvenica* was used as addition into soups and sometimes as a side dish. In different regions it was known under other names—*mrvence, dropki* or *melence.*

One of the most archaic pasta is known again under several names—*shulance, nudle, bobale* and *kokoshki.* The first name—*shulance* (rolled rods)—is connected with the method of preparation: dough is divided into several parts and then rolled on a wooden slab forming prolonged rolls, which are subsequently cut by knife into shorter pieces. Boiled potatoes were added to the dough to improve the product taste.

A very popular Slovak dish is the so-called *halushky*—small dumplings made of potato dough and boiled in hot water. Originally it was prepared by using a knife to separate and throw small pieces of thin dough from a wooden slab into boiling water. Presently a metal sieve through which is the dough pokes into boiling water is used. Dough for *halushky* was made of rye, millet and/or buckwheat flour. Similarly as with "shulance," by making "halushky" boiled potatoes were worked into the dough as well. Nowadays this approach fully prevailed in both the home-made dish in the Slovak families and in the restaurants too. *Halushky* were widespread in all regions and gained an important place in popular nutrition. People consumed this dish as the only one for lunch, sometimes together with a meat piece, but sometimes as addition to soups as well. When consumed as separate dish, it was flavoured in many ways—as sweet or salted meal. There are *halushky* with sheep cheese *bryndzove halushky*; with sour or sweet cabbage; with curds, smoked sheep cheese or soft cheese; or with scrambled eggs. The *bryndzove halushky* is the most famous specific Slovak dish. They are prepared as described above—using the combination of flour and potatoes. The genuine type of *bryndzove halushky* should be sprinkled with sheep cheese, splashed by pig fat or on surface filled with cut to small pieces and fried bacon. This process seems to be quite simple, but only some excellent cook/chefs are able to prepare such *bryndzove halushky* that everybody is satisfied. In many restaurants something is served which somehow reminds *bryndzove halushky*, but when you start to consume it, you feel that it is just an imitation (Stolicna 2003).

Noodles or vermicelli represents another group of pasta. Dough is made of wheat, rye, barley, maize or buckwheat flour. The kneaded dough is rolled into a thin form, cut to bands of similar width and then piled and sliced to thin noodles. This type of pasta is widely used for various dishes—thickening of soups, combined with other foodstuff or as a main dish with addition of vegetables, fungi, fruits or meat processed in different forms. From similar dough as used for noodles, many other types of dishes were traditionally prepared applying different final adjustments—*gagorce* (dough was rolled), *mykance* (teared by hand) or cut into squares and/or rhombus ("fliacky" and/or "shifliky").

Among the time-honoured dishes is *pirohy*, which is a favourite delicacy for many Slovaks. Kneaded and rolled dough is cut in narrow strips, and then these strips further cut into rhombus or squares. A filling—jam, cut cabbage, mashed potato, meat or any mixture of boiled vegetables—is dosed in the middle and the dough closed forming triangles. The edges should be closed with the fingers. The *pirohy* are then put into boiling water. When ready, the water is discarded and "pirohy" are flavoured by butter, fat and cream and sprinkled with ground nuts, poppy seeds or curd. *Pirohy*, also called *perky*, could be prepared from potato dough, which was kneaded and rolled as that for strudel. The flatted dough is subdivided into two equal parts. On the first half dense plum jam is dosed in regular distances and then the second half should be cautiously put over as cover. The space between the jam lumps should be pressed with the fingers and the triangles then separated with a cutting wheel that decorates them with resulting wavelike brims. *Pirohy* after cooking in boiling water should be drown by melted butter and dusted by poppy, fried breadcrumbs, nuts, curd or many other additives (Babilon 1870, Stolicna 1997, Vansova 1998, Stolicna 2002, Horecka 2007).

5.3.4 Bread

In the Slovak history bread represented not only one of the most important foodstuff but even a holy symbol. Therefore, bread production was given a unique attention in all stages of social development. Bread dough was made of several flours, usually mixed up with other additives. Rye and barley flour prevailed previously, although the south regions preferred wheat flour and mountains and poor regions the oat flour. In the case of the lack of flour, some people used hazel catkins instead, which were milled after drying, sometimes even the roots and bulbs of different plants, bark of trees or pea flour. On the whole territory of present Slovakia, potatoes were quite important parts of the bread dough. To initiate the dough rising, the so-called *kvasok*—homemade leaven—is usually applied, consisting of old bread rests or bread dough mixed up with lukewarm water and after some tempering put together with the flour. If this flour was obtained by milling germinated grains, a grated potato should be added and sprinkled with a bit of vinegar. In Slovakia it was a generally accepted practice to bake bread at weekly intervals. The preferred day was the Saturday as the bread should be freshly served on Sundays. Every family baked several loafs of bread. The housewives always engaged their daughters to assist and at the same time learn this important homework.

In most Slovak regions, the technology of bread baking was quite similar with a few minor deviations or specificities. For everyday use, the bread was made of barley, oat, rye flour or of a mixture of several flour types. For festive occasions, high-quality wheat flour was used—and the product was known as white bread. The general method was as follows: on Friday, usually in the evening hours, the leaven was prepared—salt was added to lukewarm water, raw potatoes and some flour. Next morning, boiled potatoes, caraway and the rest of the flour were added. The dough should be well kneaded and left for fermentation. In the meantime daughters prepared the oven and sons cared for the wood. It was needed to synchronize the time needed for oven heating and dough fermentation. From the fermented dough, the mother formed loafs in special wooden containers called *vahan*. The dough then was placed into a straw basket with a spread linen napkin inside powdered by flour and left to finish the fermentation process. When the oven was properly heated, mother rolled them to the oven front door forming a lump and, then using a long wooden shovel sprinkled with flour, put the formed dough and placed it into the oven. Before this step, the bread dough could be sprinkled with sweet water to be shining when ready. In some regions before baking, they put the formed dough on a cabbage leaf, which gives to the bread bottom an aesthetic appearance. Other housewives used to spread on the bread dough seeds of caraway or poppy. Bread baking lasted usually 1.5–2 h, and then the bread was cleaned—washed with water to remove the ash. It should be mentioned that this is quite important, because the contact with water strengthens the bread crust. After cooling down, the breads were stored in wooden trunks or on shelves in a cool room. When the bread seems to be too dry, it is steamed before consumption and so can be softened. There are tens of technological modifications registered in the Slovak archives that document the inventiveness and ingeniousness of housekeeping ladies in Slovak regions.

Bread baking was usually connected with its secondary product—the so-called *podplamenník* (could be translated as *under-flamer* and is also known by other regional names: *posuch*, *osuch* and *lepnik*), which is even today regarded as an excellent delicacy. It is made of fermented bread dough before the start of bread baking. The small rounded pieces of dough are flattened to approx. 10 mm thickness by a rolling pin. The surface is pricked with a fork, moistened with water, dredged with caraway, salted and baked in the oven before the bread baking started. This product consisted actually only from two crusts (the upper and lower) without any soft middle part, but it has an excellent bread taste and aroma. After baking the *podplamenník* was greased with butter and some cut garlic, fried bacon or cabbage added Sometimes—mostly for children—the upper and lower crusts were subdivided and genuine plum jam or other sweet or sour additives were spread in between (Babilon 1870, Stolicna 1997, Vansova 1998, Stolicna 2000, Demrovska 2007, Dullova 2007, Horecka 2007).

5.3.5 Legume Dishes

Legumes were cultivated in most home gardens or in vineyards and used primarily for soups, sometimes combined with noodles. Secondly, legumes were prepared in the form of a purée, which could be directly consumed or used as a side dish combined with meat. Both soups and slurries could be prepared as sweet or salted dishes. There are over 250 different recipes known—modifications of soups, slurries, purées and other dishes created in combination with pasta, various meats, vegetables and/or dairy products. In some years with poor crop yield, dry seeds of pea and bean were milled and used even for bread baking. In many families the roasted chick pea beans were used for coffee preparation.

5.3.6 Vegetable Dishes

One hundred and fifty to two hundred years ago, Slovak ancestors used a substantially lower number of vegetable species, as that time many of the presently available species were unknown. Anyway, they wished to diversify their nutrition and therefore tried to bring into the kitchen naturally occurring herbs, meadow plants and sources like leaves of nettle, wood sorrel, orache, coltsfoot and others. From boiled fresh leaves, not only medicinal concoctions were prepared but also several soups, sauces and other side dishes in combination with milk and different additives. Meadow plants were later replaced by spinach, lettuce and other plants. In several Slovak regions, fresh or dried forms of plant species like beet and turnip were consumed, but sometimes these plants were further processed, fermented and consumed in the form of very simple meals; in poor families, they were often taken as main dishes. Nearly on the whole Slovak territory, the traditional kitchen preferred cabbage, which is documented by an old lore: "When the Slovak has enough cabbage and potatoes, he cannot die of hunger." In a traditional kitchen, cabbage was usually preferred in a fermented condition. In some regions, the whole cabbage heads or their halves are fermented; in others cut form is preferred. In years of low crop yields, people fermented leaves of horseradish in wooden barrels, instead of cabbage.

There are around 200 dishes based on fresh or fermented cabbage (sauerkraut). Soups and dense sauces are the main dishes in the Slovak kitchen. Cabbage was often mixed with preboiled pasta and then prepared as a boiled, stewed or roasted dish. Fresh cabbage was a basic component of many salads and side dishes. Stewed cabbage is even presently added into various types of cakes like strudel and to the so-called *kapustníky* (cabbage pie). The very first place among the Slovak traditional meals should be assigned to thickened cabbage soup—*kapustnica*—which is known in more than one hundred modifications. In a simple way it is prepared in combination with milk, cream, legumes, different vegetables, noodles and with many known sorts of meat. An important component of *kapustnica* is dried fungi. Another very tasty dish is cabbage rolls. Boiled and beat-out cabbage leaves serve as cover of stuffing, where rice (or in past peeled barley), vegetable, meat and potato are mixed.

Different meals are prepared from pumpkin/squash in a traditional way, e.g. soups, pulps and side dishes from the squash flesh. The squash species with more delicious taste were baked mostly with cinnamon, eventually applied as stuffing of cakes or used for preparation of different concoctions with cleansing or healing effects (Stolicna 1986, Vansova 1998, Stolicna 2000, Demrovska 2007, Dullova 2007, Horecka 2007).

5.3.7   Fruit Dishes

A few hundred years ago, it was common to go to forests or meadows to collect some wildly growing fruit for everyday consumption. There were many rich offers from natural environments like apples, pears, plums, cherries, several berries and seeds of walnut, hazel, beech nut and others. These fruits were consumed mainly fresh, but sometimes they were used for preparation of soups, pulps, sauces, tea, jams, fermented beverages, spirits and other simple products. Dried plums, pears, apples and hazelnuts were stored in all Slovak regions for the traditional Christmas Eve dinner.

One of the most famous products of Slovakia is plum jam prepared by a classical technology without addition of sugar. The production technology is quite simple. The ripe plums are picked from trees, and in the evening hours, they are unstoned by the family members. Next morning an oven with a copper boiler will be prepared on the yard. The unstoned plums are gradually added into the boiler and the plum jam preparation can be started to be finished usually in late night. A wooden stirrer is used to mix up the whole content continuously to prevent the fruits from sticking to the boiler bottom. The first phase of the process results in a semi-boiling jam, the so-called *lizak* (jam for licking), which is a favourite delicacy especially for children. From time to time, the jam quality should be tested—using a wooden spoon a bit of jam is taken from the boiler—and when it is dropping down, the jam is ready, and the boiling process is properly finished. Hot jam was filled into earthen, wooden or straw vessels and allowed to solidify. The hard jam is then tipped over, cut in smaller pieces and at least in some villages packed into paper and hanged similarly to bacon on the pantry roof. In other villages jam is stored in earthen vessels overfilled with pig fat. This plum jam had to be made in such amount that the whole family was fairly supplied at least for 1 year. Well and properly boiled plum jam could be stored even for several years. It should be mentioned that in the past plum jam production by the whole family was taken as an important social event.

5.3.8 Fungal Dishes

It seems that the consumption of fungi started in ancient times. Housewives in all Slovak regions preferred cooking of fungi soups, sauces and omelettes during the fungal season. The most common species are yellow chanterelle (*Cantharellus cibarius*) and *Boletus reticulatus* which were prepared simply, just stewed on buttered flour. The procedure is as follows: on heated butter a bit of flour is added, which is then roasted up to liquid consistency; in the meantime the fungi are preboiled in salty water, and after being separated from the water, they are mixed with formerly prepared butter and then stewed in this mix until the fungi become soft.

Fresh mushrooms are often used for cooking typical dishes—the so-called *paprikash* (peppered dish) or goulash. *Paprikash* is a simple food—finely cut onion is roasted with pig fat until its colour changes to golden, and then washed and yellow chanterelle fungi preboiled in salty water are added and stewed until softened. To improve the taste and increase the dish density, some sour cream mixed with flour is added. Finally the dish is spiced with red pepper (often it is the hot type) and eventually some additional seasoning. Many gourmets are able to make over 250 different original dishes from common and less known mushrooms.

5.4 Conclusion

This survey of Slovak traditional meals provides basic information on both the species used that occur as natural resources and the species cultivated and then used. This chapter brings a mixture of old and very old knowledge, which are highly important for the present time as well. Information on more than 500 plant species used by Slovak predecessors is first class and paramount knowledge. Due to the increase of selected culture species exploited nowadays, many of the former types ceased to be utilized by the present generations. They are like a new discovery for the people that are interested in less known and forgotten species, which could be purposefully and efficiently applied in several products. Another important source of popular knowledge concerns the huge amount of original technology for foods prepared from different plant products. Many of them could be reported and registered in the frame of the EU quality policy and so contribute to the socioeconomic development of many regions.

References

Babilon J (1870) Prvá kuchárska kniha (The first cookery book), 450 p

Demrovska A (2007) Dobra slovenská kuchyna (Good Slovak kitchen). Knizne centrum, ISBN 80-85684-59-4

Dullova L (2007) Velka slovenska kucharka (Great Slovak cook). Ikar, ISBN 80-7118-783-6

Horecka J (2007) Najlepsia slovenska kuchyna (The best Slovak kitchen), ISBN 80-8064-300-8

Stolicna R (1986) Ethnocartographic research into popular food in Slovakia: some results. In: Fenton A, Kisban E (eds) Food in change. John Donald, Edinburgh, pp 147–149

Stolicna R (1997) Alternativne zdroje rastlinnej stravy v strednej Europe (Alternative plant sources of food in the Central Europe). Slovensky Narodopis 45:285–294

Stolicna R (2000) Alternative sources of food in Central Europe. In: Lysaght P (ed) Food from nature: attitudes, strategies and culinary practice. Royal Gustavus Adolphus Academy, Uppsala, pp 195–203

Stolicna R (2002) Die Slowakei als ethnokultureller Raum. In: Dröge K (ed) Altagskulturen in Grenzräumen. Peter Lang GmbH, Europäischer Verlag der Wissenschaften, Franfurt am Main, pp 187–198

Stolicna R (2003) Bryndzove halusky—Von einer "unverdaulichen Speise" zum slowakischen Nationalsymbol: Historische Sozialkunde, Nationale Mythen 3, pp 32–39

Tutka J, Vilcek J, Kovalcik M (2009) Ocenovanie verejnoprospesnych funkcii lesnych a polnohospodarskych ekosystemov a sluzieb odvetvi(Evaluation of forest and agricultural ecosystems and services). In: Aktualne otazky ekonomiky lesneho hospodarstva Slovenskej republiky: Recenzovany zbornik z odborneho seminara, Zvolen, 21–22 Oct 2009, pp 79–88, ISBN 978-80-8093-102-5

Vansova T (1998) Recepty prastarej matere(Recipes of the grand-grandmother)

Chapter 6
Traditional Foods in Turkey: General and Consumer Aspects

Semih Ötleş, Beraat Özçelik, Fahrettin Göğüş, and Ferruh Erdoğdu

6.1 Introduction

Culture takes a significant role in shaping societies, and traditions are the skeleton of a culture (Dikici 2001). In 2006, the European Commission defined "traditional" as "proven usage in the community market for a time period showing transmission between generations; this time period should be the one generally ascribed as one human generation, at least 25 years" (Guerrero et al. 2009). Each culture, ethnic group or region has specific traditions, which are formed during long periods and can hardly be changed. Traditions include eating habits, reflecting an expression of culture, history and lifestyle of nations (Weichselbaum et al. 2009).

Certain food ingredients and food preparation methods being used and transmitted from one generation to the next one are called "traditional foods", and Turkey is a very rich country in palatable traditional foods such as baklava, lokum, aşure, etc., due to the fact that the traditional Turkish cuisine has been influenced by many different cultures like Persian, Hittite and Byzantine. Moreover, since Turks have accepted Islam as religion, Arabs have also had an influence over the Turkish cuisine. In addition, the Celts, the Romans and the Turks occupied many countries over the centuries and left their culinary traces in those countries (Weichselbaum et al. 2009).

S. Ötleş (✉)
Department of Food Engineering, Ege University, Izmir, Turkey
e-mail: semih.otles@ege.edu.tr

B. Özçelik
Department of Food Engineering, Istanbul Technical University, Istanbul, Turkey

F. Göğüş
Department of Food Engineering, Gaziantep University, Gaziantep, Turkey

F. Erdoğdu
Department of Food Engineering, Ankara University, Ankara, Turkey

© Springer Science+Business Media New York 2016
K. Kristbergsson, J. Oliveira (eds.), *Traditional Foods*, Integrating Food
Science and Engineering Knowledge Into the Food Chain 10,
DOI 10.1007/978-1-4899-7648-2_6

Due to nomadic lifestyle of Turks in the past, they used to convert perishable foods into more durable ones like sucuk, pastirma, boza, yoghurt, jam, pekmez, and pickles by using different methods like fermentation, concentration and sun drying. Among these, yoghurt is a worldwide known fermented dairy product in today's world.

Turks have perhaps been more interested in home food preparation than other people in Europe. Therefore, traditional foods have been prepared at homes or in small-scale enterprises within the limited region. In spite of immigration from rural to urban regions, people do not give up consuming their traditional foods. For this reason, traditional foods are beginning to be produced industrially. However, a main problem about traditional foods is the lack of legislation and standardisation in order to protect the products as well as the producers and the consumers. This chapter highlights traditional foods, consumer aspects of Turkish traditional foods, general aspects of traditional foods and innovation in traditional foods.

6.2 Consumer Aspects of Turkish Traditional Foods

Traditional foods are perceived as well-known foods which are consumed often by consumers from childhood and which were already consumed by grandparents (Weichselbaum et al. 2009). These traditional foods are consumed regularly or related to specific dates, celebrations or seasons, produced in a specific way in a certain area, region or country (TRUEFOOD 2009). Regarding consumer aspects, traditional foods are distinguished as follows:

- *Traditional foods consumed regularly*: These are consumed without any distinctive purpose, such as black tea and Turkish coffee. Black tea is the most widely consumed non-alcoholic drink in Turkey. It is offered as a sign of friendship and hospitality especially before or after any meal. While the Chinese discovered the tea plant, Turks developed their own way of making and drinking the black tea. After boiling water in a large kettle, some of it is used to fill the smaller kettle on top and steep several spoons of loose tea leaves producing a very strong tea infusion. The remaining water is used to dilute the tea for producing weaker tea. It is served in little tulip-shaped glasses, boiling hot, with sugar, but without milk.

  The other regularly consumed Turkish traditional food is Turkish style coffee. It is less popular than black tea. In fact, there is no special variety of cocoa bean, raw material of coffee, grown in Turkey. Since the preparation method is different, it is known as Turkish coffee in the world now. It is prepared by boiling finely powdered roast coffee beans in a pot, possibly with sugar, and served into a cup.

- *Traditional foods related to specific purposes*, dates, celebrations or seasons such as Ashura or Noah's pudding (aşure in Turkish), puerperant sherbet (lohusa şerbeti in Turkish), flour-based halva (un helvası in Turkish) and Gullach (güllaç in Turkish). Ashure is a Turkish dessert which is made of a mixture consisting of various grains, fruits and nuts. According to the belief, when Noah's Ark came to rest on Mount Ararat in Northeastern Turkey, Noah's family prepared a special dish for celebration with grains, fruits and nuts since their food supplies were about to exhaust. Hence, the dessert is called Noah's pudding. Traditionally, it is prepared on the Day of Ashura, which marks the end of the Battle of Karbala. It is served to neighbours. Puerperant sherbet is a traditional cold drink prepared with rose hips, cornelian cherries, rose or liquorice and a variety of spices. Sherbet is thought to increase lactation after child birth, so it is served during celebrations after a woman gives birth. Halva is another traditional dessert. There are many types of halva such as tahini halva, summer halva, etc. Halva, prepared with flour, is cooked and served upon the death of a person. After the burial ceremony, it is served to neighbours and visitors. Gullach is another traditional dessert made with milk, pomegranate and a special kind of pastry and is consumed in Ramadan. Gullach dough is prepared with cornstarch and wheat flour. Gullach contains walnuts between the layers which are wetted with milk and some rosewater.

- *Traditional foods having a specific way of production*: Authenticity in terms of recipe, origin of raw material and production process are important to provide the distinctive trait of traditional foods. Generally fermentation and drying are the most widely used techniques in those traditional food productions. Pastirma, sucuk, boza (bosa), raki, kefir and tarhana are associated to a specific production process. Pastirma and sucuk are the most famous traditional delicatessen products. Pastirma is a dry-cured beef product whose characteristic flavour and aroma comes from the so-called cement (çemen). "Cement" is made from garlic, red pepper, paprika and ground fenugreek seeds. Sucuk is a fermented dry meat product which is produced mainly from beef meat and/or mutton and tail fat from sheep (Aksu and Kaya 2002, 2005; Batu and Kirmaci 2009; Bozkurt and Bayram 2006; Coşkuner et al. 2010; Dixit et al. 2005; Erkmen and Bozkurt 2004; Gök et al. 2008; Sagdıc et al. 2010; Soyer et al. 2005; TRUEFOOD 2009; Turkish Food Codex 2009; Yıldız-Turp and Serdaroğlu 2008).

  Tarhana, raki, bosa and kefir are popular fermented products. Tarhana is a fermented and dried soup mixture. It is prepared by mixing wheat flour, yoghurt, yeast and a variety of cooked vegetables (tomatoes, onions, green pepper), salt and spices (mint, paprika), followed by fermentation initiated by the presence of yoghurt or sour milk, for several days. Raki is a traditional and national non-sweet, anise-flavoured spirit, which is consumed as an apéritif, in particular along with mezze, red meat and seafood dishes. It has 45 % alcoholic content made from grape. It is consumed with chilled water on the side or partly mixed with chilled water. Bosa is a thick fermented wheat drink with a low alcohol content. It is generally consumed in winter and served with cinnamon and roasted chickpeas (leblebi). Kefir is a fermented milk product prepared by the incubation of kefir grains with defatted cow milk. Fermentation by lactic acid bacteria, acetic acid bacteria and yeasts has been employed. Its viscosity is similar but lower than that of stirred-type yoghurt. Nowadays, it is available with various fruit aromas at markets (Soltani et al. 2010).

- *Traditional foods associated to a certain area, region or country*: Traditional foods should be associated to a certain region. Turkish delight (lokum in Turkish), floss halva (pişmaniye in Turkish), candied chestnut (kestane şekeri in Turkish),

hamsi, raw meat ball (çiğ köfte in Turkish), kaymak, shalgam (şalgam in Turkish), butter, olive oil, baklava, pastirma and sucuk are associated to a specific region. Lokum (Turkish delight) is a sugar-based jelly-like confectionary product. Production of lokum is estimated to have begun in the fifteenth century. Since the nineteenth century, it is known as Turkish delight in Europe. The Safranbolu region is a very famous lokum producer. Lokum is a confection made from cornstarch and sugar. It is often flavoured with rosewater, lemon or some other fruit extracts. It has a soft and sticky texture and is often packaged and eaten as small cubes that are dusted with icing sugar to prevent sticking. Some recipes include small nut pieces, usually pistachio, hazelnut or walnuts (Batu 2010). Baklava is a rich, sweet pastry made of layers of filo pastry filled with chopped nuts and sweetened with syrup or honey. After mixing the ingredients, wheat flour, pistachio or other nuts, butter, wheat starch, cornstarch, water, egg, salt and lemon juice, the dough is kneaded to get baklava sheets. The sheets are baked in ovens sweetened with syrup or honey. The Gaziantep region is very famous for its baklava (Akyildiz 2010). Floss halva is a Turkish sweet made by blending flour roasted in butter into pulled sugar. The Kocaeli region is related to floss halva production. Another sweet, marron glace (kestane şekeri in Turkish) is produced specifically in the Bursa region. It is made with whole chestnuts candied in sugar syrup and then iced. Raw meatball is produced in east and south-east Anatolia. It is produced traditionally with a mixture of lean veal and bulgur, dried red pepper, salt, green or dried onion, black pepper, parsley and tomato and/or red pepper sauce. It is consumed as a raw material without heat treatment. Cream (kaymak) is a creamy dairy product, similar to clotted cream. Cream is mainly consumed today for breakfast along with the traditional Turkish breakfast. It is mainly produced in the Afyonkarahisar region. Shalgam is a fermented beverage consumed widely in southern parts of Turkey, especially with meat products like Turkish kebabs and gyros in the diets. It is produced mainly from purple carrot, bulgur and sourdough. Adana is famous for its shalgam. Hamsi is a small fish similar to the anchovy. Hamsi is used in many traditional dishes of the Black Sea region. Olive oil is the major type of oil used for cooking in Turkey. Olive trees are distributed mainly in the Aegean region of Turkey, especially in the cities of Balikesir, Aydin and Izmir (Otles 2010). Sucuk and pastirma discussed above belong to Afyonkarahisar and Kayseri regions of Turkey, respectively.
- *Traditional foods consumed since childhood*: Typical traditional foods which have been consumed since childhood are set-type yoghurt (yoğurt in Turkish), ayran, pekmez and tahin. Yoghurt is a fermented dairy product which is known worldwide. It is the Turk's gift to the world. Yoghurt-based beverage, ayran, is prepared by mixing yoghurt with water and salt. It is especially consumed in summer. Pekmez is a molasses-like syrup obtained after condensing juices of fruit must, especially grape, fig or mulberry, by boiling them with a coagulant agent. It is used as a syrup or mixed with tahini or sesame paste (tahin in Turkish) for typical Turkish breakfast. Tahini is a paste of ground sesame seeds.

There are a lot of traditional foods consumed in Turkey. The foods mentioned above are only a few famous and major e *childhood*: Typical traditional ones.

6.3   General Aspects of Traditional Products

The importance of traditional foods has significantly increased in the last couple of years. Traditional foods play an important role in maintaining diversity of foods, in serving the specific needs of the local consumers and in the presentation of the local, national cultural heritage. Although the term "traditional foods" is widely used and everybody has a rough idea of what is meant by it, defining traditional foods is not as simple as it might be presumed. Published definitions of traditional foods include temporal, territorial and cultural dimensions (Bertozzi 1998; EU 2006), the idea of a transmission from generation to generation (EU 2006; Trichopoulou et al. 2007) and, more recently, elaborative statements about traditional ingredients, traditional composition and traditional production and/or processing (Weichselbaum et al. 2009). One of these definitions has been prepared by EuroFIR. This is an elaborative definition, which includes statements about traditional ingredients, traditional composition and traditional type of production and/or processing (EuroFIR 2007; Trichopoulou et al. 2007). Traditional food is a food with a specific feature or features, which distinguish it clearly from other similar products of the same category in terms of the use of "traditional ingredients" (raw materials of primary products) or "traditional composition" or "traditional type of production and/or processing method".

The culture, lifestyle and economic conditions over a long period of time have formed the development of local food traditions. Over time, traditional foods have been influenced by many factors. One of these factors is the availability of raw materials; traditional food is thus influenced by agricultural habits and location. Geographical location has had a certain influence on the type of foods that have evolved; for instance, climatic conditions favoured the use of drying for preservation in southern European countries, while smoking was mainly used in northern countries. Regions at a lower altitude, for example, have different vegetation compared to regions at high altitudes; countries without access to the sea usually have a lower availability of fish and seafood compared to those with a large coastal area (EuroFIR 2007).

Turkey has a great number of traditional foods as a result of its long history, diversity of cultures and different climates. Different areas of Turkey have different traditions; there are local specialities which must be eaten in their home region to be fully appreciated. Thus, Kanlica in Istanbul is famous for its yoghurt, Bursa for its Iskender Kebab, Gaziantep for its pistachio nuts and baklava, the Black Sea region for hamsi (fried anchovies) and corn bread and the Syrian borderlands (Urfa and Adana) for spicy shish kebabs. Not only the location of a region but also its history has influenced the dietary patterns of its inhabitants.

Turkish food is usually not spicy, but some regions do enjoy spicy dishes. Sauces and seasonings are often used but tend to be light and simple and do not overpower the natural flavours of the foods. Most common seasonings include garlic, cinnamon, mint, dill and parsley. Yoghurt is often used in meals with both meat and vegetables. Vegetables, wheat and rice are often the basis of Turkish foods. Eggplant is the favourite vegetable of the country and zucchini is a common second in addition to cabbage, beans and artichokes that are made with olive oil.

6.4 Historical Influences on Traditional Foods

The origin of traditional foods is generally lost in ancient history. Some of these foods have been widely accepted and their popularity has spread from country to country, while the consumption of others remains limited to local areas where they originally appeared. The history of regions has influenced the dietary patterns of its inhabitants (EuroFIR 2007). At the mention of Turkish cuisine, Turkish history should come to mind, because people do not readily lose their taste in food; they do not give up foods to which they have become accustomed over thousands of years.

Turkish cuisine is often regarded as one of the greatest in the world. Its culinary traditions have successfully survived over 1300 years for several reasons, including its favourable location and Mediterranean climate. The country's position between the Far East and the Mediterranean Sea helped the Turks gain complete control of major trade routes, and an ideal environment allowed plants and animals to flourish. Such advantages helped to develop and sustain a lasting and influential cuisine.

The Turkish people are descendents of nomadic tribes from Mongolia and western Asia who moved westward and became herdsmen around A.D. 600. Early influence from the Chinese and Persians included noodles and *manti*, cheese- or meat-stuffed dumplings (similar to the Italian ravioli), often covered in a yoghurt sauce. *Manti* has often been credited with first introducing *dolma* (stuffed foods) into the Turkish cuisine. Milk and various dairy products that became staple foods for the herdsmen were nearly unused by the Chinese. This difference helped the Turks to establish their own unique diet.

By A.D. 1000, the Turks were moving westward towards richer soil where they grew crops such as wheat and barley. Thin sheets of dough called *yufka* along with crushed grains were used to create sweet pastries. The Persians introduced rice, various nuts and meat and fruit stews. In return, the Turks taught them how to cook bulgur wheat. As the Turks moved further westward into Anatolia (present-day Turkey) by 1200, they encountered chickpeas and figs, as well as Greek olive oil and an abundance of seafood.

A heavily influential Turkish cuisine was well established by the mid-1400s, the beginning of the (Turkish) Ottoman Empire's 600-year reign. Yoghurt salads, fish in olive oil and stuffed and wrapped vegetables became Turkish staples. The empire, eventually spanning from Austria to Northern Africa, used its land and water routes to import exotic ingredients from all over the world. By the end of the 1500s, the Ottoman court housed over 1400 live-in cooks and passed laws regulating the freshness of food. Since the fall of the empire in World War I (1914–1918) and the establishment of the Turkish Republic, foreign dishes such as French hollandaise sauce and Western fast food chains have made their way into the modern Turkish diet.

6.5   Traditional Products in Turkey

Turkish cuisine varies across the country. The cooking of Istanbul, Bursa, Izmir and the rest of the Aegean region inherits many elements of Ottoman court cuisine, with a lighter use of spices, a preference for rice over bulgur and a wider use of seafoods. The cuisine of the Black Sea region uses fish extensively, especially the Black Sea anchovy (*hamsi*), has been influenced by Balkan and Slavic cuisine and includes maize dishes. The cuisine of the south-east—Urfa, Gaziantep and Adana—is famous for its kebabs, *mezes* and dough-based desserts such as *baklava*, *kadayıf* and *künefe* (kanafeh). Especially in the western parts of Turkey, where olive trees grow abundantly, olive oil is the major type of oil used for cooking. The cuisines of the Aegean, Marmara and Mediterranean regions are rich in vegetables, herbs and fish. Central Anatolia is famous for specialties, such as *keşkek* (kashkak), *mantı* (especially from Kayseri) and *gözleme*. Some of the Turkish traditional foods are summarised in Table 6.1.

6.6   Innovation in Traditional Foods

Traditional foods represent a growing segment within the European food market, and they constitute a significant factor of culture, identity and heritage (Guerrero et al. 2009). Traditional foods contribute to the development and sustainability of the rural areas, protect them from depopulation, entail substantial product differentiation (Avermaete et al. 2004) and provide ample varieties for consumers (Guerrero et al. 2009).

Traditional food products are products of which the key production steps are performed in a certain area; are authentic in their recipe, origin of raw material and/ or production process; are commercially available for at least 50 years; and are part of the gastronomic heritage (Gellynck and Kuhne 2008). Even though consumers show a favourable attitude towards traditional foods (Guerrero 2001), with the developments in food processing and changes in the consumer's choices, these products now face the challenge to further improve their safety and convenience by innovations which will enable them to maintain in the market and expand their share (Guerrero et al. 2009).

Innovation, by definition, is the commercialisation of an innovative idea for commercialising and selling a product or service at profit (Karantininis et al. 2010). While innovation in the food industry combines technological innovation with social and cultural innovation through the entire process including production, harvesting, primary and secondary processing, manufacturing and distribution, the ultimate innovation is to be the new or improved consumer product and service (Earle 1997). As indicated by Earle (1997), although innovation might focus in process technology including process engineering, product formulation and consumer needs, it might also ripple through the consumer eating patterns and in general

Table 6.1 Some Turkish traditional foods in different regions

Location	Traditional food	Name in Turkish
Adana	Adana kebabı	Adana kebap
Afyon	Soudjouk	Sucuk
	Turkish delight, hashish, cream	Kaymaklı lokum
	Pastrami	Pastırma
Ankara	Carrot delight	Havuç lokumu
	Sıkma	Sıkma
	Turkish fairy floss	Pişmaniye
	Mumbar	Mumbar
	Sweet Turkish style fermented sausage	Tatlı sucuk
	A sweet made of unsalted cheese (halva with cream)	Hoşmerim
	Kapama	Kapama
	Black mulberry jam	Karadut reçeli
	Kül kömbesi	Kül kömbesi
	Rice in jug	Testi pilavı
Antalya	Locust bean molasses	Keçiboynuzu pekmezi
Aydın	Dry fig	Kuru incir
Balıkesir	Candied chestnut	Kestane şekeri
Beypazarı	80-Layer baklawa	80 katlı baklava
	Tarhana	Tarhana
Bolu	Bolçi (Bolu chocolate)	Bolçi (Bolu çikolatası)
	Hazelnut jam	Fındık reçeli
	Şakşak helva	Şakşak helvası
	Keş	Keş
	Karakışlık rice	Karakışlık pirinci
	Mengen cheese	Mengen peyniri
	"Cevizli 4 divan" helva	Cevizli 4 divan helvası
	Uhut "wheat sweet"	Uhut "buğday tatlısı"
Bolu	Hoşmerim	Hoşmerim
	Pickle "bozarmut"	Boz armut turşusu
	Cranberry soup with dried yoghurt	Kızılcık tarhanası
	Kaşık sapı	Kaşık sapı
	Keşkek	Keşkek
	Wedding soup	Düğün çorbası
	"Mudurnu saray" helva	Mudurnu saray helvası
	Çukundur hoşafı	Çukundur hoşafı
Bursa	Sliced meat on a vertical spit with round and flat bread	Pideli döner kebap
	M.Kemalpaşa semolina dessert with fresh cheese	M.Kemalpaşa peynir tatlısı
	Chestnut sweet	Kestane şekeri
	İnegöl meatball	İnegöl köftesi

(continued)

Table 6.1 (continued)

Location	Traditional food	Name in Turkish
Çorum	Çorum roasted chickpea	Çorum leblebisi
	Iskilip stuffed green peppers	İskilip dolması
	Gül burma baklawa	Gül burma baklava
	Hingal	Hingal
	Leblebi (roasted chick pea)	Leblebi
Denizli, Çal	Grape molasses	Üzüm pekmezi
Elazığ	Orcik candy	Orcik şekeri
	Mulberry flour	Dut unu
	Çedene coffee	Çedene kahvesi
Erzincan	Bryzna	Tulum peyniri
	Cottonseed	Çiğit
	Povik pickle	Povik turşusu
	Avrenç (special dry cottage cheese)	Avrenç (has çökelek)
	Şavak cheese	Şavak peyniri
	Kesme soup	Kesme çorbası
Erzurum	Curd dolma	Lor dolması
	Haşıl	Haşıl
	Kaysefe	Kaysefe
	Kuymak	Kuymak
	Thick soup of chickens, wheat, meat, tomato and onion	Halim aşı
Gaziantep	Pomegranate syrup	Nar ekşisi
	Coarse bulgur/coarse cracked wheat	Pilavlık bulgur
	Baklava	Baklava
Gediz	Soup with dried yoghurt	Kuru tarhana
Hatay	Pomegranate syrup	Nar ekşis
	Kunefe	Künefe
Iğdır	Aubergine jam	Patlıcan reçeli
	Lepe	Lepe (kırık nohut)
Isparta	Rose jam	Gül reçeli
Izmir	Stuffed artichoke with rice, minced meat	Enginar dolması
	Rolled pastry with nettle	Isırgan otlu kol böreği
	Black-eyed bean salad	Börülce salatası
	Stuffed fig	İncir salatası
	Dry fig	Kuru incir
Karaman	Coarse bulgur	Pilavlık bulgur
Kayseri	Evlik soudjouk (irişkrik)	Evlik sucuk (irişkrik)
	String cheese	Tel peyniri
	Soudjouk	Sucuk
	Pastırma	Pastırma
	Meat pasty	Mantı
Kırşehir	Grape molasses	Üzüm pekmezi

(continued)

social and cultural areas. Based on this, innovation strategies in the food industry

Table 6.1 (continued)

Location	Traditional food	Name in Turkish
Manisa	Mesir paste	Mesir macunu
Kahramanmaraş	Soup with dried yoghurt	Kuru tarhana
	Maraş ice cream	Maraş dondurması
Mersin, Mut	Coarse bulgur	Pilavlık bulgur
Mersin	Pomegranate syrup/dip roman	Nar ekşisi
	Tantuni	Tantuni
	Cezerye	Cezerye
Safranbolu	Turkish delight	Lokum
South East Anatolia	Dried layers of grape pulp	Üzüm pestili
	Grape molasses	Üzüm pekmezi
Şanlıurfa	Bastık	Bastık
	Gün pekmezi (a thick syrup made by boiling down grape juice)	Gün pekmezi
	Küncülü akıt	Küncülü akıt
	Mırra (bitter coffee with cardamon)	Mırra
	Pomegranate molasses	Nar pekmezi
	Isot jam (preserve of urfa red peppers used as spread)	Isot reçeli
	Urfa isot	Urfa isotu
	Çiğköfte (steak tartar a la turca)	Çiğköfte
	Şıllık	Şıllık
	Lahmacun (very thin Turkish pizza covered with seasoned minced meat and onions)	Lahmacun
	Çekçek	Çekçek
Uşak	Soup with dried yoghurt	Kuru tarhana
Ürgüp	Dried layers of grape pulp	Üzüm pestili
Van	Ayran aşı	Ayran aşı
	Bulgur aşı	Bulgur aşı
	Kurutlu erişte aşı	Kurutlu erişte aşı
	Borani	Borani
Van	Quince meal	Ayva yemeği
	"Kurutlu" minced meat ball	Kurutlu köfte
	Semgeser	Semgeser
	Şille	Şille
	Roasted plum	Erik kızartması
	Şor fish	Şor balık
	Herby cheese	Otlu peynir

(continued)

Table 6.1 (continued)

Location	Traditional food	Name in Turkish
Various	Turkish delight	Lokum
	Pişmaniye—Turkish fairy floss	Pişmaniye
	Grape molasses, *Juniperus drupacea*	Andız pekmezi
	Turkish raki	Raki
	Almond paste	Badem ezmesi
	Yoghurt	Yoğurt
	Ayran (yoghurt mix)—buttermilk	Ayran
	Turkish coffee	Türk kahvesi
	Fermented pickle	Turşu
	Tel kadayıf (shredded dough baked in syrup topped with crushed nuts)	Tel kadayıf
	Tahin (crushed sesame seeds)	Tahin
	Turnip juice	Şalgam suyu
	Turkish bagels	Simit
	Sahlep	Salep
	Dried layers of mulberry pulp	Dut pestili
	Semolina dessert/semolina helva	İrmik helva
	Güllaç (rice wafers stuffed with nuts/cooked in milk)	Güllaç
	Fenugreek	Çemen
Various	Boza	Boza
	Kefir/kephir	Kefir
	Curd	Çökelek
	Brynza	Tulum peyniri
	White cheese	Beyaz peynir
	Aşure (Noah's pudding/wheat pudding with dried nuts)	Aşure
	Herby cheese	Otlu peynir
	Fig molasses	İncir pekmezi
	Stuffed grape leaves with olive oil	Zeytinyağlı sarma
	Rice pudding	Sütlaç
	Pudding/duff	Muhallebi
	Keşkül (a milk pudding with coconut)	Keşkül
	Kazandibi (pudding with a caramel base)	Kazandibi
	Döner	Döner
Zile	Zile grape molasses	Zile pekmezi

and also in traditional foods should concern both the production and social-environmental changes to satisfy the personal, nutritional and social needs with the safety issue.

Innovation in the production of traditional foods might be a novelty by replacing a traditional technology with an innovative process, an improvement in any step of the processing or a fundamental change without affecting any traditional aspect of the food product. In addition, the innovations might include the packaging and changes in products size and shape for possible new ways of using the product (Kuhne et al. 2010). All these approaches might be based on the consumers' needs, behaviour and attitudes with the changes in the social, economical and cultural environment. Incorporation of the opinion of the consumers supports the realisation of competitive advantage through the implementation of innovation (Earle 1997). Hence, a consumer-based research prior to developing and introducing innovation is generally acknowledged to be useful (Verbeke et al. 2007).

The food industry is characterised by a larger number of small- and medium-sized enterprises, and this applies in particular to the traditional food sector (Kuhne et al. 2010). In the current globalising food market, innovation is the strategic tool for these enterprises to achieve a competitive advantage especially by traditional foods despite the controversy between innovation and tradition (EC 2007) as innovations in the traditional foods bring the chance to strengthen and widen the market to increase the profits.

6.7 Conclusion

Culture takes a significant role in shaping societies and its traditions. The European Commission defined the "traditional" as "proven usage in the community market for a time period showing transmission between generations ascribing a human generation for at least 25 years". Due to the effect of many different cultures that the Turkish people faced during their history, Turkey is one rich country in palatable traditional foods. Traditional foods constitute a significant factor of culture, identity and heritage and contribute to the development and sustainability of the rural areas, protect them from depopulation, entail substantial product differentiation and provide ample varieties for consumers. With the developments in food processing and changes in the consumer's choices, these products now face the challenge to further improve their safety and convenience by innovations to maintain in the market and expand their share, as in today's globalising food market, innovation became the strategic tool to achieve a competitive advantage.

References

Aksu MI, Kaya M (2002) Effect of commercial starter cultures on the fatty acid composition of pastirma (Turkish dry meat product). J Food Sci 67(6):2342–2345
Aksu MI, Kaya M (2005) Effect of modified atmosphere packaging and temperature on the shelf life of sliced pastirma produced from frozen/thawed meat. J Muscle Foods 16:192–206
Akyildiz E (2010) Geleneksel Türk Tatlisi "Baklava". In: 1st international symposium on traditional foods from Adriatic to Caucasus. Tekirdag, 15–17 April

Avermaete T, Viaene J, Morgan EJ, Pitts E, Crawford N, Mahon D (2004) Determinants of product and process innovation in small food manufacturing firms. Trends Food Sci Technol 15:474–493

Batu A, Kirmaci B (2009) Production of Turkish delight (lokum). Food Res Int 42:1–7

Batu A (2010) Liquid and pasty date pekmez production. African J Food Sci Technol; Vol.1,No.4, pp. 82–89

Bertozzi L (1998) Tipicidad alimentaria y dieta mediterránea. In: Medina A, Medina F, Colesanti G (eds) El color de la alimentación mediterránea. Elementos sensoriales y culturales de la nutrición. Icaria, Barcelona, pp 15–41

Bozkurt H, Bayram M (2006) Colour and textural attributes of sucuk during ripening. Meat Sci 73:344–350

Coşkuner Ö, Ertaş AH, Soyer A (2010) The effect of processing method and storage time on constituents of Turkish sausages (sucuk). J Food Process Preserv 34:125–135

Dikici A (2001) The place and importance of tradition in a society. Fırat Univ J Soc Sci 11(2):251–258

Dixit P, Ghaskadbi S, Mohan H, Devasagayam TPA (2005) Antioxidant properties of germinated fenugreek seeds. Phytother Res 19:977–983

Earle MD (1997) Innovation in the food industry. Trends Food Sci Technol 8:166–175

EC (2007) European research on traditional foods—project examples. DG Research, European Commission, Brussels

Erkmen O, Bozkurt H (2004) Quality characteristics of retailed sucuk (Turkish dry-fermented sausage). Food Technol Biotechnol 42(1):63–69

EU (2006) Council Regulation (EC) No. 509/2006 of 20 March 2006 on agricultural products and foodstuffs as traditional specialties guaranteed. Off J Eur Union L 93/1

EuroFIR (2007) FOOD-CT-2005-513944, EU 6th Framework Food Quality and Safety Programme. http://www.eurofir.net. Accessed June 2007

Gellynck X, Kuhne B (2008) Innovation collaboration in traditional food chain networks. J Chain Netw Sci 8:121–129

Gök V, Obuz E, Akkaya L (2008) Effects of packaging method and storage time on the chemical, microbiological, and sensory properties of Turkish pastirma—a dry cured beef product. Meat Sci 80:335–344

Guerrero L (2001) Marketing PDO (products with dominations of origin) and PGI (products with geographical identities). In: Frewer I, Risvik E, Shifferstein H (eds) Food, people and society: a European perspective of consumers' food choices. Springer, Berlin, pp 281–296

Guerrero L, Guardia MD, Xicola J, Verbeke W, Vanhonacker F, Zakowska-Biemans S, Sajdakowska M, Sulmont-Rosse C, Issanchou S, Contel M, Scalvedi ML, Granli BS, Hersleth M (2009) Consumer-driven definition of traditional food products and innovation in traditional foods. A qualitative cross-cultural study. Appetite 52:345–354

Karantininis K, Sauer J, Furtan WH (2010) Innovation and integration in the agri-food industry. Food Policy 35:112–120

Kuhne B, Vanhonacker F, Gellynck X, Verbeke W (2010) Innovation in traditional food products in Europe: do sector innovation activities match consumers' acceptance? Food Qual Prefer 21:629–638

Otles S (2010) Olive Oil-Culture & Health. In: 1st international symposium on traditional foods from Adriatic to Caucasus, Tekirdag, 15–17 April

Sagdıc O, Yilmaz MT, Ozturk İ (2010) Functional properties of traditional foods produced in Turkey. In: 1st international symposium on traditional foods from Adriatic to Caucasus, Tekirdag, 15–17 April

Soltani M, Saydam Bİ, Guzeller N (2010) Kefir. In: 1st international symposium on traditional foods from Adriatic to Caucasus, Tekirdag, 15–17 April

Soyer A, Ertaş AH, Uzumcuoglu Ü (2005) Effect of processing conditions on the quality of naturally fermented Turkish sausages (sucuks). Meat Sci 69:135–141

Trichopoulou A, Soukara S, Vasilopoulou E (2007) Traditional foods: a science and society perspective. Trends Food Sci Technol 18(8):420–427

TRUEFOOD Traditional United Europe Food (2009) Report on the similarities and differences of views of producers, consumers and support organisations on the bottlenecks and success factors

Turkish Food Codex (2009) No. 2000/4 of 2 February 2009 on meat products. Off J Turk 27133

Verbeke W, Frewer LJ, Scholderer J, De Brabender HF (2007) Why consumers behave as they do with respect to food safety and risk information. Anal Chim Acta 586:2–7

Weichselbaum E, Benelam B, Soares Costa H (2009) Traditional foods in Europe. In: European Food Information Resource (EuroFIR) consortium, EU 6th framework food quality and safety thematic priority, Contract FOOD-CT-2005-513944. http://www.eurofir.net/. Accessed 2015

Yıldız-Turp G, Serdaroğlu M (2008) Fatty acid composition and cholesterol content of Turkish fermented sausage (sucuk) made with corn oil

Further details could also be retrieved on June 2015 from the following websites

http://en.wikipedia.org/wiki/Turkish_cuisine

http://www.foodbycountry.com/Spain-to-Zimbabwe-Cumulative-Index/Turkey.html

Part II
Traditional Dairy Products

Chapter 7
Indian Traditional Fermented Dairy Products

Narender Raju Panjagari, Ram Ran Bijoy Singh, and Ashish Kumar Singh

7.1 Introduction

The requirement to store and preserve foods has been recognized from the time immemorial. Along with salting, cooking, smoking and sun drying, fermentation is one of the earliest preservation methods developed to extend the possible storage time of foods. Fermented foods were very likely among the first foods consumed by human beings. This was not because early humans had actually planned or had intended to make a particular fermented food, but rather because fermentation was simply an inevitable outcome that resulted when raw food materials were left in an otherwise unpreserved state. Fermented foods were developed simultaneously by many cultures across the globe for two main reasons: (a) to preserve harvested or slaughtered products, which were abundant at certain times and scarce at others, and (b) to improve the sensory properties of an abundant or unappealing produce. Fermentations use a combination of principles of food preservation, viz. minimizing the level of microbial contamination onto the food (asepsis), inhibiting the growth of the contaminating microbiota and killing the contaminating microorganisms. Fermentation could be described as a process in which microorganisms change the sensory (flavour, odour, etc.) and functional properties of a food to produce an end product that is desirable to the consumer (Guizani and Mothershaw

N.R. Panjagari • A.K. Singh
Dairy Technology Division, National Dairy Research Institute,
Karnal, Haryana 132 001, India

R.R.B. Singh (✉)
Dairy Technology Division, National Dairy Research Institute,
Karnal, Haryana 132 001, India

Faculty of Dairy Technology, Sanjay Gandhi Institute of Dairy Technology,
BVC Campus, Jagdeopath, Patna, Bihar 800 014, India
e-mail: rrb_ndri@rediffmail.com

© Springer Science+Business Media New York 2016
K. Kristbergsson, J. Oliveira (eds.), *Traditional Foods*, Integrating Food
Science and Engineering Knowledge Into the Food Chain 10,
DOI 10.1007/978-1-4899-7648-2_7

2007). The acceptability of a food to the consumer, in general, is based mainly on its sensory properties. The sought-after sensory properties of fermented foods are brought by the biochemical activity of microorganisms. Fermented foods are classified in a number of different ways. They may be grouped based on the microorganisms, the biochemistry or the product type. Based on the biochemical products produced, they can be classified into lactic acid, acetic acid, ethanol and carbon dioxide fermentations. Based on product type, they are classified into beverages, cereal products, dairy products, fish products, fruit and vegetable products, legumes and meat products (Campbell-Platt 1987).

Milk is particularly suitable as a fermentation substrate owing to its carbohydrate-rich, nutrient-dense composition. Fresh bovine milk contains about 5 % lactose and 3.3 % protein and has a water activity of 0.93 and a pH of 6.6–6.7, perfect conditions for most microorganisms. Lactic acid bacteria are saccharolytic and fermentative and, therefore, are ideally suited for growth in milk. In general, they will outcompete other microorganisms for lactose and, by virtue of acidification, will produce an inhospitable environment for would-be competitors. Therefore, when properly made, cultured dairy products have long shelf lives, and, although growth of acid-tolerant yeast and moulds is possible, growth of pathogens rarely occurs. Given the early recognition of the importance of milk in human nutrition and its widespread consumption around the world, it is not surprising that fermented dairy products, also called as cultured dairy products, have evolved on every continent. Their manufacture was already well established thousands of years ago, based on their mention in the Old Testament as well as other ancient religious texts and writings. The manufacturing procedures, the sources of milk, organoleptic quality and the names of many of the cultured dairy products produced across the world vary considerably. However, they share many common characteristics in terms of the therapeutic properties that they confer. In the present paper, the types of cultured dairy products which are native to the Indian subcontinent (including India, Pakistan and Bangladesh), their manufacturing procedures and desirable sensory attributes are outlined.

7.2 Dahi

Dahi, also called simply as curd, is a popular Indian fermented milk product. The use of *dahi* has been prevalent since Vedic times, and it is mentioned in ancient scriptures like the *Vedas* (1500 BC), *Upanishads* (1000 BC) and various hymns. It is popular with consumers due to its distinctive flavour, and it is believed to have good nutritional and therapeutic value. It is utilized in various forms in many Indian culinary preparations (Pal and Raju 2007). Although *dahi* is quite analogous to plain yoghurt in appearance and consistency, the flavour and other biochemical aspects of the product are different owing to the starter cultures used in its preparation. *Dahi* is also traditionally used as an article in rituals and an ingredient of *panchamrut* (five nectars). *Ayurveda*, the traditional scientific system of Indian medicine, in its treatises, *Charaka Samhita* and *Sushruta Samhita*, discusses various properties of cow and buffalo milk *dahi* and emphasizes its therapeutic characteristics. Ayurveda also describes the properties of various types of *chhash* (stirred diluted *dahi*) and their role in the control of intestinal disorders. *Dahi*, which came into use as a means of preserving milk nutrients, was probably used by Aryans in their daily diet, as it reduced putrefactive changes and provided an acidic, refreshing taste (Prajapati and Nair 2008). *Dahi* (curd) is consumed throughout India, either as a part of the daily diet along with the meal or as a refreshing beverage, or it may be converted into *raita* (seasoned with onion, spices, etc.). It is mostly consumed with rice in South India and with wheat preparations in North India; it is also used as a beverage or dessert. According to the Prevention of Food Adulteration Act (1955) of India, *dahi* or curd is defined as "the product obtained from pasteurized or boiled milk by souring, natural or otherwise, by a harmless lactic acid or other bacterial culture. It may contain added cane sugar and shall contain the same percentage of fat and solids-not-fat as the milk from which it is prepared". The Act also permits the use of milk solids in the preparation of *dahi*. The Bureau of Indian Standards (BIS) specifications for fermented milk products are based on the type of culture used in their preparation (IS: 9617 1980). Mild *dahi* is made from mesophilic lactococci such as *Lactococcus lactis*, *L. diacetylactis* and *L. cremoris* in single or in combination. Leuconostocs may be adjunct organisms for added buttery odour and flavour. Sour *dahi* contains additional cultures belonging to the thermophilic group, which are generally employed in the manufacture of yoghurt. These thermophilic organisms grow rapidly at 37–45 °C, producing *dahi* in less than 4 h.

Dahi is still largely made by local confectionery (*halwais*) shops and restaurants and in homes by traditional methods involving milk of buffalos, cows and goats. However, buffalo milk is best suited for *dahi* having better sensory quality, particularly body and texture because of its inherent properties. At the consumer's household or *halwais'* level, milk is boiled, cooled to room temperature, inoculated with 0.5–1.0 % starter (previous day's *dahi* or *chhash*) and then incubated undisturbed for setting for about overnight. In cold weather, the *dahi* setting vessel is usually wrapped up with woollen cloth to maintain appropriate temperature (Singh 2007). However, with the recent lifestyle-related changes, the branded packaged *dahi* has increasingly become popular among consumers. As a consequence, many organized dairies are now preparing *dahi* adopting the standardized method. In the method adopted by organized sector for preparing mild *dahi*, fresh, good quality milk is preheated and subjected to filtration and clarification. The milk is standardized to 4–5 % fat and 10–12 % milk solids-not-fat (SNF), homogenized and heat treated followed by cooling to incubation temperature and inoculated with specific *dahi* starter culture. It is then filled in suitable containers (plastic cups) of the appropriate size, heat sealed and incubated at 40–42 °C for 3–4 h. When a firm curd is formed and the acidity reaches to about 0.7 %, *dahi* cups are transferred to cold room maintained at about 4–5 °C and stored at that temperature till consumption (Fig. 7.1). However, some manufacturers are using thermophilic cultures to reduce the incubation time and obtain the desired product in short time.

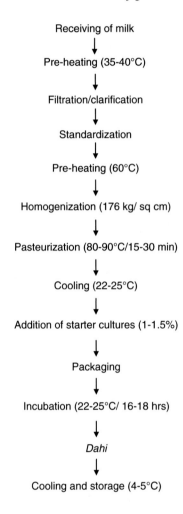

Fig. 7.1 Flow diagram for the manufacture of *dahi*

Receiving of milk
↓
Pre-heating (35-40°C)
↓
Filtration/clarification
↓
Standardization
↓
Pre-heating (60°C)
↓
Homogenization (176 kg/ sq cm)
↓
Pasteurization (80-90°C/15-30 min)
↓
Cooling (22-25°C)
↓
Addition of starter cultures (1-1.5%)
↓
Packaging
↓
Incubation (22-25°C/ 16-18 hrs)
↓
Dahi
↓
Cooling and storage (4-5°C)

The desirable attributes that are important for evaluating *dahi* are colour and appearance, flavour, body and texture and acidity (Ranganadham and Gupta 1987).

Colour and appearance: The colour of *dahi* should be pleasing, attractive and uniform without any signs of visible foreign matter. The colour of *dahi* ranges from creamy yellow for cow milk to creamy white for buffalo. It should be free from browning. *Dahi* should have smooth and glossy surface without appearance of any free whey on top.

Flavour: Flavour of *dahi* is the most important quality attribute. A pleasant sweetish aroma and a mild clean acid taste are looked for in the product. It should be free from any off flavours. A good pleasant diacetyl flavour is desired in *dahi*. It should not show any signs of bitterness, saltiness or other off flavours.

Body and texture: Good *dahi* is a weak gel-like junket, when whole milk is used. It has a creamy layer on top, the rest being made up of a homogenous body of curd.

The surface should be smooth and glossy while the cut surface is trim and free from cracks and gas bubbles.

Acidity: Generally, an acidity of 0.75–0.85 % lactic acid is appropriate for good *dahi*. Excessive acidity gives the product a sour, biting taste. However, in sour *dahi*, the acidity can go up to 1.0 %.

7.2.1 Technological Developments in the Manufacture of Dahi

With the continuous developments in dairy science and technology, new technologies, viz. improved functionality starter cultures, probiotic cultures, membrane technology, etc., that have emerged have been successfully adopted for the manufacture of *dahi*. Whey separation is the most common defect found in set-type fermented dairy products such as *dahi*. Certain strains of lactic acid bacteria are able to synthesize exopolysaccharides (EPS) that are secreted into their environment such as milk. Such EPS-producing lactic acid bacteria have been found to contribute to the texture, mouthfeel, taste perception and stability of the final cultured dairy products such as yoghurt, drinking yoghurt, cheese, fermented cream, etc. (Duboc and Mollet 2001). Vijayendra et al. (2008) isolated EPS-producing non-ropy strains of lactic acid bacteria (*Leuconostoc* sp. CFR 2181) consisting mainly of glucose (91 %) with minor quantities of rhamnose (1.8 %) and arabinose (1.8 %) from *dahi*. Behare et al. (2009) reported that *dahi* prepared with EPS-producing mesophilic strains, namely, *Lactococcus lactis* subsp. *lactis* B-6 and *Lc. lactis* subsp. *lactis* KT-24, was found to have lower susceptibility to whey separation and received higher sensory scores than *dahi* prepared with non-EPS-producing culture. Kumar and Pal (1994) studied the suitability of reverse osmosis (RO) concentrates for the manufacture of *dahi* and reported that the quality of *dahi* made from 1.5-fold RO concentrates was highly satisfactory. Further, it was reported that the use of highly concentrated RO milk (more than 1.5-fold) for *dahi* making failed to bring the pH down to the desirable level and yielded a product that was extremely thick, lumpy and that lacked a clean pleasant flavour.

With the growing interest in functional foods and being a widely consumed dairy product, *dahi* was chosen as a vehicle to develop functional foods and combat chronic and non-communicable diseases in India. Fortification with minerals (Singh et al. 2005; Ranjan et al. 2006), incorporation of dietary fibre in the form of fruits (Pandya 2002; Kanawjia et al. 2010) and incorporation of probiotic organisms (Yadav et al. 2005) are some of the applications reported. Studies conducted on calcium fortification of cow and buffalo milks for *dahi* making revealed that among the three salts studied, viz. calcium chloride, calcium lactate and calcium gluconate, the quality of *dahi* made from cow milk enriched with calcium gluconate (Singh et al. 2005) and buffalo milk enriched with calcium gluconate and calcium lactate (Ranjan et al. 2006) was comparable with the *dahi* made from the non-calcium fortified milk. Human feeding trials with *dahi* containing *Lactobacillus acidophilus* exhibited no allergic reaction (Patil et al. 2003), and the appearance of probiotic organisms in the urine of 20 % female volunteers indicated their colonization in the vaginal tract (Patil et al. 2005). Probiotic *dahi* has been developed using probiotic lactobacilli, viz. *L. acidophilus* and *L. casei* either alone or in combination with mesophilic *dahi* culture (NCDC-167) or *Lactococcus lactic* subsp. *lactis* biovar. *diacetylactis* (NCDC-60) (Sinha 2004). Recently, Yadav et al. (2007) reported that the probiotic *dahi*-supplemented diet containing *Lb. acidophilus* and *Lb. casei* significantly lowered the risk of diabetes and its complications such as delay in the onset of glucose intolerance, hyperglycaemia, hyperinsulinaemia, dyslipidaemia and oxidative stress in high-fructose-induced diabetic rats. In another study, Rajpal and Kansal (2008) reported that probiotic *dahi* containing *L. acidophilus* and *Bifidobacterium bifidum* was found to attenuate carcinogenesis in gastrointestinal tract.

7.3 Lassi

Lassi is made by blending *dahi* with water, sugar or salt and spices until frothy. The consistency of *lassi* depends on the ratio of *dahi* to water. Thick *lassi* is made with four parts *dahi* to one part water and/or crushed ice. It can be flavoured in various ways with salt, mint, cumin, sugar, fruit or fruit juice and even spicy additions such as ground chillies, fresh ginger or garlic. The ingredients are all placed in a blender and processed until the mixture is light and frothy. Sometimes, a little milk is used to reduce the acid tinge and is topped with a thin layer of *malai* or clotted cream. *Lassi* is chilled and served as a refreshing beverage during extreme summers (Sabikhi 2006; Kanawjia et al. 2010). While sweetened *lassi* is popular mainly in North India, its salted version is widely relished in the southern parts of the country. Various varieties of salted *lassi* include buttermilk, *chhach* and *mattha*. Ancient Indian literature reports that regular use of buttermilk has therapeutic advantages, being beneficial for haemorrhoids (piles), swelling and duodenal disorders. Buttermilk warmed with curry and/or coriander leaves, turmeric, ginger and salt is a therapy for obesity and indigestion as per the Indian medicinal science of *Ayurveda* (Sabikhi and Mathur 2004).

7.3.1 Technological Developments in the Manufacture of Lassi

The keeping quality of *lassi* is extended considerably under refrigeration. Although further extension of shelf life of *lassi* is achieved by ultra-high temperature (UHT) processing of product after fermentation and packaging it aseptically, the sensory quality is adversely affected due to wheying off. To overcome this problem, Aneja et al. (1989) developed a method for manufacture of long-life lassi that does not settle down over extended storage in aseptic packs. Now, UHT-processed *lassi* and spiced buttermilk are commercially manufactured and marketed by different dairies in India. Kumar (2000) developed *lassi* for calorie-conscious and diabetic people using an artificial sweetener and reported that aspartame at a rate of 0.08 % on *dahi* basis was the most acceptable level to prepare low-calorie *lassi*. Recently, George et al. (2010a) successfully replaced sucrose in lassi with blends of artificial sweeteners and reported that aspartame and acesulfame-K obtained highest sensory scores when compared with the best optimized single sweetener aspartame. The storage studies of artificially sweetened lassi revealed that the product resembled the control (15 % sucrose sweetened) in all the sensory attributes up to 5 days of storage (George et al. 2010b). Recently, Khurana (2006) developed suitable technologies for the manufacture of mango, banana and pineapple*lassi* along with their low-calorie counterparts using artificial sweeteners.

7.4 Shrikhand

Shrikhand is another Indian indigenous fermented and sweetened milk product having a typical pleasant sweet–sour taste. It is prepared by blending *chakka* (quarg-like product), a semisolid mass obtained after draining whey from *dahi*, with sugar, cream and other ingredients like fruit pulp, nut, flavour, spices and colour to achieve the finished product of desired composition, consistency and sensory attributes. It is referred to as *shikhrini* in old *Sanskrit* literature and has been a very popular dessert in Western India for several hundred years (Prajapati and Nair 2008). *Shrikhand* has a typical semisolid consistency with a characteristic smoothness, firmness and pliability that makes it suitable for consumption directly after meal or with *poori* (made of a dough of whole-meal wheat, rolled out and deep-fried) or bread. Although largely produced on a small scale adopting age-old traditional methods, *shrikhand* is now commercially manufactured in organized dairy sector to cater to the growing demand. The traditional method of making *shrikhand* involves the preparation of curd or *dahi* by culturing milk (preferably buffalo milk) with a natural starter (curd of the previous batch). After a firm curd is formed, it is transferred in a muslin cloth and hung for 12–18 h to remove free whey. The *chakka* obtained is mixed with the required amount of sugar, colour, flavouring materials and spices and blended to smooth and homogenous consistency (Upadhyay and Dave 1977; Pal and Raju 2007). *Shrikhand* is stored and served in chilled form. The batch-to-batch large variation in the quality and poor shelf life of *shrikhand* are the serious drawbacks of the traditional method. Generally, the recovery of solids in *chakka* is also low. With a view to overcome the limitations of the traditional method, Aneja et al. (1977) developed an industrial process for the manufacture of *shrikhand*. Normally, skim milk is used for making *dahi* for the manufacture of *shrikhand* in this method (Fig. 7.2). By using skim milk, not only fat losses are eliminated, but also faster moisture expulsion and less moisture retention in the curd are achieved (Patel 1982).

The use of the right type of culture is an essential prerequisite for the manufacture of *shrikhand*. Among different starter cultures recommended by various

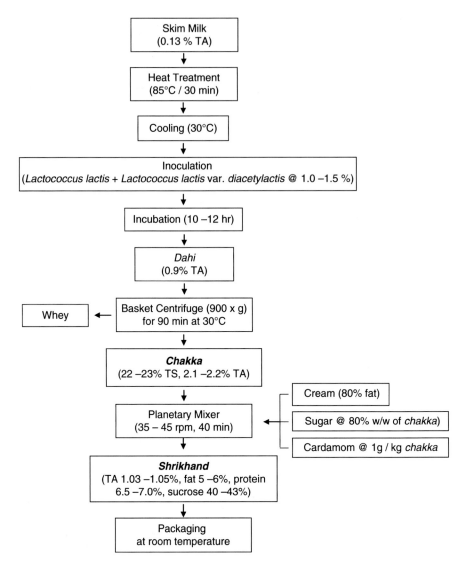

Fig. 7.2 Industrial method of making *shrikhand*

workers, LF-40, a culture containing *Lactococcus lactis* subsp. *lactis* and *Lactococcus lactis* var. *diacetylactis*, has received wide acceptance by many *shrikhand* manufacturers. The LF-40 culture at 1–1.5 % is added to milk, and the milk is incubated at 30 °C for 10–12 h in order to get 0.9 % of titratable acidity in the curd. *Dahi*, so obtained, is centrifuged in a basket centrifuge to get *chakka*. Measured quantities of *chakka*, sugar, cream and additives are mixed in a planetary mixer to get *shrikhand* (Patel and Chakraborty 1985). With a view to extend the shelf life of *shrikhand*, many dairies normally practice thermization. *Shrikhand*, being a semisolid product, is packed in heat-sealable polystyrene containers of various sizes ranging from 100 g to 1.0 kg and stored at refrigerated temperature.

7.4.1 Technological Developments in the Manufacture of Shrikhand

Successful attempts were made by Sharma and Reuter (1998) to adopt ultrafiltartion (UF) technology for making *chakka*, the base material for *shrikhand*. The objective was to recover all the whey proteins and increase the yield of the final product while automating the process. The process involved heating skim milk in a double-jacketed vat with slow agitation up to a temperature of 95 °C for 5 min and then cooling it to 21–22 °C (Fig. 7.3). This was followed by inoculating with mixed starter culture (*Lactococcus lactis*, *Lactococcus lactis* var. *diacetylactis* and *Lactococcus cremoris*) at a rate of 0.1–0.15 %. It was incubated at 21–22 °C for16–18 h so as to get curd with pH of 4.6–4.5 and with a pleasant diacetyl aroma. The coagulum obtained was agitated slowly and subjected to ultrafiltration. Whey was removed in the form of permeate. The *chakka* thus obtained was mixed with 70 % fat cream and sugar so as to manufacture *shrikhand* that contained 6 % fat, 41 % sugar and 40 % moisture. The mixture was then kneaded in a planetary mixer at 25–26 °C in order to get a smooth paste-like semisolid consistency with no feeling of sugar grains. It was reported that there was practically no difference between traditional and UF *shrikhand*. Recently, Md-Ansari et al. (2006) also developed shrikhand using UF pre-concentrated skim milk.

Several attempts have been made to incorporate different additives into *shrikhand* to address the growing interest in the diversification of food products to attract a wider range of consumers. The pulp of fruits such as apple, mango, papaya, banana, guava and sapota (Bardale et al. 1986; Dadarwal et al. 2005), cocoa powder with and without papaya pulp (Vagdalkar et al. 2002) and incorporation of probiotic organisms (Geetha et al. 2003) have been tried in *shrikhand*.

7.5 Misti Dahi

Misti dahi, also called as *mishti doi* or *lal dahi* or *payodhi*, is a sweetened variety of *dahi* popular in Eastern India (De 1980; Aneja et al. 2002; Gupta et al. 2000). It is characterized by a creamy to light brown colour, firm consistency, smooth texture and pleasant aroma. Traditionally, *misti dahi* is prepared from cow or buffalo or mixed milk. It is first boiled with a required amount of sugar and partially concentrated over a low heat during which milk develops a distinctive light cream to light brown caramel colour and flavour. This is then cooled to ambient temperature and cultured with sour milk or previous day's *dahi* (culture). It is then poured into consumer- or bulk-size earthen vessels and left undisturbed overnight for fermentation. When a firm body curd has set, it is shifted to a cooler place or preferably refrigerated. Till recently, *misti dahi* preparation was mainly confined to domestic- or cottage-scale operations (Pal and Raju 2007). With the development of technology for the industrial manufacture of *misti dahi* (Ghosh and Rajorhia 1990), it is now being

Fig. 7.3 Method of making *shrikhand* using UF *chakka*

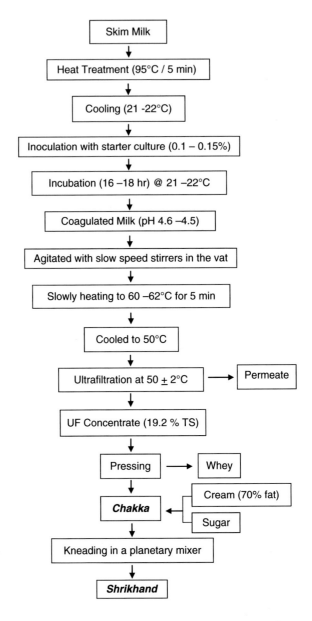

produced and marketed on industrial scale also. The process involves standardization of buffalo milk (5 % fat and 13 % SNF) followed by homogenization at
5.49 MPa pressure at 65 °C, sweetening with cane sugar (14 %) and heating mix to
85 °C for 10 min. Then, the mix is cooled to incubation temperature and inoculated
with suitable starter culture and subsequently incubating to obtain a firm curd. The
firm curd is transferred to cold storage (4 °C) and served chilled.

7.5.1 Technological Developments in the Manufacture of Misti Dahi

Besides several useful virtues as a fermented dairy product, *misti dahi* contains high milk fat and high sugar, which are causes of concern for health-conscious consumers. The growing awareness of impact of high-fat and high-sugar foods on health among consumers may hamper the successful marketing of *misti dahi* in other parts of the country for which fat reduction may be a solution. Successful attempts were made by Raju and Pal (2009) to develop a technology for the production of reduced-fat *misti dahi*. It was reported that highly acceptable reduced-fat *misti dahi* can be produced from buffalo milk standardized to 3 % milk fat and 15.0 % milk solids-not-fat (MSNF) and addition of 14 % cane sugar and 0.1 % caramel powder (Fig. 7.4).

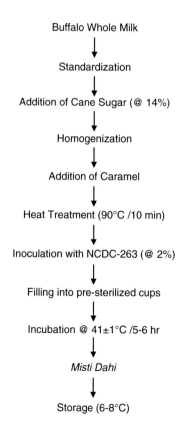

Fig. 7.4 Flow diagram for the manufacture of reduced-fat *misti dahi*

Buffalo Whole Milk
↓
Standardization
↓
Addition of Cane Sugar (@ 14%)
↓
Homogenization
↓
Addition of Caramel
↓
Heat Treatment (90°C /10 min)
↓
Inoculation with NCDC-263 (@ 2%)
↓
Filling into pre-sterilized cups
↓
Incubation @ 41±1°C /5-6 hr
↓
Misti Dahi
↓
Storage (6-8°C)

7.6 Conclusion

Traditional fermented dairy products are an integral part of Indian heritage and have great social, religious, cultural, medicinal and economic importance and have been developed over a long period with the exceptional culinary skills of homemakers. Each of these products has its unique flavour, texture and appearance. With the developments in the processing equipment, packaging materials and discovery of novel food additives, the quality of these fermented dairy products have considerably improved over a period of time. Also, with the changing lifestyle and growing awareness of the link between the diet and health among the consumers, the demand for fermented dairy products with enhanced health benefits is increasing.

References

Aneja RP, Vyas MN, Nanda K, Thareja VK (1977) Development of an industrial process for the manufacture of *Shrikhand*. J Food Sci Technol 14(4):159–163

Aneja RP, Vyas MN, Sharma D, Samal SK (1989) A method for manufacturing of *lassi*, Indian Patent 17374

Aneja RP, Mathur BN, Chandan RC, Banerjee AK (2002) Technology of Indian milk products. Dairy India, New Delhi

Bardale PS, Waghmare PS, Zanjad PN, Khedkar DM (1986) The preparation of *Shrikhand*-like product from skimmed milk *chakka* by fortifying with fruit pulps. Indian J Dairy Sci 39(4):480–483

Behare P, Singh R, Singh RP (2009) Exopolysaccharide-producing mesophilic lactic cultures for preparation of fat-free *dahi*—an Indian fermented milk. J Dairy Res 76:90–97

Campbell-Platt G (1987) Fermented foods of the World: a dictionary and guide. Butterworths, London, p 290

Dadarwal R, Beniwal BS, Singh R (2005) Process standardization for preparation of fruit flavored *Shrikhand*. J Food Sci Technol 42(1):22–26

De S (1980) Outlines of dairy technology. Oxford University Press, New Delhi

Duboc P, Mollet B (2001) Applications of exopolysaccharides in dairy industry. Int Dairy J 11:759–768

Geetha VV, Sarma KS, Reddy VP, Reddy YK, Moorthy PRS, Kumar S (2003) Physico-chemical properties and sensory attributes of probiotic *shrikhand*. Indian J Dairy Biosci 14(1):58–60

George V, Arora S, Sharma V, Wadhwa BK, Singh AK (2010a) Stability, physico-chemical, microbial and sensory properties of sweetener/sweetener blends in lassi during storage. Food Bioprocess Technol 5(1):323–330. doi:10.1007/s11947-009-0315-7

George V, Arora S, Wadhwa BK, Singh AK, Sharma GS (2010b) Optimization of sweetener blends for the preparation of *lassi*. Int J Dairy Technol 63(2):256–261

Ghosh J, Rajorhia GS (1990) Technology for production of *misti dahi*—a traditional fermented milk product. Indian J Dairy Sci 43(2):239–246

Guizani N, Mothershaw A (2007) Fermentation as a method for food preservation. In: Rahman MS (ed) Handbook of food preservation, 2nd edn. CRC Press, Boca Raton, pp 215–236

Gupta RC, Mann B, Joshi VK, Prasad DN (2000) Microbiological, chemical and ultrastructural characteristics of *mishti doi* (sweetened *dahi*). J Food Sci Technol 37(1):54–57

IS: 9617 (1980) Indian Standard Specification for *Dahi*. Indian Standards Institution, Manak Bhavan, New Delhi

Kanawjia SK, Chatterjee A, Raju PN (2010) Developments in buffalo milk cheeses and fermented milks. In: Proceedings of international buffalo conference, New Delhi, 1–4 Feb, pp 147–153

Khurana HK (2006) Development of technology for extended shelf life fruit *lassi*. Ph.D. Thesis, National Dairy Research Institute (Deemed University), Karnal

Kumar S, Pal D (1994) Quality of *dahi* (curd) manufactured from buffalo milk concentrated by reverse osmosis. Indian J Dairy Sci 47(9):766–769

Kumar M (2000) Physico-chemical characteristics of low-calorie lassi and flavored dairy drink using fat replacer and artificial sweetener. M.Sc. Thesis, National Dairy Research Institute(Deemed University), Karnal

Md-Ansari IA, Rai P, Sahoo PK, Datta AK (2006) Manufacture of *shrikhand* from ultrafiltered skim milk retentates. J Food Sci Technol 43(1):49–52

Pal D, Raju PN (2007) Traditional Indian dairy products-an overview (Theme paper). In: Souvenir of the international conference on traditional dairy products, N.D.R.I., Karnal, 14–17 Nov, pp i–xxvii

Pandya CN (2002) Development of technology for fruit *dahi*. M.Tech. Thesis, National Dairy Research Institute (Deemed University), Karnal

Patel RS (1982) Process alterations in *shrikhand* technology. Ph.D. Thesis, Kurukshetra University, Kurukshetra

Patel RS, Chakraborty BK (1985) Factors affecting the consistency and sensory properties of *shrikhand*. Egypt J Dairy Sci 13(1):73–78

Patil AM, Khedkar CD, Chavan BR (2003) Studies on safety aspects of probiotic *dahi*. Seminar Report on fermented foods, health status and social wellbeing, Anand, 13–14 Nov

Patil AM, Desosarkar SS, Kalyankar SD, Khedkar CD (2005) Is a probiotic *dahi* safe for human consumption? In: XXXIV dairy ind. conference, Bangalore, DP No. 49, 23–25 Nov

Prajapati JB, Nair BM (2008) The history of fermented foods. In: Farnworth ER (ed) Handbook of fermented functional foods, 2nd edn. CRC Press, Boca Raton, pp 1–24

Rajpal S, Kansal VK (2008) Buffalo milk probiotic *dahi* containing *Lactobacillus acidophilus*, *Bifidobacterium bifidum* and *Lactococcus lactis* reduces gastrointestinal cancer induced by dimethylhydrazine dihydrochloride in rats. Milchwissenschaft 63(2):122–125

Raju PN, Pal D (2009) The physico-chemical, sensory and textural properties of *misti dahi* prepared from reduced fat buffalo milk. Food Bioprocess Technol 2:101–108

Ranganadham M, Gupta SK (1987) Sensory evaluation of *dahi* and yoghurt. Indian Dairyman 39(10):493–497

Ranjan P, Arora S, Sharma GS, Sindhu JS, Singh G (2006) Sensory and textural profile of curd (*dahi*) from calcium enriched buffalo milk. J Food Sci Technol 43(1):38–40

Sabikhi L (2006) Developments in the manufacture of *lassi*. In: Lecture compendium of the short course on developments in traditional dairy products. Centre of Advanced Studies, NDRI, Karnal, pp 64–67

Sabikhi L, Mathur BN (2004) Dairy products in human health: traditional beliefs vs. established evidence. Indian Dairyman 56(8):61–66

Sharma DK, Reuter H (1998) Ultrafiltration technique for *shrikhand* manufacture. Indian J Dairy Sci 45(4):209–213

Singh R (2007) Characteristics and technology of traditional Indian cultured dairy products. Bull Int Dairy Fed 415:11–20

Singh G, Arora S, Sharma GS, Sindhu JS, Rajan P, Kansal VK (2005) Sensory and textural profile of *dahi* from calcium enriched cow milk. Indian J Dairy Sci 58(1):23–26

Sinha PR (2004) Utilization of probiotic organisms in the preparation of *dahi* and *lassi*. Annual report 2004-2005, National Dairy Research Institute, Karnal, p 18

Upadhyay KG, Dave JM (1977) *Shrikhand* and its technology. Indian Dairyman 28(9):487–490

Vagdalkar AA, Chavan BR, Mokile VM, Thalkari BT, Landage SN (2002) A study on preparation of *shrikhand* by using cocoa powder and papaya pulp. Indian Dairyman 54(4):49–51

Vijayendra SVN, Palanivel G, Mahadevamma S, Tharanathan RN (2008) Physico-chemical characterization of an exopolysaccharide produced by a non-ropy strain *Leuconostoc* sp. CFR 2181 isolated from *dahi*, an Indian traditional lactic fermented milk product. Carbohydr Polym 72(2):300–307

Yadav H, Jain S, Sinha PR (2005) Preparation of low fat probiotic *dahi*. J Dairying Foods & Home Sci 24(3/4):172–177

Yadav H, Jain S, Sinha PR (2007) Antidiabetic effect of probiotic *dahi* containing *Lactobacillus acidophilus* and *Lactobacillus casei* in high fructose fed rats. Nutrition 23:62–68

Chapter 8
Traditional Bulgarian Dairy Food

Anna Aladjadjiyan, Ivanka Zheleva, and Yordanka Kartalska

8.1 Introduction

Traditional Bulgarian food is mostly based on the diets of the stockbreeders and land tillers (farmers who cultivate the land) from the historical and cultural regions of the Balkans, the Caucasus, and Hither Asia, as well as from the territories inhabited by the Slavs. The familiarity is associated with the use of different fermentation agents (for dough, milk, fruit, and vegetable manufacturing).

As described by Markova (2006), the food of the Eastern Slavs—typical land tillers—is based on yeast-leavened dough products, flour porridges, fermented fruit, vegetables and drinks, and pork. They used two kinds of milk yeast (ferment and rennet) as well as the ferment for wine preparation. The usual menu of the typical stockbreeders (the population of Central Asia and Hither Asia) was based on the yeastless dough and fermented dairy foods.

Less known or used were fermented vegetables and fruit (dried rather than fermented fruit is preferred).

Most of these food products are spread over Bulgarian traditional diet. In the past the bigger part of Bulgarian people used to be tillers and stockbreeders, which explains the availability of raw materials like milk, meat, fruits, and vegetables.

One of the most popular Bulgarian traditional foods is milk-based products.

As many sociological and ethnological studies underline (Markova 2006; Keliyan 2004; Murdzheva-Pancheva 2007), the fermented dairy foods take an important place in Bulgarian diet.

A. Aladjadjiyan, Professor, DSc,PhD • Y. Kartalska
Agricultural University, 12 Mendeleev Str., Plovdiv, Bulgaria

I. Zheleva (✉)
Rousse University, 8 Studentska Str., Rousse, Bulgaria
e-mail: vzh@abv.bg

© Springer Science+Business Media New York 2016
K. Kristbergsson, J. Oliveira (eds.), *Traditional Foods*, Integrating Food
Science and Engineering Knowledge Into the Food Chain 10,
DOI 10.1007/978-1-4899-7648-2_8

There are a number of fermented milk products: yogurt, katyk, cheese, kashkaval, ayran, etc. The most popular recipes for preparation vary depending on the sources and locations.

8.2   Bulgarian Yogurt

Kiselo mlyako is a Bulgarian yogurt.

The Bulgarian yogurt is a unique and famous worldwide lactic acid product. For many years it has been considered a food that helps keep people's life longer and healthier.

When and where it came from is still unknown. One of the theories for its origin is connected to the Thracians. Ancient Thracia had rich soil, vegetation, and pastures which contributed to the development of sheep farming, so the basic domestic animal of Thracians was sheep. They discovered that the milk "spoiled" from lactic acid lasted longer than the row milk. By adding spoiled yogurt to boiled milk, they produced the well-known fermented milk referred to as "prokish."

Another theory is connected to the proto-Bulgarians. It is assumed that the Bulgarian-type lactic acid milks originate from the lactic acid drink "kumis," which the proto-Bulgarians prepared from mare's milk. When they settled into current Bulgarian territories and hoarding sheep, they began to make "kumis" from sheep milk. Proto-Bulgarians produced lactic acid milk by the name of "katuk" as they curdled raw sheep milk with cheese whey. This product usually was produced at the end of the summer when the milk contains higher content of dry substances. The katuk was considered delicious and they lasted the whole winter.

The uygurits who lived in the Sindgan region in Northwest China still call the lactic acid milk "kuthuk." They make it from both mare's milk and sheep's milk. The Bulgarian name suggests the proto-Bulgarian origin of yogurt. Neither one of these theories is confirmed.

Bulgarian acid milk is first mentioned in the literature during the eighth century, by the Turkish name yogurt. Genghis Khan (1206–1227) used yogurt as food for his army and as a means to preserve meat. The milk was stored in sheep's stomachs, and the microflora present in milk undergoes lactic acid fermentation, resulting in acid milk.

In Western Europe yogurt owes its fame to the French king François I. He suffered from an incurable diarrhea and asked for help from his alley, the Ottoman sultan Suleiman the Great. He sent him a doctor who managed to cure him with a diet of yogurt. In his gratitude the French king spread the information about the food that helped him all over Europe.

The first Bulgarian ever to study the microflora of Bulgarian yogurt was Stamen Grigorov (1878–1945), who was a student of medicine in Geneva. In 1905 he describes it as a one rod- and one round-shaped bacteria. In 1907 the rod bacteria is called *Lactobacillus bulgaricus*. In 1917 Orla-Jensen proved that in the production of the Bulgarian milk, not only *Lactobacillus bulgaricus* participates, but *Streptococcus thermophilus* also has a role.

The unique characteristic of yogurt is that it contains cultures of *Streptococcus thermophilus* and *Lactobacillus delbrueckii* ssp. *bulgaricus*.

All dairy fermentations use lactic acid bacteria for acidification and flavor production.

Although lactic acid bacteria are genetically diverse, common characteristics of this group include being gram-positive, non-spore forming, non-pigmented, and unable to produce iron-containing porphyrin compounds (catalase and cytochrome); growing anaerobically but being aerotolerant; and fermenting sugar with lactic acid as a major end product. Lactic acid bacteria tend to be nutritionally fastidious, often requiring specific amino acids, B vitamins, and other growth factors but unable to utilize complex carbohydrates.

There are currently 11 genera of lactic acid bacteria, of which four—*Lactobacillus*, *Streptococcus*, *Lactococcus*, and *Leuconostoc*—are commonly found in dairy starter cultures. A fifth genus, *Enterococcus*, is occasionally found in mixed-strain (undefined) starter cultures. Important phenotypic taxonomic criteria include morphological appearance (rod or coccus), fermentation end products (homofermentative or heterofermentative), carbohydrate fermentation, growth temperature range, optical configuration of lactic acid produced, and salt tolerance.

Lactic acid bacteria are generally associated with nutrient-rich habitats containing simple sugars. These include raw milk, meat, fruits, and vegetables. They grow with yeast in wine, beer, and bread fermentations. In nature, they are found in the dairy farm environment and in decomposing vegetation, including silage.

The only *Streptococcus* sp. useful in dairy fermentation is *S. thermophilus*. This microorganism is genetically similar to oral streptococci (*S. salivarius*) but can still be considered a separate species. *S. thermophilus* is differentiated from other streptococci (and lactococci) by its heat resistance, ability to grow at 52 °C, and ability to ferment only a limited number of carbohydrates (Axelsson 1993).

The *Lactobacillus* genus consists of a genetically and physiologically diverse group of rod-shaped lactic acid bacteria. The genus can be divided into three groups based on fermentation end products. Homofermentative lactobacilli exclusively ferment hexose sugars to lactic acid by the Embden-Meyerhof pathway. They do not ferment pentose sugars or gluconate. These are the lactobacilli (*Lb. delbrueckii* subsp. *bulgaricus*, *Lb. delbrueckii* subsp. *lactis*, and *Lb. helveticus*) commonly found in starter cultures. They grow at higher temperatures (45 °C) than lactobacilli in the other groups and are thermoduric. Another member of this group, *Lb. acidophilus*, is not a starter culture organism, but it is added to dairy foods for its nutritional benefits.

Facultative heterofermentative lactobacilli ferment hexose sugars either only to lactic acid or to lactic acid, acetic acid, ethanol, and formic acid when glucose is limited. Pentose sugars are fermented to lactic and acetic acid via the phosphoketolase pathway. Yogurt is made using a combination of *S. thermophilus* and *Lb. delbrueckii* subsp. *bulgaricus*. These organisms grow in a cooperative relationship, resulting in rapid acidification. The presence of lactobacilli stimulates growth of the more weakly proteolytic *S. thermophilus*, because lactobacilli liberate free amino acids and peptides from casein. *S. thermophilus*, in turn, stimulates growth of *Lb. delbrueckii* subsp. *bulgaricus*, possibly by removing oxygen, lowering pH, and producing formic acid and pyruvate. Yogurt may also contain *Lb. acidophilus* or other nutritionally beneficial cultures. The most important characteristics for yogurt cultures are rapid acidification, production of characteristic balanced flavor, and ability to produce the desired texture. The ideal yogurt flavor is a balanced blend of acidity and acetaldehyde. This is achieved through culture selection, balance of rod to coccus ratio, and fermentation control.

The western world calls it Bulgarian yogurt, but in its homeland, Bulgaria, it's called sour (kiselo) milk (mlyako).

Bulgarian yogurt (Kiselo mlyako) can be produced from fresh milk from cows, sheep, goats, and buffalo; the lowest fat content is found in cows' yogurt, while buffalo has the highest fat content. It is used as part of a number of main courses as well as being consumed as a dessert.

Bulgarian yogurt may be organic if the animals are kept organically, and it is an excellent diet food and source of "good" bacteria and proteins that are needed in the digestive tract, which are necessary for the body to function at its best. These good bacteria prevent the growth of harmful bacteria that cause bacterial infections and diseases. They also promote digestive health and boost the immune system. It helps to prevent osteoporosis and reduces the risk of high blood pressure. Active cultures help certain gastrointestinal conditions including lactose intolerance, constipation, diarrhea, colon cancer, inflammatory bowel disease, pylori infection, and many others. Yogurt is nutritionally sound and it also makes you feel full faster.

8.2.1   Home Preparation of Bulgarian Yogurt

Bring the milk to boil, put in jars or other large containers, and wait until it cools down. It has to be hot but cold enough so you can put your finger in it and hold it there for a few seconds without burning yourself.

Dilute the starter bacteria in some lukewarm milk and add it to the jars. Stir well. Cover the jars with blankets and tuck them in good. Do not put the lids on the jars so the milk can breathe. Let them stay overnight or for about 8 h. After this the yogurt can be stored in a refrigerator.

8.2.2   Yogurt Fair

Bulgarians are proud of their yogurt and even have a special yogurt fair. If you come to Bulgaria, to the town of Razgrad in July, you can feel the unique atmosphere of the traditional yogurt fair—the only one of its kind in the world. There are competitions for homemade yogurt and traditional dairy dishes and international folk concerts in the open. Before performing on the stage, all participants march along the main boulevard in Razgrad in colorful costumes performing music and dances. Every year at the bazaar in the central town square, producers and distributors offer a wide range of different trademarks of yogurt and milk products.

8.3 Katyk

Katyk or *krotmach* (*krutmach, kurtmach*) is a specific national dairy product with a salty-sour taste, typically prepared in Bulgaria. The name actually is used for several products with similar taste but produced in different ways.

Katyk has a special reserved place on the table of the contemporary Bulgarian family. It is an object of scientific investigations related to its influence on health, together with some other traditional dairy products. In these products the species composition of lactic acid bacteria has more variability and inconstancy compared to commercially traded products. An investigation by Tserovska et al. (2002) has shown that 18 lactic acid bacterial strains can be isolated from homemade katyk. Nine of them were reported to belong to lactic acid cocci, and others were referred to genus *Lactobacillus*. The lactic acid fermentation, which these bacteria perform, has long been known and applied by people for making different foodstuffs. It plays an essential role in the production of all dairy products and is involved in the production of many other Bulgarian foods and drinks—sausages, pickles, boza, etc.

Traditionally, katyk is a dairy product that was obtained by inoculation of sheep's milk, with cheese used for yeast. The resulting product is durable and can be kept for several months. To obtain a thicker and denser product, the sheep's milk needs to be boiled on slow fire for several hours prior to inoculation.

An old recipe for katyk preparation is described in the ethnographic investigation of village Svoboda by Ivanka Kancheva Murdzheva-Pancheva. The starting material must be sheep's milk, collected in August, when milk is most dense. It is additionally thickened by boiling in a water bath. The milk is placed in a smaller container and immersed in a larger water tank. The small container containing the milk should not touch the outer tank. Milk must be boiled continuously until thickened to the desired consistency and form a thick surface cream. It needs to be stirred periodically with a wooden spoon. Once cooked, milk is removed from the water bath and allowed to cool while being stirred continuously. Once cooled, salt is added to taste. It is then continuously stirred every day to thicken the product.

Katyk can also be made with uncooked milk, but the cooked one is considered more delicious. In the case of uncooked milk, it is initially salted to taste and stirred every day, but it thickens more slowly.

After its preparation katyk must be stored in appropriate containers. In the past, it was usually stored in sheep skins—sheep skin that had been well washed with soap. After the katyk has been poured in the sheep skin pouch, the air is expelled, and it is stored in a cool, ventilated place. Today it may be stored in glass or plastic containers in a cool place. When katyk is stored in sheep skin, the residual liquid is released through the pores of the skin, which makes katyk more delicious.

Nowadays the name katyk often used for a more or less homogeneous mixture of yogurt and sheep's milk cheese with a possible addition of butter. This, of course, is not the original product.

A contemporary recipe from the Rodopi mountain region is as follows: a 5 L fresh sheep's milk is boiled in a water bath, then salted, and, when cool, mixed with 250 g of sheep cheese and 125 g melted butter. This mixture is then poured into a clay pot and stored.

There is a similar recipe given in the blog of Bulgarian lady which gives the following as necessary ingredients:

5 L fresh sheep milk
2 full spoons homemade sheep yogurt
6–7 guts cheese yeast
250–300 g homemade sheep cheese
3 spoons salt

The way of preparation is as follows:

The milk is boiled in a water bath for approximately 2 h. Then, it is cooled so that it may be touched. In another bowl salt, yogurt, yeast, and some fresh milk are stirred well until salt is melted. Then, this mixture is poured in the container with the fresh milk and stirred well. Then, the crumbled cheese is added and the mixture is stirred again. The container is closed and set aside for 12 h. After this period it is stirred again. Then, it is allowed to drain well. Once drained, katyk is placed in jars and put in a refrigerator to cool, and then it is ready for consumption.

8.4 Sirene

Bulgarian white cheese is a variety of feta cheese but can only be produced in Bulgaria. This is due to specific lactose-tolerant bacteria which convert milk into yogurt and then sirene. The bacteria are only found in this part of the world; hence, it is named *Lactobacillus bulgaricus*. Bulgarian white cheese is a brined goat, sheep, or cow cheese.

8.5 Kashkaval

Kashkaval is the typical yellow cheese of Bulgaria. Very similar to cheddar types of cheese, kashkaval is made of goat, sheep, or cow's milk. It is then aged for a certain period of time (about 6 months) to develop its very particular and defining flavor.

8.5.1 Homemade Kashkaval

Fresh milk (at least 3.5 % fat) without any additional ingredients needs to be pasteurized—heated to almost boiling temperature and then cooled to 37 °C. Ready-pasteurized milk can be used too. The process of pasteurization kills any harmful bacteria that would alter the taste of the cheese.

After cooling the protein of milk is coagulated with rennet–renin—derived from lamb ventricles or from some plants or synthetic. The process is referred to as curdling. Milk can also be "curdled" with adding a weak acid, for example, salt of lemon or vinegar but then missing bacteria that ultimately determine the taste of cheese during maturing, and the cheese ripens more slowly.

The cheese mass (white flakes) is then salted, leavened, and placed under a load to remove excess water and to increase density. Salting also helps to exude the water and increase tightness of cheese portions.

If the cheese mass is heated to 50–70 °C and kneaded, the result will be hard cheese, called kashkaval in Bulgaria.

After removal from the press, the cheese lump is salted periodically. Then, it is stored at around 10–12 °C for 20–45 days until it reaches "maturity."

The ready cheese may be coated with paraffin or wax to prevent drying (immersed in the molten defensive wax).

8.5.2 Another Recipe

A popular in Bulgaria recipe says:

Five liters of homemade milk (non-pasteurized) is boiled. Pour in it three soup spoons crushed sea salt; stir well to dissolve the salt. Then, add one not-full soup spoon salt of lemon and stir the mixture for 10 min on weak fire. The resulting slurry is placed in a cloth to drain. After 12 h the slurry is pressed with a burden and put in the refrigerator.

After 2 days the cheese is ready for eating.

8.6 Ayran

Ayran or airan is a yogurt drink, very popular in Bulgaria and many other countries in the Balkan region and further east. In general, it is a mixture of yogurt, water, and salt. It originated as a way of preserving yogurt by adding salt. If you want to add some twist to it, you can try it with some finely chopped mint leaves mixed into the ayran. Ingredients are four cups yogurt, two cups water, and pinch of salt. Preparation is very easy—mix well, cool lightly, and the drink is ready for consumption.

8.7 Tarator

Tarator is typical Bulgarian milk product—it is a kind of a cold soup made of yogurt and cucumber (dill, garlic, walnuts, and sunflower oil are sometimes added). Tarator is very popular in Bulgaria, especially during the summer time. Home preparation of tarator is very easy: Ingredients are 1 long cucumber, chopped or grated (we prefer it peeled); 1 garlic clove, minced or mashed; 4 cups yogurt; 1 cup water; 1 teaspoon salt (we like it saltier); 1 tablespoon dill, finely chopped; 4 big pecans, well crushed; and 3 teaspoons olive oil. Preparation—put all ingredients together and mix well. When ready garnish with olive oil (or other favorite oil). Best when chilled. All Bulgarians and tourists like this famous summer soup tarator very much.

References

Axelsson LT (1993) Lactic acid bacteria: classification and physiology. In: Salminen S, von Wright A (eds) Lactic acid bacteria. Marcel Dekker, New York, p 1

Kancheva Murdzheva-Pancheva I (2007) Ethnografic investigation of village Svoboda (Alipashinovo), Chirpan region (Bulgarian: Иванка Кънчева Мурджева–Панчева, Етнографско изследване на с. Свобода (Алипашиново), Чирпанско, ИК Флъорир)

Keliyan M (2004) The intelligentsia, consumption patterns and middle strata self-identification in two peripheral communities (Bulgarian). Sociological problems (1–2/2004), pp 169–182

Markova M (2006) Traditional food and typology farmers/stockbreeders (Bulgarian). Bulgarian Ethnology (2/2006), pp 161–173

Tserovska L, Stefanova S, Yordanova T (2002) Identification of lactic acid bacteria isolated from katyk, goat's milk and cheese. J Cult Collect 3:48–52

Chapter 9
Dulce de Leche: Technology, Quality, and Consumer Aspects of the Traditional Milk Caramel of South America

María Cecilia Penci and María Andrea Marín

9.1 Introduction

Dulce de leche, also called milk caramel, is a soft cream dairy-based confectionery product of a brown color, similar in many ways to sweetened condensed milk. Widely consumed in Argentina and in other Latin American countries for more than a century, *dulce de leche* is now also appreciated in other countries for household and industrial uses (Malec et al. 2005). Traditionally, it is made by a heat concentration of whole milk with sucrose, until a thick, creamy product with 70 % (w/w) total solids is obtained. *Dulce de leche* can be eaten alone or spread on bread or toast (Fig. 9.1); it can be used as filling for crepes, cookies, cakes, and *alfajores*, product consisting of two or more cookies, crackers, or baked dough, separated by fillings such as jams, jellies, or other sweets, having a bath or outer covering (Código Alimentario Argentino, Argentine Food Code 2009, Article 132) (Fig. 9.2), or as topping for ice cream and fruits. It can also be used for industrial purposes to produce candies, ice cream, and chocolate, among others.

The origin of *dulce de leche* is still under discussion, since Argentina, Uruguay, Chile, and other Latin American countries claim it as a "national product."

Buenos Aires' (Argentina) oral tradition states that in 1829, during the signing of the Pact of Cañuelas, a peace treaty between Juan Manuel de Rosas and Juan

M.C. Penci (✉)
Departamento de Química Industrial y Aplicada, Facultad de Ciencias Exactas, Físicas y Naturales Universidad Nacional de Córdoba, Córdoba, Argentina
e-mail: cpenci@efn.uncor.edu; cpenci@gmail.com

M.A. Marín
Departamento de Química Industrial y Aplicada, Facultad de Ciencias Exactas, Instituto de Ciencia y Tecnología de los Alimentos, Físicas y Naturales Universidad Nacional de Córdoba, Córdoba, Argentina
e-mail: ma.andrea.marin@gmail.com

© Springer Science+Business Media New York 2016
K. Kristbergsson, J. Oliveira (eds.), *Traditional Foods*, Integrating Food Science and Engineering Knowledge Into the Food Chain 10,
DOI 10.1007/978-1-4899-7648-2_9

Fig. 9.1 Dulce de leche spread on toast, usually as part of breakfast

Fig. 9.2 Alfajores filled with dulce de leche, a common sweet snack in Argentina (household)

Lavalle, a servant of Juan Manuel de Rosas was in charge of heating the milk with sugar for breakfast. When Lavalle arrived, the maid was distracted and forgot the mixture on the cooker; it began to boil and when she came back to the kitchen, she found this caramel color paste in the pot. Juan Manuel de Rosas tried it out and found it very good and tasteful. This casual event turned into the traditional recipe of homemade *dulce de leche*.

In 2003, Argentina claimed at the United Nations that *dulce de leche* should be declared a gastronomic and cultural heritage of the country, while Uruguay claimed that it should be called as a product from "Rio de la Plata" (the river between Argentina and Uruguay), instead.

Furthermore, in 2006, the Secretariat of Agriculture, Livestock, Fisheries and Food of Argentina (SAGPyA) defined a quality protocol (Code: SAA012 Version: 06) for *dulce de leche*. This protocol describes the attributes of quality expected of a milk caramel to bear the stamp of quality "Argentine Food, a Natural Choice," both in Spanish and in English (Resolution SAGPyA N° 798/2006).

In each Latin American country, *dulce de leche* has different names, and its recipe is slightly diverse, except in Argentina and Uruguay, where both the recipe and the name are the same. In Mexico it is called *cajeta*, and it is made of goats' milk. In Colombia, Venezuela, and Panama, it is called *arequipe*; in Bolivia, Ecuador, Chile, and Peru, *manjar blanco* (sometimes only *manjar*); in Brazil, *doce de leite*; and in Cuba it is called *dulce de leche cortada* (cut, as they cook it differently, and it is not as soft a paste as in other countries).

Argentine regulations (Argentine Food Code 2009, Chapter VIII, Article 592, Secretary for Policy, Regulation and Sanitation (SPRyRS) and SAGPyA Joint Resolution No. 33/2006 and No. 563/2006) establish several types of this product: *dulce de leche*, *dulce de leche* for confectionary, *dulce de leche* with aggregates (cocoa, chocolate, almonds, peanuts, dried fruits, cereals, and/or other food products alone or in combinations), and even a low calorie content *dulce de leche* is admitted (Argentincan Food Code 2009, Chapter XVII, Article 1339—Joint Resolution Secretary of Policy, Regulation and Institutes (SPReI) and SAGPyA No. 94/08 and No. 357/08).

In 2010, Argentina produced 129,000 ton of *dulce de leche*. About 6,800 ton has been exported to different countries like Chile, Canada, Brazil, Syria, and the United States, among others. Argentine production is commercialized around the world in glass, tin, or cardboard containers. Five kilogram cardboard packages constitute the largest exports, around 50 %, and tins, glass, or cardboard of a net content below 1 kg represents around 42 % of total exports (National Health Service and Food Quality SENASA 2010). The typical Argentine consumes approximately 3 kg of *dulce de leche* a year, in different desserts. In Argentina, *dulce de leche* has an average price of 1.49 US$/kg.

9.2 *Dulce de Leche* Production

9.2.1 *Manufacturing Process*

Dulce de leche (DDL) is subjected to a heat concentration process in which water is removed by evaporation while milk protein reacts with reducing sugars and the emulsion of milk fat stabilizes in the sugar phase. Milk helps to develop color, by Maillard reactions, and flavor while holding moisture within the sugary matrix.

Although it seems to be a very simple process, there are particular requirements specified in the *Grupo Mercado Común* (GMC, Common Market Group) Resolution No. 137/96, MERCOSUR Technical Regulation "Identity and quality of *Dulce de leche*." This resolution defines some basic raw materials that can be used. Among them, milk and sucrose are considered as mandatory, and cream, solids of dairy origin, mono- and disaccharides to replace sucrose in a maximum of 40 % w/w, starch, and others are optional. Allowed additives are also established, and it is stated that the name *dulce de leche* is reserved for the product in which the dairy base contains only fat and protein from this source.

There are also safety and quality requirements for raw materials, and production inputs as well as basic fixed operations, equipment, and process control parameters are involved in the manufacturing process of DDL.

The above description is condensed in a single tool, a provision and production matrix, as shown in Fig. 9.3, including:

(a) Raw materials or ingredients and production inputs
(b) Quality and safety requirements for ingredients and inputs
(c) Manufacturing stages of *dulce de leche*
(d) Equipment and process control parameters

The industrial manufacturing stages listed in the matrix refer to a standard process. The main ingredient supplying the solids is normally either whole or skimmed milk, but concentrated milk may be used to reduce process time. Milk acidity must be adjusted to 13 °D (°D, Dornic degree, acidity expressed in lactic acid) adding sodium bicarbonate (0.04–0.06 % w/w) as a pH drop produces milk protein coagulation and also contributes to lactose crystallization resulting in a DDL with sandy texture which tends to reduce consumer acceptance. An acidity excess may also be detrimental to the Maillard reaction. In contrast, if more than the required sodium bicarbonate is used, DDL will have a strong coloration and a rubbery texture.

Sucrose is the other main ingredient traditionally used to produce DDL (max 30 kg 100 L^{-1} milk). Alternative sugars such as glucose are generally used to replace a proportion of the sucrose in confectionary products in order to modify the sweetness and/or textural properties. For DDL, the replacement of 5–15 g 100 g^{-1} with glucose syrup will have the effect of lowering the overall crystal size and/or smoothing the confection (Pepper 1990).

Raw materials are kept in accordance with safety requirements. The neutralized milk is preheated in a tubular heat exchanger until 30 °C. After that, sugar (sucrose/glucose) is added according to the recipe used in a mixing chamber at 60–70 °C.

The concentration of solids in this mixture to produce DDL can be achieved with different type of industrial equipment, in a continuous or a mixed process (Fig. 9.4).

For a batch heat concentration process, a jam pail is used (Fig. 9.5). In general, a typical jam pail for industrial purposes has a 1,000 L capacity (500 kg of DDL) (Fig. 9.6). In small-scale manufacturing processes of DDL, milk neutralization, preheating, and sugar addition can be done at once in the jam pail according to the experience of the operator. Usually, when solids achieve 55°Brix–60°Brix, glucose syrup is added. Starch or other texture modifiers are added too. The solids are concentrated until 70°Brix (sugar content, one degree Brix corresponds to 1 g of sucrose in 100 g of solution). In general, flavors (vanilla) are the latest ingredients to be incorporated.

A continuous heat concentration process can be achieved employing a heating (85 °C) and pre-concentration heating stage (110–150 °C, 1 min) before mixing neutralized milk and sugar. The milk acquires a characteristically brown color of Maillard reactions. After that, the current is cooled (50–55 °C) and fed to a multiple-effect evaporator (triple).

Raw materials/Ingredients and production inputs	Quality and safety requirements for ingredients-inputs	MANUFACTURING STAGES OF DULCE DE LECHE	Equipment and process control parameters
- **MILK:** whole, low-fat or skimmed; fluid or evaporated.	T< 5 °C **Acidity:** 0,14 to 0,18 (g lactic acid . 100ml⁻¹) **Density** at 15°C: 1,028-1,034 **Milk solids not fat content** >8,2 g.100g⁻¹ **Total protein** >2,9 g.100g⁻¹ **Freezing point** < −0,512 °C **Alcohol testing:** stable **Boiling test:** stable **GMP at dairy manufacturing plant**	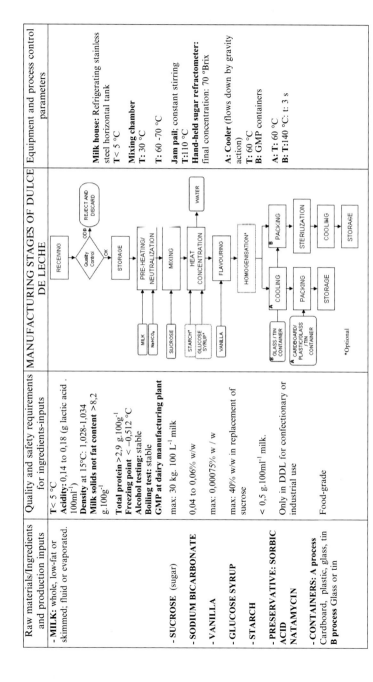	**Milk house:** Refrigerating stainless steel horizontal tank T< 5 °C **Mixing chamber** T: 30 °C T: 60 -70 °C
- **SUCROSE** (sugar)	max: 30 kg. 100 L⁻¹ milk		**Jam pail:** constant stirring T:110 °C **Hand-held sugar refractometer:** final concentration: 70 °Brix
- **SODIUM BICARBONATE**	0,04 to 0,06% w/w		
- **VANILLA**	max: 0,00075% w / w		
- **GLUCOSE SYRUP**	max: 40% w/w in replacement of sucrose		**A: Cooler** (flows down by gravity action) T: 60 °C **B:** GMP containers
- **STARCH**	< 0,5 .100ml⁻¹ milk.		
- **PRESERVATIVE: SORBIC ACID** **NATAMYCIN**	Only in DDL for confectionary or industrial use		**A: T:** 60 °C **B: T:**140 °C: t: 3 s
- **CONTAINERS: A process** Cardboard, plastic, glass, tin **B process** Glass or tin	Food-grade		

Fig. 9.3 Provision and process matrix summarizing safety and quality requirements as well as equipment and process parameters for the traditional *dulce de leche* manufacture, both with and without sterilization

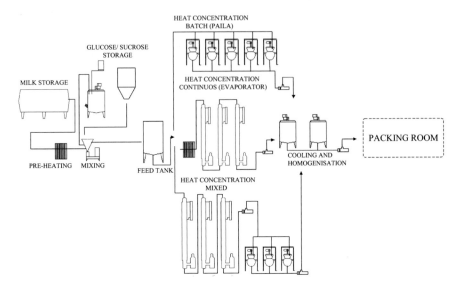

Fig. 9.4 Processing line for *dulce de leche* using traditional jam pails (at the *top*), a continuous heat concentration process (in the *middle line*), and a mixed process (at the *bottom*)

Fig. 9.5 Stainless steel jam pail for most traditional dulce de leche manufacturing process

Fig. 9.6 Internal view of jam pail and its stirring system while producing DDL

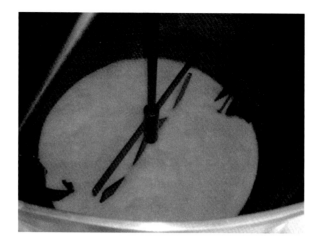

For a mixed heat concentration process, double- or triple-effect evaporators are used to pre-concentrate the mixture of milk, sugar, and additives (Oliveira et al. 2009). After that, the DDL is fed to a jam pail to reach 70°Brix. The advantage of this method is to spend less process time in pails.

DDL is ready when solid content reaches 70°Brix and the product develops its typical brown color. Although nowadays process control instruments are very common in the food industry (hand refractometer), the heating concentration stage finishes when the *dulce de leche master* decides the sensory profiling is correct regarding color, texture, and consistency.

After heat concentration, DDL is immediately cooled to 60 °C. This stage can be done in the jam pail (batch process) replacing steam for cold water, or an auxiliary tank can be employed. Cooling rate is important; big crystals can be formed if temperature goes down slowly (Hynes and Zalazar 2009).

DDL requires a homogenization stage to acquire smoothness and brightness. Again, the expertise of *dulce de leche master* is important in order to decide when to finish this process stage.

Packing equipment and operations differ according to containers, but the process has to be done in conditioned rooms, and DDL temperature must be kept around 60 °C. For glass and tin containers, sterilization can also be performed to make shelf life longer. Once it is packed, DDL is stable at room temperature during shelf life, but it must be refrigerated when it becomes opened.

As DDL is a very common product in some countries in Latin America, innovations in the traditional recipe have been proposed regarding manufacturing process and consumer preferences. Lactose crystals cause "sandiness" in DDL and are probably the worst defect reported for this product. Studies have been conducted in order to control this defect by partial hydrolysis of lactose with beta-D-galactosidase (Sabioni et al. 1984; Giménez et al. 2008). Martínez et al. (1991) studied the lactose seeding effect on consumer acceptance with a sensory panel, showing that a seeding procedure was effective in eliminating sandiness in *dulce de leche* over prolonged storage.

Ultrafiltration technology is widely applied to separate milk components. Low-lactose-content milk could be an alternative source to DDL manufacturing, but this alternative is until under research.

9.2.2 Dulce de Leche *Formulation and Its Implication on the Final Product*

While the traditional Argentine recipe of *dulce de leche* includes only milk and sucrose as ingredients, the industrialization of this product has led to the introduction of some additives during the manufacturing process, in order to extend product shelf life. According to Argentine regulation, whole or skimmed milk and sugar are mandatory ingredients, and preservatives, flavors, stabilizers, and thickeners are commonly added (Argentine Food Code 2009). Table 9.1 shows the main additives for *dulce de leche* formulation.

As *dulce de leche* is a creamy and viscous product prepared by heat concentration, its rheological behavior is an interesting factor for food technologists to consider for product innovation. Different types of DDL (household, for confectionery) have been studied in rotational viscometers or by dynamic oscillatory tests (Rovedo et al. 1991) showing thixotropic and pseudoplastic characteristics (Navarro et al. 1999). Casson's model, Herschel and Bulkley's model, and Michaels-Bolger's model were used to characterize the flow (Pauletti et al. 1990).

Navarro et al. (1999) studied the rheological behavior of DDL, low-calorie DDL, and confectionery DDL types and found that it was intermediate between a concentrated solution and a gel. The DDL confectionery type was closer to a gel-like behavior than the other two. The presence or absence of thickening agents can be related with differences observed (Verbeken et al. 2006).

A texture analyzer can be used to characterize DDL (Pauletti et al. 1992; Ares et al. 2006; Valencia García and Millán Cardona 2009). Hardness, spreadability, adherence, cohesiveness, and elasticity are the characteristics usually evaluated.

Like texture, *dulce de leche* color is one of the most appreciated attributes. Different conditions of processing, formulation, or storage time can vary the color obtained in the final product widely (Garitta et al. 2006; Rozycki et al. 2010). Sucrose replacement by glucose tends to produce a darker product because Maillard reactions are favored. Color measurements of DDL employing CIE $L^*a^*b^*$ system show that in general, L^* (lightness) is the parameter with more differences (Pauletti et al. 1996).

The main Latin American production of DDL is based in cow milk. However, goat and ewe are a very common source for dairy products in many countries. Because their milk has a steady production, its further processing into *dulce de leche* and cheese contributes to the development of some rural economies. Goat milk is recommended for childhood feeding and in cases in which cow milk is not tolerated. Furthermore, goat milk and their products have importance in human nutrition. Developing countries are used to consume more goat milk than cow milk

Table 9.1   Additives for *dulce de leche* formulation

Function	Name	INS number	Maximum concentration[a]
Preservative	Sorbic acid or Na, K, Ca salts[b]	200, 201, 202, 203	600 mg kg^{-1} (sorbic acid) and 1000 mg kg^{-1} (sorbic acid) for industrial confectionary DDL
	Natamicin[b]	235	1 mg (dm^2)$^{-1}$
Humectants	Sorbitol	420	50 g kg^{-1}
Flavor	Vanilla		GMP[c]
Food coloring	Caramel I, II, III, IV	150a, 150b, 150c, 150d	GMP, only admitted for industrial ice cream DDL
Stabilizer	Calcium lactate	327	GMP
Texturizer	Sodium citrate	331	GMP
	Alginic acid, sodium, potassium, ammonium, calcium, salts, and propylene glycol alginate	400, 401, 402, 403, 404, 405	5000 mg kg^{-1}
	Carrageenan	407	5000 mg kg^{-1}
	Agar	406	5000 mg kg^{-1}
	Carob bean gum	410	5000 mg kg^{-1}
Thickener[c]	Tragacanth gum	413	5000 mg kg^{-1}
	Gum arabic (Acacia gum)	414	5000 mg kg^{-1}
	Xanthan gum	415	5000 mg kg^{-1}
	Karaya gum	416	5000 mg kg^{-1}
	Gellan gum	418	5000 mg kg^{-1}
	Konjac flour	425	5000 mg kg^{-1}
	Pectins	440	5000 mg kg^{-1}
	Microcrystalline cellulose (cellulose gel)	460i	5000 mg kg^{-1}
	Methyl cellulose	461	5000 mg kg^{-1}
	Hydroxypropyl cellulose	463	5000 mg kg^{-1}
	Methyl ethyl cellulose	465	5000 mg kg^{-1}
	Sodium carboxymethyl cellulose (cellulose gum)	466	5000 mg kg^{-1}

[a]Final product, Argentine Food Code 2009
[b]Surface sprinkled before packing close
[c]For confectionary or ice cream DDL mixture, thickeners are allowed up to 20,000 mg kg^{-1} (final product)
GMP good manufacturing process

in order to improve nutritional properties of food for malnourished people (Costa et al. 2005). Regarding the Argentine Food Code (2009), a *dulce de leche* produced with any other milk source different from cow milk, like goat milk, must be called *dulce de leche from goat milk*. The formulation and manufacturing process for this product are, in general, similar to *cow's dulce de leche*. Sensory profile and final product are rather different because of milk composition. Ewe's milk contains more fat and protein than goat or cow milk (Pandya and Ghodke 2007). In general, DDL from goat milk is less brown, sweet, and viscous than *dulce de leche* from cow milk.

9.3 Quality and Consumer Aspects

There are economic, physiological, social, cultural, and environmental factors influencing food choices and consumption of particular products. Countries have specific legislation governing food, food markets, and merchandising techniques in order to educate and protect consumers from misleading advertising and moral hazard. Consumers have the right to be informed and to choose regarding their welfare. Nowadays, consumers are more concerned with safety and health, as well as ethical considerations when they refer to food quality.

All of these also have implications for food researchers who have to bridge the gap between consumer demand, technological problems, and developments of new products.

Dulce de leche, particularly, has to achieve attributes related to physicochemical and biological parameters, macroscopic and microscopic criteria, sensory characteristics, organic and inorganic contaminants, and containers and labeling, defined in the Argentine Food Code, to be marketed. General considerations of hygiene practices for product development must also comply with Chapter II of the same Code, in keeping with Codex Alimentarius CAC/RCP 1-1969, Rev. 4-2003 "Recommended international code of practice: general principles of food hygiene," regarding sanitary conditions and good manufacturing practices (GMPs) which are established to satisfy regulatory requirements as well as to maintain the high-quality standards of products for consumer acceptance.

9.3.1 General Requirements for Dulce de Leche

The chemical composition for traditional DDL made in Argentina according to Argentine regulations is shown in Table 9.2. DDL must also meet some requirements regarding factors affecting safety. Microbiological criteria are shown in Table 9.3.

Table 9.2 Chemical composition for dulce de leche made in Argentina

Component	Dulce de leche (traditional) (g 100 g^{-1})
Water	Max. 30.0
Proteins	Min. 5.0
Carbohydrates	Min. 56
Fat content	6.0 to 9.0
Ash	Max. 2.0

Table 9.3 DDL regulation safety requirements

Criteria		Requirement
Microbiological	Coagulase positive staphylococci	$n=5, c=2, m=10, M=100$
	Yeast and molds	$n=5, c=2, m=50, M=100$
Impurities and foreign bodies		0 (absence)
Contaminants		Limit established for each particular in code

n: number of sample units tested

c: maximum number of sample units and the results can range from m (quality) and M (provisionally acceptable quality)

m: maximum level of the organism in the food to an acceptable quality

M: maximum level of the organism in the food to an acceptable quality temporarily

Due to the high sugar concentration, DDL is a safe product, but yeast and molds may grow, in most cases, due to external contamination because of reckless handling during the manufacturing process (Hentges et al. 2010).

DDL has to be packed in suitable food-use containers to protect the product, considering materials for the intended conditions of storage and shelf life.

When labeling the product, other ingredients besides milk, sucrose, glucose syrup, sodium bicarbonate, and vanilla are declared not only on nutritional tables but by adding to the name *dulce de leche* a denomination that expresses the additional ingredient. If starch or modified starch is added, it is called "DDL for confectionary."

The sensory profile of DDL is one of the most important aspects related to consumer preferences. Consistency, color, taste, and flavor are the most specific and relevant attributes described. DDL has to be creamy and spreadable, without perceptible crystals. The consistency may be stronger in the case of DDL for confectionary, bakery, or ice cream. It could also be semisolid or solid and partially crystallized when humidity exceeds 20 % w/w.

Regarding color, its brown caramel color comes only from the Maillard reaction except in DDL for ice cream, in which case the color may correspond to the dye added.

DDL flavor, taste, and even aftertaste are sweet but unique.

9.3.2 Argentine Quality Protocol for Dulce de Leche

Regarding quality and consumer aspects, for a DDL to obtain the quality certification "Argentine Food, a Natural Choice," it has to comply with differential attributes referred in the quality protocol, related to product, process, and eventually to the container.

The protocol states that the unique ingredients for manufacture are raw cow milk; "cane" sugar, described as common type and/or higher grades (Argentine Food Code, Art. 768bis); glucose; sodium bicarbonate; and possibly vanillin.

Table 9.4 Particular requirements for access to the seal of quality "Argentine Food, a Natural Choice"

Distinguishing attributes, quality protocol SAA012
Raw material
Milk
• Cows fed mainly pastures
• *Good Livestock Practices* (GLPs) at dairy farm
• Code of hygienic practice for milk and milk products CAC/RCP 57-2004 developed by the Codex Committee on Food Hygiene
• Time between milking and processing <1 h
• Storage T during transport <6 °C
• Coming from farms free of brucellosis and tuberculosis, with official certification
• Milk fat not less than 3.2 % w/w
• Somatic cell count: not more than 400,000 cells/mL
• Mesophilic aerobic bacteria count: not more than 100,000 CFU/mL
• Absence of antibiotic residues
• pH: 6.55–6.75
Process
Assurance safety system
• Standards of hygiene and safety: *GMP* and *HACCP* from receiving raw materials to final product sale
Packaging
• Filling: under strict hygienic standards, automatically or semiautomatically in enabled environments
• Packaging rooms separated from the processing room
Packaging control
• Visual observation and package inspection
• Treatment with filtered air (filter sterile) pressure
• UV light
• Metal detector
• Washing with chlorinated water: 0.4 ppm minimum
• Potable water treatment at temperatures above 80 °C
Storage and transport characteristics
• Storage: fresh, dry, closed, and free of contaminants
• Transport: hygienic conditions of the vehicle, enabled and used only for food transport
• T <30 °C and relative humidity <80 %, protected from solar light
Package
• Transparent glass primary packaging, cap containing safety button, outer plastic coating or sash
• Tin or other innovative material approved by the competent authority acceptable to the market
• PET (polyethylene terephthalate) or cardboard: not allowed
Final product
• Total milk solids: minimum 24 % w/w
• Preservatives: not allowed
• Salmonella: absence in 25 g
• Listeria monocytogenes: absence in 25 g

More strict safety requirements have to be accomplished such as Good Livestock Practices (GLPs) at dairy farm, GMP, and Hazard Analysis and Critical Control Points (HACCP) from receiving raw materials to final product sale.

The distinguishing attributes stated in quality protocol, SAA012 Version 06, regarding raw materials, process, package, and final product are summarized in Table 9.4.

9.4 Conclusions

Dulce de leche is dairy product basically made by a heat concentration of milk and sugar. Although cow milk is the most used raw material, goat and ewe milk are also used.

Further research has to be done in order to find out sources to replace milk protein and sugar. Sensory profile and consumer aspects have to be improved when nontraditional ingredients are used in the formulation.

Since the ingredients and technology are simple, DDL has the advantage of being a product that can be manufactured worldwide.

References

Ares G, Giménez A, Gámbaro A (2006) Instrumental methods to characterize nonoral texture of dulce de leche. J Texture Stud 37:553–567

Código Alimentario Argentino, Argentine Food Code (2009) Administración Nacional de Medicamentos, Alimentos y Tecnología Médica. Secretaría de Políticas, regulación y relaciones sanitarias, Ministerio de Salud, República Argentina. http://www.anmat.gov.ar/webanmat/normativas_alimentos_cuerpo.asp. Accessed 25 Feb 2011

Costa AM, Bonomo P, Lima PM, Solares LS, Bonomo RC, Leal CS, Veloso CM, Wanderley MA, Carneiro JC, Fontan RC (2005) Effect of the goat milk in sensory quality and processing time of dolce de leite. Cienc Tecnol Aliment 4(5):315–318

Garitta L, Hough G, Sánchez R (2006) Sensory shelf life of dulce de leche. J Dairy Sci 87:1601–1607

Giménez A, Ares G, Gámbaro A (2008) Consumer reaction to changes in sensory profile of dulce de leche due to lactose hydrolysis. Int Dairy J 18:951–955

Hentges D, Teixeira da Silva D, Alves Dias P, Conceição RCS, Nunes Zonta M, Dias Timm C (2010) Pathogenic microorganism survival in dulce de leche. Food Control 21:1291–1293

Hynes E, Zalazar C (2009) Lactosa in dulce de leche. In: MacSweeney PLH, Fox PF (eds) Advanced dairy chemistry, vol 3, Lactose, water, salts and minor constituents. Springer, New York, pp 58–67

Malec LS, Allosa R, Naranjo GB, Vigo MS (2005) Loss of available lysine during processing of different dulce de leche formulations. Int J Dairy Technol 58(3):164–168

Martínez E, Hough G, Contarini A (1991) Sandiness prevention in dulce de leche by seeding with lactose microcrystals. J Diary Sci 73:612–616

Navarro AS, Ferrero C, Zaritski NE (1999) Rheological characterization of dulce de leche by dynamic and steady shear measurements. J Texture Stud 30(1):43–58

Oliveira MN, Penna ALB, García Nevarez H (2009) Production of evaporated milk, sweetened condensed milk and dulce de leche. In: Tamine AY (ed) Dairy powders and concentrated products. Wiley-Blackwell, Oxford, pp 158–177

Pandya AJ, Ghodke KM (2007) Goat and sheep milk products other than cheeses and yoghurt. Small Rumin Res 68(1–2):193–206

Pauletti M, Venier A, Sabbag N, Stechina D (1990) Rheological characterization of dulce de leche, a confectionery dairy product. J Dairy Sci 73(3):601–603

Pauletti M, Calvo C, Izquierdo L, Costell E (1992) Color y textura de dulce de leche. Selección de métodos instrumentales para el control de calidad. Rev Española Cienc Tecnol Aliment 32(3):291–305

Pauletti MS, Castelao EL, Sánchez R (1996) Influence of soluble solids, acidity and sugar on the color of dulce de leche. Food Sci Technol Int 2(1):45–49

Pepper T (1990) Alternative bulk sweeteners. In: Jackson EB (ed) Sugar confectionery and manufacture. Blackie Academia and Professional, Glasgow, pp 15–33

Raynal-Ljutovac K, Lagriffoul G, Paccard P, Guillet I, Chilliard Y (2008) Composition of goat and sheep milk products: an update. Small Rumin Res 79:57–72

Resolution Secretariat of Agriculture, Livestock, Fishing and Food (SAGPyA) N° 798/2006 Quality protocol (Code: SAA012). http://www.alimentosargentinos.gov.ar/programa.../dulce-deleche_ingles.pdf. Accessed 25 Feb 2011

Rovedo CO, Viollaz PE, Suarez C (1991) The effect of pH and temperature on the rheological behavior of dulce de leche, a typical dairy argentine product. J Dairy Sci 74(5):1497–1502

Rozycki SD, Buera MP, Pauletti MS (2010) Heat-induced changes in dairy products containing sucrose. Food Chem 118:67–73

Sabioni JG, Rezende Pinhero ADO, Olzani SD, Paes Chavez JB, Chaer BA (1984) Control of lactose crystallization in "dulce de leche" by beta-D-galactosidase activity from permeabilized *Kluyveromyces lactis* cells. J Dairy Sci 67:2210–2215

SENASA (2010) Oficina de Estadisticas de Comercio Exterior, Lacteos/Dairy. http://www.senasa.gov.ar/estadistica.php. Accessed 25 Feb 2011

Valencia García FE, Millán Cardona LDJ (2009) Estimación de la vida útil de un arequipe bajo en calorías. Rev Lasallista Invest 6(1):9–15

Verbeken D, Bael K, Thas O, Dewettinck K (2006) Interactions between k-carrageenan, milk proteins and modified starch in sterilized dairy desserts. Int Dairy J 16:482–488

Part III
Traditional Cereal Based Products

Chapter 10
Austrian Dumplings

Josefa Friedel, Helmut Glattes, and Gerhard Schleining

10.1 History

Dumplings are a typical and traditional Austrian dish and are served there in a great variety, in soups and as appetizers, main dishes or side dishes with meat or vegetables, or even as sweat desserts. Therefore, dumplings are variegated and diversified. They can be made out of semolina, wheat, bread, potatoes, curd cheese, or yeast dough. There are also dumplings that are stuffed or unfilled, boiled, baked, or au gratin. They can be small, big, round, oval, flat, or in some other shape pressed loosely together or even longish. The dumpling traditionally came from Upper Austria, Salzburg, Tyrol, and Bavaria. Outside of the alpine cuisine "Knödel" are called "Klöße."

Dumplings played a significant role in the historical development of Austrian cooking culture. The exact origin of the dumpling is unknown. The dumpling is not an achievement from a singular imaginative invention. It is a logical consequence of cooking and eating conditions. The history of the dumpling goes far back to ancient times. The primary, ancestral shape was not exactly like the dumplings we know today. It was a simple presentation of chopped meat mixtures. In the region around the "Mondsee" (Austria), archaeological excavations of Neolithic settlements and lake dwellings of the Bronze Age showed remnants of dough. Small round flat dough cakes were found. These flat cakes were made out of barley and wheat flour and they had an unusual curvature. In these flat cakes, meat and fruits were wrapped. The earliest written evidence is dated back to Roman times. It can be found in an old "cooking book," "Notes from Gavius Apicius" which is from the occidental cultural

J. Friedel • H. Glattes • G. Schleining (✉)
Department of Food Science and Technology, BOKU, Vienna, Austria
e-mail: gerhard.schleining@boku.ac.at

© Springer Science+Business Media New York 2016   139
K. Kristbergsson, J. Oliveira (eds.), *Traditional Foods*, Integrating Food
Science and Engineering Knowledge Into the Food Chain 10,
DOI 10.1007/978-1-4899-7648-2_10

area and was written in the time of the Roman imperator Tiberius around the time when Jesus was born. There existed dumpling-like foods because the Roman people ate while they were lying, and so it was easier for them to eat the food when it was formed in dumplings. The food was taken with the hands and pushed slightly balled into the mouth. In a fresco from twelfth century (1280) found in the chapel of the castle, Hocheppan in Meran, South Tyrol near Bozen/Bolzano (Italy), a woman is depictured, maybe a nun or the midwife, with Maria next to the child's crib. The woman is holding in one hand a cookware over the fire with five round dumplings and in the other hand a pierced dumpling which she is leading to her mouth. The dumpling is skewered on a dumpling knife. This is the oldest illustration of dumplings outside of a book. Was there a dumpling tradition?

The oldest cooking and parboiling methods were smoking the food, barbecuing on a spit, cooking in a stock, or steaming and stewing in a pot. As far as traditional cooking cultures go, the so called boiler cultures (Kesselkulturen) can be found from the Northern Alps to the Baltic Sea. In these regions there were mainly boiled foods like boiled and cooked dumplings. In the Southern and Eastern Alps, there was the "pan culture" (Pfannenkultur). Here the food was mainly fried so they had lard pastries, strudel, "Teigfleck," "Nockerl," and "Schmarrn."

Upper and Lower Austria were the "dumpling stronghold" and were influenced from the Bavarian "boiler culture."

Until the seventeenth century, there existed no flatware like we know it today. The people ate the food with their hands. Only the kitchens of the upper class used knives as well as a few men that needed them for hunting. The foods were prepared like a ragout, or the raw meat, fish, or other ingredients were mixed with spices and binding agents, like cereals or bread, from which round balls were formed. This was an easier way of cooking and barbecuing and it was easier to eat with your bare hands. Therefore, in old cooking books, instructions can be found for this kind of cooking method. These were the prototypes of our dumplings today. The dumplings had their beginnings in meat dishes. There are many various types of dumplings today which can be traced back to this original kind of meat dumpling, for example, the liver, bacon, ham, chicken, brain, and various game dumplings as well as the fish, crab, and ray dumplings.

The name came from the Romans: "nodus = der Knoten" and "nodulus = das Knötchen." The Bavarian-Austrian ideogram for the dumpling—Knödel—is related etymologically to the Old High German chnodo, chnoto; Middle High German knode, knote; and New High German Knoten (a knot or small lump) (Helmreich and Staudinger 1993). The word "Knoten" can be translated into "Knötlein," "Knötelein," or "Knödchen" (Horn 1976). The dumpling has a great history of development since its existence. The name of this everyday food has changed a lot in the different dialects over periods of time.

Beginning in the eighteenth century, there were better technical possibilities. In the middle of the sixteenth century, producing firm and consistent dough became possible because of developments in milling technology. The equipment for sifting and sieving improved so the flour could be better separated into its finer and coarser compounds. By then the dumplings was only known regionally, but it soon became the most common food of people in many places in Bavaria, Habsburg Austria, here in much greater variety. Habsburg Austria also included Bohemia and Hungary.

Since the 30 Years' War (1618–1648) and in the following two centuries, meat was a luxury good. So the people used alternatives like vegetables and especially dumplings made out of brown or white bread before the potato was brought to Europe. Since meat was expensive, not everyone could always afford it. As a result people developed different types of meals with ingredients like bread, oats, millet, barley, corn, peas, beans, and mushrooms. Some meat and giblets were supplemented to these ingredients and then formed into dumplings. These were either cooked in water or lard, steamed, or fried. After cooking the dumplings, they were often put in a stock. The dough was always made of the same raw consistency that was chopped and after that pressed together. There was practically no kind of food that was not used to make some kind of dumpling. Dumplings can include almost every kind of meat, fish, vegetables, herbs, mushrooms, and cheeses. So there is a great variety of recipes.

Around the dumpling there developed gradually fixed traditional cooking and eating habits and something like a myth. The larger filled meat dumplings or also napkin dumplings were usually eaten on Sundays. Thursday was dumpling day and in some southern region Tuesday was another dumpling day as well.

Before the dumplings are cooked, a test dumpling should be made to make sure the dough is suitable and the meal will be a success. The ability of a woman to cook dumplings was a qualification to marry because it meant she had a good judgment with her eyes and hands to form a perfect dumpling. Dumplings were served in special bowls and plates. These were especially beautiful bowls with pot handles and arched tops that had steam vents. These bowls were made out of tin or heavy stone. Only when dumplings were being served did everyone have his own plate. Otherwise the people ate together from the pan or bowl. There is a legend that the dumpling was invented in the town hall in Tyrol. The legend tells the story of a landlady and her tavern which was invaded by a mob of wild journeymen. They demanded a meal from her. But the pantry of the landlady was empty. Therefore, she decided to take old leftovers like bread, bacon, sausage, eggs, milk, salt, and onions that she had and cut these ingredients into small cubes, mixed them, and formed the first Tyrolean bacon dumplings. There is another special eating habit in regard to eating dumplings. They should never be cut with a knife but only ripped open, for example, with a fork so that they stay fluffy and the sauce can soak in.

10.1.1   The Dumpling Knife

In the medieval times, specially designed dagger knives were developed to eat dumplings. The dumpling knife is a serving knife, "Vorlegemesser," with a large blade and an extension that looks like a dagger. It was used to spear meat pieces and dumplings. The earliest description of the dumpling knife is from the eleventh century, but it is believed to be much older than that and was used for a long time after that. It was called "Knödelwürger" or "Knödelhenker" (dumpling executioner). A slightly changed version of the dumpling knife with a smaller knife point was still known in recent times. Then the fork adopted this function in the eighteenth century and has kept it until today.

10.2 Types of Dumplings

An overview of typical dumplings sorted by Austrian regions is given in Table 10.1.

10.2.1 Bread Dumplings

As meat became too expensive for the ordinary population, the quantity of meat in meals decreased and instead more bread was used. So the bread dumplingwas developed. This was the prototype of the bread dumpling we know today. It evolved from the old meat dumpling. Meat was more and more replaced by other foods and the bread ratio was increasing until a pure bread dumpling was invented.

Dumplings were increasingly made out of leftover bread, to such an extent that the Semmelknödel (dumpling made from breakfast roll bread cubes) became one of the most typical forms—isolated mentions can be found in the eighteenth century. And we should not forget that it is old and dry rolls or other white bread which provides the cubes which can be used as the material for dumplings but especially to bind a considerable number of meat, fish, vegetable, and other dumplings and when fried in butter provides a cushion for sweet dumplings (curd cheese, apricot, plum, etc.) and by contrast enriches the flavor (Helmreich and Staudinger 1993).

Basic Recipe for Bread Dumplings (Neuhold and Svoboda 2001)

250 g bread cubes
20 g butter
2 tablespoons chopped onion
1 tablespoon chopped parsley
2 eggs
250 mL milk
2 tablespoons flour
Salt

Whisk the eggs, milk, and salt and then pour the mixture over the bread cubes. Pour in the butter-fried onions and parsley over the dumpling dough and stir everything well so all bread cubes can soak in liquid. Let it stand for about 10–15 min. During this, loosen up, stir, and compress the dough several times. The dumpling dough should be moist and sticky after this but not dripping wet. If it looks too wet, flour can be added to bind the spare moisture. With hands dipped in warm water,

Table 10.1 Typical Austrian dumplings by region

Region	As garnish or main dish	In a soup	Sweet dumplings
Vienna	• Flour dumplings	• Sweetbread soup with bread crumb dumplings	• Filled pear dumplings
	• Leek dumplings	• Liver dumplings	• Apple dumplings
	• Semolina dumplings		• Fruit dumplings made with potato dough
	• "Klosterneuburger" dumplings		• Almond dumplings
	• Folded dumplings		
	• Bandit dumplings		
	• Cabbage dumplings		
	• "Schwarzplentene" dumplings		
	• Bread crumb dumplings		
	• Bone marrow dumplings		
	• Bread dumplings		
	• Napkin dumplings		
Lower Austria	• Cream cheese dumplings	• Bacon semolina dumplings in a tasty beef soup	• Mostviertler apple pear dumplings
	• Vinegar dumplings	• Horseradish dumplings in a beet soup	• Apricot dumplings with almond paste filling
	• "Strabazi" dumplings	• Fowls soup with mock crumb dumplings	• Flambéed ice cream dumplings
	• Lent dumplings	• Plain vegetable soup with nut cheese dumplings	• Curd cheese dumplings
	• Oat flake dumplings	• Fried pressed cheese dumplings in plain herbal soup	• Amaretto dumplings
	• Fried glazed dough dumplings		
	• Potato dumplings escalloped with cheese sauce		
	• Half silken dumplings		
	• Crackling leek dumplings		
	• Potato bread dumplings		
	• Filled meat dumplings with cheese		
	• Wieselburger balls		
	• Colored pumpkin dumplings		
	• Waldviertler dumplings		

(continued)

Table 10.1 (continued)

Region	As garnish or main dish	In a soup	Sweet dumplings
Upper Austria	• Baked veal dumplings	• Turkey hen soup with bread crumb dumplings	• Apple poppy seed dumplings
	• Tuna dumplings	• Polenta dumplings in mashed squash soup	• Curd cheese dumplings coated with poppy seed
	• Spinach dumplings	• Beef soup with liver dumplings	• Apricot dumplings
	• Potato dumplings coated with sunflower seed		• Plum dumplings
	• Wild garlic dumplings with ham		• Nougat-nut dumplings
	• Baked potato-lamb dumplings		• Curd cheese dumplings
	• Hash dumplings		• Flour dumplings
	• Blood sausage (Blunze) dumplings in a napkin		
	• Green yeast dumplings with wild garlic filling		
	• Tomato dumplings		
	• Vegetables napkin dumplings		
	• Steyrian wedding dumpling (Steyrische Hochzeitsknödel)		
	• Game dumplings with mushrooms		
	• Ham-leek dumplings		
	• Baked yeast bacon dumplings		
	• Escalloped crackling dumplings		
	• "Mühlviertler" potato dumplings		
	• Mixed up dumplings		
	• Meat dumplings with sauerkraut		
	• Bacon dumplings, baked		
Styria	• Potato balls	• Pearl barley dumplings	• Apple dumplings
	• Brown bread dumplings	• Spelt semolina dumplings	• Fried plum dumplings
	• Cabbage dumplings	• Pheasant dumplings	• Strawberry dumplings made from glazed dough
	• Pumpkin seed dumplings	• Polenta dumplings	• "Steirische" ice cream dumplings
			• Chestnut dumplings

Tyrol	• Spinach dumplings	• Classical Tyrolean dumplings in a beef soup	• Fruit dumpling variations made out of glazed dough
	• Fried cream cheese dumplings	• Fried sesame-liver dumplings in a root soup	• Mozart dumplings
	• Potato dumplings in sage butter with tomato sauce	• Carrot-spelt-semolina dumplings in a root soup	• Millet curd cheese dumplings
	• Dumpling lasagna	• Chamois soup with cep mushrooms dumplings	• Poppy seed-pear dumplings
	• Zucchini dumplings with lentil ragout	• Beef soup with bacon dumplings or with Tyrolean dumplings	• Curd cheese dumplings
	• Buckwheat dumplings with gorgonzola cream		
	• Beet "Rohne" dumplings		
	• Chestnut dumplings coated in potatoes		
	• "Graukäse" cheese dumplings		
	• Polenta dumplings		
	• Fried cheese dumplings		
	• Innviertler bacon dumplings		
	• Bacon dumplings		
	• Cep mushroom dumplings		
	• Tyrolean dumplings		
Carinthia	• Sheep cheese dumplings	• Wild garlic balls	• Apple yeast dumplings
	• Hunter's dumplings	• Chard roll	• Prune-almond paste dumplings
	• Sauerkraut dumplings	• Bacon semolina dumplings	• Pear chestnut dumplings
	• Curd cheese dumplings in "Kärntner" style	• Fried cheese balls in tomato soup	• Fine curd cheese semolina dumplings
	• "Gurktaler" bacon dumplings	• Milt roll	• Sweet bread dumplings
	• Pumpkin seed dumplings	• Brown bread dumplings	
	• "Sterz" bacon dumplings	• "Nigalan"	
	• Minced meat dumplings		

(continued)

Table 10.1 (continued)

Region	As garnish or main dish	In a soup	Sweet dumplings
Salzburg	• Spinach dumplings • Hash dumplings • Wild herbs curd cheese dumplings • Chanterelle dumplings • "Stinker" dumplings • Crackling dumplings • Mozzarella curd cheese dumplings • Mozart dumplings • Liver dumplings	• Baked liver dumplings in a beef soup • Semolina dumplings • Bread crumb dumplings • Fried pressed cheese dumplings	• Fried apple dumplings • Strawberry dumplings • Curd cheese dumplings • Yeast dumplings • Fried curd cheese dumplings on apple sauce
Vorarlberg	• Silbertaler dumplings • Walser cheese dumplings • Bregenzerwälder semolina dumplings • Cep mushroom dumplings • Tomato-potato dumplings • Pretzel napkin dumplings		• Blackberry dumplings • Rice dumplings • Snowballs
Burgenland	• Vegetable dumplings • Polenta dumplings • Smoked meat dumplings • Spinach dumplings • Filled crackling dumplings • Pumpkin seed dumplings • Fried bacon dumplings • Herbal dumplings • Zesty curd cheese dumplings • Hungarian cabbage dumplings • Potato dumplings	• Fried bread crumb dumplings • Cheese dumplings • Bread crumb dumplings in a chicken broth • Liver dumplings • Fine semolina dumplings • Rice dumplings	• Curd cheese semolina dumplings • Nougat dumplings • Apricot dumplings • Spelt curd cheese dumplings • Fruit dumplings with curd cheese dough • Yeast dumplings

form dumplings and put them in plenty heated salted water for about 10 min. The dumpling water should be steaming and slowly boiling. When they are ready, they tend to float up to the surface. The finished dumplings are lifted out of the dumpling water with a slotted spoon. At this moment the odor says more than a thousand words.

10.2.2 Potato Dumplings

The potato was originally cultivated in Peru and Chile, but after the discovery of America, the potato was brought to Europe, where at first the potato was hardly noticed and not accepted other than a mire botanical interest. At first they were very expensive and only the fine middle-class cuisine used them. In the rural areas the potato was discovered much later and was used especially for the fruit dumplings which are still very popular today. As the cultivation of wheat, rye, oat, and barley declined in the eighteenth century, the potato had its triumphal march. The first recipe in Germany was published by the reformer of the cuisine, the cook Marcus Rumport in his work *Ein new Kochbuch*, published in 1581. He called the potato "Erdepfel" which much later found its way into German. Until the end of the eighteenth century, the potato was only occasionally mentioned in the literature. Later potatoes were found in the modern cookbooks of those times. In 1826 a special potato cookbook was published in Kaschau (Bohemia). It included three recipes for potato dumplings. About 1800 potato dumplings were mentioned in various cookbooks with variegated types and shapes. The potato dough was much different than the dumplings that were developed previously.

Potato dumplings, as unfilled dumplings, are favored to eat as a side dish with meat and sauce. In addition they are used as doughy wrapping for many filled dumplings with such stuffing as sausages, vegetables, mushrooms, crackling, etc. The good taste of the potato dough is very enhancing when used as wrapping for filled dumplings and makes the production fast and easy.

Basic Recipe for Filled Potato Dumplings (Neuhold and Svoboda 2001)

Ingredients for the potato dough:
800 g floury potatoes
200 g flour
2 eggs
20 g butter
Salt and nutmeg
Flour for preparing
Stuffing depends on the recipe

The hot, cooked, and peeled potatoes which are chased through the potato ricer are mixed together with flour, eggs, melted butter, salt, and nutmeg. Everything is put into a bowl and kneaded to a smooth dough. Cut off the amount of dough for one dumpling with floured hands stretch and flatten it in your hands to a circular disk. Put a little bit of the stuffing in the middle of the dough disk and press up the edges of the dough and seal the dumpling carefully. Roll the dumplings between the palms of your hand until they are round and cover them in flour. The dumplings are put in boiling salted water and simmered with medium heat for 20 min without the lid.

The rural cuisine in Austria has its typical Austrian specialties. Well-known classics are pressed cheese dumplings, Tyrolean dumplings, spinach dumplings, and yeast dumplings filled with plum sauce. The eccentric ones are zucchini dumplings, pheasant dumplings, dumpling lasagna, sweet bread dumplings, and prune-almond paste dumplings. An extended list of typical Austrian dumpling types sorted by regions is given in Table 10.1 in the annex of this paper.

Small dumplings are used for the soups, middle-sized ones with vegetables, and large dumplings as main dishes called napkin dumpling. The dumplings are filled, breaded, backed, spiced, or sweetened, and they can be eaten with meat, fish, or fruits as side or main dishes. You can also use dumplings as leftovers as a salad with oil and vinegar or by frying them, "G'röste Knöd'l," "Knödelschmarrn," etc. Dumplings could be a full meal with a soup or steamed vegetables. Today they are mainly served as a side dish or dessert. In the rural and middle-class cuisine, there are mainly semolina and potato dumplings, and in the fine cuisine there are fish and fruit dumplings.

10.2.3   Soup Dumplings

Soup dumplings in general are less voluminous than dumplings that are a garnish or a course for themselves and therefore should be called "Knöderln." Almost all soup dumplings depend for the development of their full and true flavor on an accompaniment or availability of an appropriate soup. There is no limit to the list of ingredients that may be used for soup dumplings, so they are means of making good use of leftovers.

Pressed Cheese Dumplings (Helmreich and Staudinger 1993)

200 g cheese
4 rolls
Flour as necessary
2 medium-sized boiled potatoes
250 mL milk
2 eggs
Onion
Salt
Pepper
Nutmeg
1 tablespoon fat
Oil for frying

Fig. 10.1 Smoked ham semolina dumplings in beef soup

Whisk the milk, eggs, and spices and pour over the diced rolls. Mix the chopped browned onions, cover, and let rest. Thicken the thinly cut cheese and mashed potatoes if necessary with flour and mix into the dumpling mass. Form the dumplings, press it flat, and brown it in floating hot oil. These dumplings taste delicious in all clear soups (Fig. 10.1).

Smoked Ham Semolina Dumplings: Speckgriessknödel (Neuhold and Svoboda 2001)

125 mL milk
50 g wheat semolina
50 g smoked ham
20 g butter
20 g stripped white bread
1 egg
Salt

Boil up the milk with the salt, add the wheat semolina, and stir constantly so that it can thicken by boiling. Take it off the oven and let it cool down. Bark the white bread and cut it into small cubes like the bacon. Heat the butter in a pan, add the bacon, and briefly roast it gently. Then add the white bread cubes and roast both ingredients until the white bread cubes have a gently brown color. Separate the egg. Stir the egg yolk under the semolina pudding. Stir the egg whites until a stiff foam and fold it together with the roasted bacon and white bread cubes under the semolina pudding. Mix gently and let it rest for ten minutes. Form small dumplings from the mass with wet hands. The quantity of ingredients should be enough for eight dumplings. Add the dumplings in boiling salt water and steep them for 15 min by medium heat. Lift the dumplings out of the water with a slotted spoon, allow the water to drip off, and serve them in a hot beef broth sprinkled with freshly chopped parsley.

Liver Dumplings (Pernkopf and Wagner 2010)

40 g creamy butter or butter lard
250 g beef or pork liver (alternatively lamb, canard, fowl or game liver)
50 g onion cubes
2 eggs
220 g stuffing bread or bread cubes
1 tablespoon chopped parsley
Salt
Marjoram
Pepper
Optional: garlic, rosemary, sage, allspice, pimento, coriander, lemon zest or juice,
   or Worcester sauce

Mix the minced liver and butter-sweated onion cubes with the other ingredients
when they are cooled down. The dough is spiced and then rest for 30 min. Form
dumplings out of this dough with wet or oily hands. Boil them in salt water or beef
soup. Another possibility is to roll them in bread crumbs and fry them in hot fat until
they are golden brown. Serve the liver dumplings with a beef soup or with sauerkraut or with a mixed salad. As filler for canard, game, or fowl liver dumplings,
caramelized apple pieces can be optionally mixed in. For game liver dumplings, use
deer or venison and season it with marjoram, thyme, ground juniper, garlic, and
coriander.

Semolina Dumplings (Gössl 2003)

50 g butter
1 egg
100 g semolina (durum wheat)
Salt
Nutmeg
Some oil

Stir the butter until it is creamy; add the egg, semolina, and salt; make it spicy
with nutmeg; and let the mixture stand for about 10–15 min. Form "Nockerln" with
a wet teaspoon and drop them in boiling salt water. Let them simmer for 10 min.
Strain them and chill them, covered for another 10 min. Semolina dumplings are
served in a beef soup.

10.2.4 Dumplings as Garnish or Main Dish

When dumplings are served as garnish or main dish, they usually take up more
space. They cannot really be distinguished in this function as main or subsidiary
matter and the capability of bread crumb or potato dumplings to serve as main dish.
When they are of top quality, they stand in, with absolutely no need of roast pork,

beef goulash, mushroom sauce, etc. They can be served with a refreshing salad as contrast demotes these to mere garnishes. Bread dumplings or dumplings in a napkin are a typical form of dumpling served as garnish or main dish. The basic dough for both of them is similar to the basic recipe for bread dumplings which is mentioned in Sect. 10.2.1. But there is a small difference concerning the "Klosterneuburger" and Tyrolean bacon dumplings (Fig. 10.2). Both of them are additionally made with meat. The "Klosterneuburger" dumpling is a special kind of napkin dumpling.

Tyrolean Bacon Dumplings (Hierl 2004)

150 g (5½ oz) bacon, cut in cubes
6 plain bread rolls
3 eggs
250 mL milk
Salt
50 g wheat flour or grit
150 g fatless chopped ham
½ bunch of parsley
1 onion
40 g butter

The diced bacon and bread rolls are fried in a hot pan until the bacon gets glassy and the bread rolls are soaked full with fat. Pour the whisked eggs, milk, and salt over the cooled bread rolls and bacon. Stir in the flour or semolina and let it rest for 15 min. Then the chopped parsley and ham cubes are mixed together. Take a piece as big as a spoon from the dough and push it flat. Put a little bit of the ham parsley mixture in the middle of the dough and form dumplings out of it. Put them into salt water and boil them for 15 min. The peeled onions are cut into rings and fried with butter in a pan. Break the dumplings with a fork in two pieces and spill the roasted onions over the dumplings.

Fig. 10.2 Tyrolean bacon dumplings

"Klosterneuburger" Dumplings (Obermann and Kröpfl 2009)

5 traditional Austrian bread rolls
250 mL milk
250 g cooked and smoked meat
4 eggs
140 g margarine
Fresh parsley
30 g bread crumbs
Salt

Cut the bread rolls into cubes, add the milk, and let it soak. Slowly mix the eggs, the cubed meat, and cut parsley into the bread rolls and add salt. Stir the margarine until it is creamy and mix it to the bread roll dough. If the dough needs to be more compact, add bread crumbs. Then grease a cloth napkin, fold the bread mass into the napkin, and tie it up with a string. Cook it in salted water for 1 h. The dumplings can serve with salad and a mushroom sauce (Fig. 10.3).

Spinach Dumplings (Pernkopf and Wagner 2010)

200 g mashed, well-drained spinach
200 g boiled, cold passed through potatoes
100 g very fine bread crumb
25 g melted butter
1 egg
1 egg yolk
Salt
Pepper
Garlic

Fig. 10.3 "Klosterneuburger" dumplings with salad

Mix the egg and spinach together and then add the other ingredients. Let the dough rest for a short time. If the dough is too smooth, add more bread crumb. Form dumplings and place them in almost boiling salt water for about 10–12 min. Either arrange them with a cream sauce or sprinkle with Parmesan. Another option is to escallop them with ham stripes and cheese and serve them with chives butter.

10.2.5 Sweet Dumplings

In 200 B.C. the consul Cato mentioned sweet dumplings. These sweet dumplings were similar to our curd cheese dumplings today. The dumplings were a compound made of a special kind of barley—barley groats—and fresh sheep cheese (curd cheese). The formed dumplings were fried in fat and served with honey and poppy seeds.

In the medieval times, apple dumplings were popular, as well as rice dumplings with cinnamon and ginger. But there were also dumplings made out of white bread and filled with fruit. There were also sweet dumplings with chicken meat. Later on plum, cherry, and apricot combined with potato dough were included. Fruit dumplings of those days were much different from the fruit dumplings which are known today. They were made from chopped, steamed, or boiled apples and stewed sour cherries or pears. These ingredients were mixed with butter or lard and roasted bread crumbs and were seasoned with sugar and cinnamon. From this dough small dumplings were formed, dusted with flour, and fried in lard. Sweet dumplings are esteemed mostly as dessert but also can be enjoyed as main dish. Curd cheese dough, potato dough, glazed dough, strudel dough, and noodle dough are used for wrapping (Fig. 10.4).

Fig. 10.4 Fluffy curd cheese dumplings with vanilla sauce and strawberries

Fluffy Curd Cheese Semolina Dumplings (Obermann and Kröpfl 2009)

100 g butter
2 egg yolks
2 egg whites
500 g curd cheese
100 g semolina
100 g flour
1 tablespoon salt
50 g butter
100 g bread crumbs

Whisk butter and egg yolks until foamy and add in the curd cheese, semolina, flour, and salt. Beat the egg whites stiff and fold them slowly into the mixture. Form small dumplings and simmer in salted water for ca. 15 min. Melt 50 g butter in a pan and fry the bread crumbs carefully in it. Put the cooked dumplings in the pan and fry them in the browned bread crumbs. You can serve them with strawberries and vanilla sauce.

Apricot Dumplings (Hohenlohe 2005)

1 kg mealy potatoes
1 pinch of salt
3 egg yolks
Flour
750 g equally sized apricots
Lump sugar
150 g butter
100 g bread crumbs
Sugar for sprinkling

The apricots are skinned, sliced, and pitted after blanching in boiling water. Cook the potatoes in their peel, peel them, and pass them through a strainer or potato press. Mix salt with the egg yolks and the flour to a smooth dough. Cut the dough into squares and put an apricot which is prepared with lump sugar into the middle of each square, fold the dough around the fruit, and form a dumpling. Simmer the dumplings in plenty of boiling salt water for 10 min. Meanwhile roast the bread crumbs in butter until they are golden brown. Lift the dumplings out of the water with a slotted spoon, allow the water to drip off, and roll the dumplings in the bread crumbs. And then they are sprinkled with icing sugar and served.

Yeast Dumplings (Gössl 2003)

Yeast dough:

    25 g fresh yeast
    250 mL milk
    100 g sugar
    1 pinch salt

1 pinch vanilla sugar
A little bit grated lemon zest
1 tablespoon rum
4 egg yolks or 2 eggs
500 g flour
80 g butter

Filling: Plum sauce
For basting:

Butter
Powdered sugar
150 g poppy seed

Crumble yeast in lukewarm milk; add sugar, spices, egg yolks, sieved flour, and then the gently heated butter. Beat to a smooth dough until it can be lifted easily from the dish in one piece.

Let the yeast dough rise for about 60 min. Fill palm-sized pieces with plum sauce and close it carefully and form dumplings. Leave the dumplings to rest on a floured board. Put them in boiling salt water for about 10 min. Remove the dumplings from the water, douse them with hot, golden brown butter and sprinkle them with powdered sugar and poppy seeds.

Yeast dumplings also can be steamed: fill the pot half full with water, clamp a clean cloth over the pot rim, lay the dumplings on it, close the lid, and let rise in the steam.

References

Gössl H (2003) Das Grosse Österreichische Kochbuch. Thomas Bauer, Bad Wiesse-Holz
Helmreich F, Staudinger A (1993) Nur Knödel: the ultimate dumpling book from Austria, Bavaria & Bohemia. Christian Brandstätter, Wien
Hierl T (2004) Hausgemacht: Knödel. Österreichischer Agrarverlag Druck- und Verlagsges.m.b.H, Leopoldsdorf
Hohenlohe M (2005) Viennese cuisine, cook and enjoy. Pichler, Wien
Horn E (1976) Von Knötelein, Knödchen und Knödeln. Kulturhistorische Betrachtungen rund um die Knödel und Klöße. Heimeran, München
Neuhold M, Svoboda M (2001) Das Buch vom Knödel. Steirische Verlagsgesellschaft m.b.H, Graz
Obermann I, Kröpfl B (2009) Österreichische Bäuerinnen kochen Knödel. Die besten Rezepte aus allen Bundesländern. Loewenzahn in der Studienverlag Ges.m.b.H., Innsbruck
Pernkopf I, Wagner C (2010) Knödel Küche. Die 250 besten Rezepte von pikant bis süß. Pichler Verlag in der Verlagsgruppe Styria GmbH & Co KG, Wien

Chapter 11
French Bread Baking

Alain Sommier, Yannick Anguy, Imen Douiri, and Elisabeth Dumoulin

11.1　Introduction

Bread has been a basic foodstuff for thousands of years.

Before discovering how to make bread, man ate cereals in the form of a paste and then as small cakes. The Egyptians and the Babylonians knew how to use yeast and to apply bread-making techniques. The Greeks improved the bread maker's art and spread it throughout their conquests.

Although wheat was harvested in China, in the Middle East, and in Egypt several thousands of years BC, the people used to eat bread made from rye, buckwheat, and maslin (a mixture of wheat and rye). It was only in the nineteenth century that the production of bread made from wheat flour developed in Europe using yeast.

Although bread consumption in France fell considerably in the twentieth century, it has recently started to increase once more due mainly to the diversity of products available to consumers.

The production of bread requires energy, time, and a very precise dexterity. Only a few simple ingredients are used. The bread maker seems to be a kind of magician (Fig. 11.1).

Here, we describe briefly its current state of the art and then present the materials used and the methods of measurement. We highlight the links between the kinetics of mass loss from a French loaf of bread, the variation in the relative humidity of the air as it comes out of the oven, and the thickness of the product, which is linked with the internal pressure.

A. Sommier (✉) • Y. Anguy
I2M-TREFLE UMR CNRS 5295, Esplanade des Arts et Métiers,
Talence Cedex 33405, France
e-mail: a.sommier@i2m.u-bordeaux1.fr

I. Douiri • E. Dumoulin
AgroParisTech-site de MASSY, Avenue des Olympiades, Massy Cedex 91744, France

© Springer Science+Business Media New York 2016　　　　　　　　　　　　157
K. Kristbergsson, J. Oliveira (eds.), *Traditional Foods*, Integrating Food
Science and Engineering Knowledge Into the Food Chain 10,
DOI 10.1007/978-1-4899-7648-2_11

Fig. 11.1 Different types of French bread

11.2 Technology of Bread Production

Bread is the result of very complex physical transformations, chemical reactions, and biological activity that occur within a mixture of flour, water, salt, and yeast (only *Saccharomyces cerevisiae* yeast is used) and sometimes a few other ingredients (Feillet 2000) (ascorbic acid, bean flour, exogenous enzymes, emulsifiers, etc.), with the controlled addition of mechanical and heat energy. The exact formula differs, depending on the type of bread; traditional bread contains no sugar, milk, or fat. Traditional French bread has the "pain de tradition française" appellation, created in an article of the Bread Decree of 1993. It is produced by baking a dough which contains no additives and which has undergone no freezing during production (Feillet 2000). French bread typically has a thin, golden crust which is shiny and crisp (Roussel and Chiron 2002) unlike the white bread that is called "English bread" (with no crust, dense in texture, and which is baked in a mold).

Bread making takes about 5 h and consists of the following stages:

11.2.1 Kneading in the Mixer

This carries out two essential functions: it ensures that the dough mixture is homogenous (with flour, water, yeast, and salt), smooth, firm, and viscoelastic based on the two main constituents, flour and water, and it incorporates tiny air pockets into the dough whose walls become impermeable to gas to a certain extent, and it is from these that the cell structure of the bread will develop. In addition, fermentable sugars start to be released which enable the yeasts to multiply and grow later in the process.

The dough is subjected to intense forces of stretching, compression, and shearing, depending on the geometry of the constituent parts of the mixer, the size and shape of the blades, the speed at which they rotate, and also the rheological properties of the dough (Feillet 2000).

If mechanical kneading is carried out, this consists of two phases:

- **Mixing** the ingredients at a first speed (40 rev/min) for 3–5 min.
- **Working** the dough, and for a better result this is carried out at a higher speed (80 rev/min). Hydration continues, the dough is worked (blade), and the gluten is stretched and softened. The components bind together even further and the dough becomes smooth, elastic, and supple.

Thanks to the rapid mixing, air is incorporated into the dough, which gradually comes away from the sides of the mixer bowl, and becomes smooth, dry, and elastic (Guinet 1982) (Fig. 11.2).

Finally, 5 min before the end of kneading, salt is added to firm up the gluten structure and hold the shape of the dough better. When salt is added at the end of the kneading stage, this whitens the dough and hence the bread (Guinet 1982); it also helps tighten the gluten structure.

11.2.2 Tank Fermentation

This technique (duration, temperature, phases of resting the dough, hygrometry of the proofer) has evolved in order to simplify the work of the baker (Feillet 2000). The total time that the dough ferments (from 2 h 30 min to 5 h) is divided into several stages based on transforming the kneaded dough into small units of a given shape.

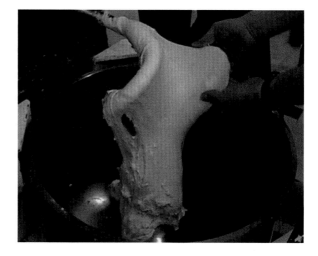

Fig. 11.2 Dough obtained at the end of the kneading process, with a smooth and flexible structure that can form a veil-like film once the gluten network is in place

There are generally considered to be three distinct phases; the first phase of fermentation is the **first rise** which starts when the yeast is added to the dough and lasts until the dough is divided up. The aim of this stage is to allow the fermentation agent the time needed to adapt to the environment and produce the carbon dioxide necessary for the dough to rise, ethyl alcohol, and a number of other products from the breakdown of sugars.

11.2.3 Weighing, Dividing, Shaping, and Resting

The dough is divided into regular individual pieces by weight (Fig. 11.3a).

Forming into balls (Fig. 11.3b) is optional, depending on the baker or regional customs, and it may consist of gentle handling or it may be more energetic kneading. This is a restructuring action to firm up the dough and consists of turning the ball around on itself, to give it a fairly regular round shape (Guinet 1982).

The second phase of fermentation is **resting**, which follows dividing and precedes shaping. This gives the dough time to relax and prepares it to pass to the shaping stage.

11.2.4 Shaping

Shaping gives the dough the form of the finished product (Fig. 11.3c) (Guinet 1982). Finally, the third and last phase of fermentation or gas retention will take place in a climate control proofer at 27 °C and 85 %RH. This stage is called the **final proof** and during this stage the dough piece will expand with the force of the gas from the fermentation process. The gluten network stretches and forms gas-filled pockets of varying sizes depending on the bread-making method. This is the final fermentation phase: this determines the future volume of the bread.

Fig. 11.3 (**a**) Division of the dough, (**b**) forming into balls, (**c**) dough is shaped before being placed in the proofer for the fermentation stage

11.2.5 Baking

During this last operation, the fermented dough is transformed into bread as a result of the effect of heat. It is the result of a heat exchange between the oven and the product being baked and consists of the expansion and physicochemical transformation of the dough. These changes ensure the bread's organoleptic and preservation qualities (Roussel and Chiron 2002). We describe this stage, which is crucial in order to obtain a quality product.

11.3 Baking in Ovens

11.3.1 Description of the Ovens

While remaining true to its traditional origins, the oven has undergone various modifications through the centuries and has contributed to changing the quality of bread. It has adapted to the different types of bread devised by the baker and offered to consumers.

Whether it is old or more recent, an oven is always made up of the following parts (Roussel and Chiron 2002):

– The cement or metal **housing** with one or several **baking chambers**: the bottom or the "sole" and the top or the "vault." They contain outlets to evacuate water vapor and/or flue gases.

The soles are made of natural stone or mineral elements with refractory cements. The thickness of the soles, 2–3 cm, determines the time it takes for the ovens to get up to temperature and whether or not there are temperature peaks.

In modern "convection" ovens, the baking chamber consists of a compartment into which the trolley(s) is introduced.

– The **burner**, where combustion occurs, is the heat source.
– **Insulation** is made up of a variety of materials (glass wool, sand, fire clay, etc.).
– There may be various **accessories**: boiler, steam device, chimney, pyrometer, and thermostat.

There are traditionally two categories of oven:

– Direct combustion ovens: the burner and the products to be baked are in the same chamber.
– Indirect combustion ovens: the heat source heats up an intermediate fluid (water vapor or air), which carries the heat to the baking chamber.

In each case, there are different types of oven:

- The sole may be fixed or mobile.
- Heating may be by combustion or by electrical resistance.
- There may be different heat transfer methods.

11.3.2 Baking

11.3.2.1 Loading

The end of the final proof marks the beginning of the loading phase. This consists of taking the products out of the proofer when fermentation is complete and placing them on a conveyor belt or oven peel, without damaging the dough pieces. The pieces have previously been scored using a knife blade (or **scarification**), where an incision is made on the surface. The purpose is to reduce the dough's resistance and thus facilitate expansion at the beginning of the baking process. If there is no scoring, then this can lead to a high pressure buildup in the dough, which may restrict expansion and also create uneven shapes (e.g., bursting) in some less-resistant areas. The slits channel the force of the gas toward the cut areas and also improve the aesthetic appearance of the product (Roussel and Chiron 2002).

11.3.2.2 The Role of Steam

Before and/or after loading the products, steam is injected into the oven. This condenses on the surface of the dough pieces, forming a thin film of water, which softens the dough, helping the carbon dioxide to expand and the pieces to rise. For a short time, this film of water protects the surface of the dough pieces against the effects of heat and delays the formation of the crust, which is thus reduced in thickness. In addition, it helps reactions, which color the bread crust, giving it a golden yellow color and a shiny appearance. Without the steam, the bread is dull and the surface is rough.

Finally, the steam also limits evaporation of the water in the dough to some extent, thus improving the production yield (Guinet 1982). If baking is done without steam, there may be a lot of cavities in the crust associated with the porosity of the crumb, which thus forms a series of chimneys through which the water vapor in the dough can escape (Fig. 11.4a). In Fig. 11.4b steam injection helps to close the porosity of the dough. Condensation happens at the surface of the bread immediately (loading at 27 °C) where the starch grains then have an excess of water and swell rapidly, which tends to fill in/close up the cells that were originally in contact with air. This closed porosity in the crust (away from areas of scoring) then limits steam transfer between the very moist crumb and the baking chamber by acting as a barrier.

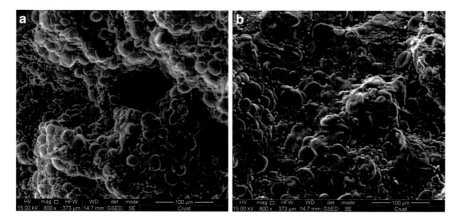

Fig. 11.4 SEM image of bread dough crust (×400). (**a**) Crust baked without steam, many cavities. (**b**) Crust has smoother appearance when baked with steam injection + starch grains are larger

11.3.3 Baking Conditions

11.3.3.1 Baking Temperature

This depends on the mass of the dough pieces to be baked, the level of color, the stability of the dough, and its level of hydration.

If the oven is too hot, the dough is browned, the crust forms more quickly, the bread is less able to rise, and the slits are more irregular. Since the color reactions occur earlier, the product has to be taken out of the oven earlier, so the bread is moister; this is a way of increasing yield.

If the oven temperature is too low, even if the crust is formed at a later stage, expansion is also hindered because the gases expand less; the bread has a tendency to extend outward and thus has a flatter cross section, and the slits are more torn. Problems with color often require a longer cooking time, resulting in the top part drying out, a drop in yield, and also a thicker crust.

In industrial ovens, with the possibility of regulating temperature at the top and bottom and at different levels between the products going into the oven and coming out, baking conditions are flexible and can be optimized. As a general rule:

– When the sole is at a higher temperature at the beginning of baking, this promotes expansion, which then decreases.
– The temperature in the vault increases gradually so that the crust does not form too rapidly and the level of color is set at the end of the baking process (Roussel and Chiron 2002).

11.3.3.2   Baking Time

This varies according to the mass, the shape, and the type of bread. However, it is generally accepted that it takes 1 h to bake 2-kg loaves and half an hour for 500-g loaves. Baking is assessed according to the degree of resistance and the color of the crust and the resonance of the base of the loaf when tapped with a finger (Ammann 1925).

11.3.4   Changes in the Dough During Baking

When the dough pieces are put into the oven, the outside, the part that will form the crust, is subjected to a high temperature. Inside, in the part that will form the crumb, the heat is transmitted slowly.

11.3.4.1   In the Crust

The water that evaporates from the crust via the surface prevents the surface temperature from rising above about 100 °C which in turn prevents a real crust from forming. There is only a film of crust. By releasing the water vapor, this limits heat penetration into the loaf (Guinet 1982). The layer of water that is deposited on the bread, or that covers the bread, transforms the starch into soluble starch and dextrin, which gives the crust its beautiful golden color. The crust dries out by gradually solidifying as the temperature at the surface of the bread comes closer to that of the oven: there is almost total desiccation of all the outside parts of the bread; next the gluten browns, and the starch itself can begin to be roasted. This phenomenon occurs more deeply into the mass of the bread the higher the oven temperature, and we obtain a crust that is fairly thick, cracked, and fairly dark in color.

11.3.4.2   In the Crumb

Inside the bread, the temperature rises gradually: the volume of the bread increases quickly as the gases in the cells dilate and then more gradually due to the acceleration of fermentation until the yeasts are rendered inactive by the heat (at about 55 °C). At the same time, amylolysis continues at a faster rate, and then this also stops but a little later as amylases are not destroyed until about 70 °C. During amylolysis, the stiffening of the starch and/or gelatinization starts at about 55 °C. This phenomenon comes to an end at about 83 °C. The gluten starts to coagulate around 70 °C and continues until about 100 °C. This is the end of changes inside the oven as the glutinous structure is now "set."

Above 70 °C there can be considerable dilation of the cells filled with air which is saturated with water vapor. At the end of baking, water evaporation can cause a final expansion (100 °C) even though the crumb is starting to set (Sommier et al. 2005). The alcohol formed during fermentation vaporizes in the ambient air.

Experiments carried out by Zanoni et al. (1993) show that the variation in temperature and water content during bread baking is determined by the formation of a vaporization front at 100 °C. As this front gradually advances into the product, the result is that two distinct zones are formed: the crust, where the water content is very low and the temperature tends asymptotically toward the oven temperature, and the crumb, where the water content is constant and the temperature tends asymptotically toward 100 °C. This result was previously found by Pyler (1973) (cited by Yin and Walker 1995) who confirmed this and found that the temperature of the crumb cannot reach the boiling temperature of water because the amount of heat evacuated as the water evaporates is greater than that absorbed by the crumb.

On the other hand, after carrying out experiments on German bread, Seibel (1984) believed that to obtain a good quality of bread at the end of the baking process, the temperature of the crust has to reach up to 102–105 °C.

11.3.4.3 Expansion in Volume

The expansion of the volume of the dough, which starts in the first minutes after being placed in the oven, is estimated to be one third of its original size (Pyler 1973). Although this phenomenon is generally attributed to the effect of heat penetration, it is currently considered to be the result of a series of reactions.

One of the physical effects of heat on a gas is to increase its pressure. If it is in an elastic compartment, then we observe that it expands in volume. Another physical effect of heat is to reduce the solubility of gases. A large amount of carbon dioxide formed during fermentation is present in solution in the liquid that makes up the dough. When the dough temperature reaches 49 °C, the CO_2 in solution is released, causing the pressure inside the dough to increase. Another effect of the heat is to transform liquids with a low boiling point into vapor. Thus, the alcohols present in the dough evaporate at 79 °C and also contribute to increasing pressure inside the dough.

11.4 Microscopic and Macroscopic Approach to Bread Baking

In this example we use the following ingredients to produce the bread: flour, yeast, water, and salt. The flour and yeast are kept at 4 °C. A summary of the procedure is shown in Table 11.1, which gives the detailed formula and the equipment and the time required for the mixing and shaping operations. Before the bread is placed in the oven, a single slit is cut into the surface lengthways. The dough piece is placed directly onto the sole of the oven.

Table 11.1 Summary of standard conditions used

Formula	1.5-kg Corde Noire special flour
	945 g demineralized water (63 %)
	37.5-g yeast (2.5 %) CAPPA
	33-g salt (added 5 min before kneading complete) (2.2 %)
Kneading with Artofex mixer	4 min at speed 1 (50 rpm)
	17 min at speed 2 (70 rpm)
First rise	20 min at 27 °C and 85 % RH
Division and molding	10 min. 7–10 portions of 250 or 350 g
Resting	20 min at ambient temperature
Shaping	10 min
Fermentation or final proof	1 h 30 min at 27 °C and 85 %RH
Baking	27 min (sole 250–260 °C, vault 235–245 °C)

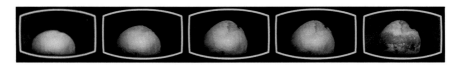

Fig. 11.5 Baking without steam at Tsole = 260 °C, Tvault = 245 °C at five observation times (1, 4, 6, 8, 27 min)

As soon as the dough is loaded into the oven, the baker injects steam (for a few seconds), which condenses on the surface of the product (the dough pieces leaving the proofing chamber at 27 °C). This deposited water then acts as a plasticizer and facilitates the distortion of the product macroscopically, which gives rise to a considerable increase in mass (Sommier and Douiri 2006). In microscopic terms, for very short time the starch grains have an excess of water, and this, combined with the heat given off by the vapor (enthalpy of vaporization), causes them to swell, and this dilation closes up the orifices of the cell structure, transforming the surface into a continuous film (Fig. 11.4a, b). The dough piece then becomes a closed cell network (apart from the scored area), which limits the phenomena of mass loss by evaporation.

Because of the reactions that it produces, steam injection is the reason for one of the organoleptic characteristics typical of French bread, i.e., a thin, crisp, golden crust. If there is no steam injection, this will result in a lesser increase in volume (absence of plasticizer) and a more rapid drying out of the product surface, leading to a thick and crunchy brown crust (cf. Figs. 11.5 and 11.6 at $t = 27$ min).

In Figs. 11.5 and 11.6, we can compare two different types of baking carried out in similar thermal conditions with the same temperatures for the sole (260 °C) and the vault (245 °C) in the same oven, both without (Fig. 11.5) and with (Fig. 11.6) steam injection. With steam injection, the product expands more quickly. Its surface

Fig. 11.6 Baking with steam at Tsole = 260 °C, Tvault = 245 °C at five observation times (1, 4, 6, 8, 27 min)

dehydrates less quickly, slowing the formation of the crust, which means that its volume can continue to increase. This product reaches its maximum volume at $t = 8$ min, whereas the product without steam sets more quickly in the areas that are not very moist (to the right and the left). The crust forms quickly, thus setting/locking the structure and preventing it from increasing in volume any further. The dough piece expands only around the moist area where the knife blade was used to score the surface; maximum volume is reached at 6 min.

At the same time, inside the product, heat transfers have a direct influence on transfers of matter, and at the heart of the product, there is still a core of dough that evolves as the product expands (cf. Fig. 11.10).

This doughy core remains cold, while the surface temperature increases rapidly, thus leading to a gradual transformation of the dough into crumb, from the outside walls toward the core of the product. The initial water content of the dough is fairly uniform (when it is placed in the oven), and this is subjected to successive waves of evaporation and condensation at the level of each pore. As evaporation takes place in the hottest cell wall (i.e., close to the surface), vapor circulates through the cell network at great speed before condensing on a cold cell wall.

As Wagner et al. (2007) have shown, we then see variations in the water content in the product. The dough core becomes the area where the water content is not only greatest but also higher than the mean water content of the product at the beginning of baking.

At the same time, the product expands rapidly, the cells dilate, the cell walls become thinner and tear, and we observe that the pockets coalesce, forming a foam which has a greater porosity. The original scoring ruptures, exposing the cell network to the air; maximum volume is reached after 8–10 min (Fig. 11.6) and the porous network is now opened up. This phenomenon can be clearly seen in Fig. 11.7 where the increase in the internal pressure of the product causes the structure to expand in the first few moments, but after 10 min, although the pressure continues to increase, this is no longer sufficient to expand the product, which still remains moldable: even though it is under pressure, the structure starts to sag (collapse also visible in Fig. 11.10).

Figure 11.8 shows a piece of bread crumb (after baking) imaged using X-ray computerized tomography. At the spatial scale and resolution (30 μm/*voxel*) assumed in Fig. 11.8, the porous space of the crumb is clearly irregular and heterogeneous (as opposed to sandwich loaf). The large disparity in the size of air cells expresses mainly the bread-making stages prior to the baking step, namely, the fermentation and the shaping technique. As shown in the higher resolution Fig. 11.9

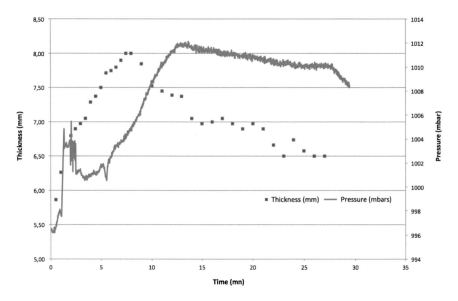

Fig. 11.7 Example of typical variation of internal pressure and thickness of bread during baking

Fig. 11.8 3D tomography image of a sample of bread crumb (in *red*). The porosity cells are transparent. Image size is $630 \times 630 \times 440\,voxels$, i.e., $18.9 \times 18.9 \times 13.2$ mm. Spatial resolution: 30 μm per *voxel* (a *voxel* is the 3D counterpart of a 2D TV pixel). Porosity is 93 %

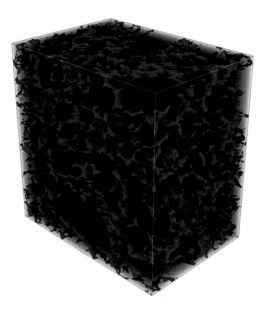

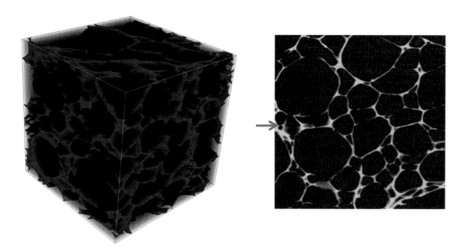

Fig. 11.9 (*Left view*) 3D tomography image of a sample of bread at a higher spatial resolution of 12 μm/*voxel*. Image size is 550×550×500 *voxels*, i.e., 7.5×7.5×7.5 mm. Porosity is 89 %. (*Right view*) 2D section taken from the 3D image (*left view*). The *arrow* shows a closed bubble of porosity (whose diameter is about 180 μm) trapped in a local zone where the crumb is thicker and denser. In this display, *light gray* levels represent the crumb

(12 μm/*voxel*), the working and shaping of the dough by hand can result in a marked elongation of the air cells along a preferential direction: see in Fig. 11.9 how the cells are stretched in a horizontal plane. In Figs. 11.8 and 11.9, it can be seen clearly that most of the air cells are connected, forming a continuous and open network bounded by very thin cell walls. Yet, the higher spatial resolution that typifies (Fig. 11.9) also stresses that *locally*, the crumb can be thicker and can include small *closed* air cells that remain disconnected from the main continuous porous network. This closed fraction of porosity represent *early times bubbles* whose expansion was blocked at some point of the process. In all respects, *the whole art of the baker is to ensure that this unwanted fraction of closed porosity remains small or negligible.*

The change in the water content in the product and the hygrometry of the air in the oven are of course not the only factors responsible for the expansion of the product. Minor ingredients in the product mass (such as yeast, starter, ascorbic acid), the mechanical energy used during kneading which can influence the quality of the gluten network, and the temperature of the dough at the end of kneading (which is itself closely linked with the initial temperature of each ingredient and with the intensity with which the dough is worked) are all levers with which to adjust/guide the final quality of the product in terms of flavor, volume, and uniform cell structure which are not the subject under discussion here. In the case of baking, it would seem more interesting to highlight the interaction between the product and the oven, looking especially at the temperatures used in the baking stage.

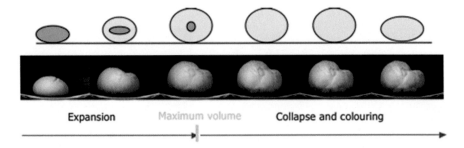

Fig. 11.10 Baking using steam injection at Tsole = 250 °C, Tvault = 235 °C at five observation times (1, 4, 6, 8, 27 min)

Figure 11.10 shows baking a dough piece with initial steam injection using the same experimental protocol as in Fig. 11.7 but with temperatures of 250 °C for the sole and 235 °C for the vault. The product appears to expand less rapidly than was the case in Fig. 11.6 (10 °C hotter), and the volume at 6 min is less. As we have seen, the temperature of the oven sole is one of the key factors in expansion for many cereal products. This heat input enables the gases held in the dough to dilate, which leads in turn to an increase in the internal pressure of the product and contributes to expansion. However, it is in competition with the coalescence of the cells which gradually form a network opening onto the outside via the only porous area in contact with the air in the oven and the initial scoring by the knife blade, which tends to extend and thus makes a large contribution to exposing the crumb to the air by promoting evaporation of the water in the form of steam (loss of mass). The second factor responsible for the acceleration in the loss of product mass is the increase in exchange surface area (direct consequence of the product increasing in volume). The expansion seen in Fig. 11.6 brings the upper part of the product closer to the heating elements, and thus the radiation received by the product is greater than in the case shown in Fig. 11.10, which accelerates the drying of the crust and therefore sets the structure and reduces the collapse phenomena that are more clearly visible when baking at a lower temperature.

Baking is therefore not only a tool used to set the structure of the product, but it is also clearly a means of guiding the final quality of the product in terms of expansion, color, or sensory perception. A thin, crisp crust that has also helped ensure that a higher water content is retained within the product will give the consumer a crumb that is softer and more aerated as it is less dense.

Acknowledgments The environmental scanning electron microscope used in this work was cofinanced by the European Union (EU). The EU supports Aquitaine via the European Regional Development Fund.

Bibliography

Ammann L (1925) Meunerie et Boulangerie, 2ème édn. Edition J.-B. Bailliere et Fils, 528 p

Feillet P (2000) Le grain de blé. Edition INRA, Paris, 308 p

Guinet R (1982) Technologie du pain français. Edition B.P.I, Paris, 181 p

Pyler EJ (1973) Baking science and technology, vol 2. Siebel, Chicago, 1240 p

Roussel P, Chiron H (2002) Les pains français : évolution, qualité, production. Edition Maé, 433 p

Seibel W (1984) Research needs of the European bread industry, state of the art of baking and recommendation for further work. In: Zeuthen et al (eds) Thermal processing and quality of foods—concluding seminar of cost 14–18 November 1983 Athène Grèce. Elsevier, Londre, pp 362–367

Sommier A, Douiri I (2006) Approche et compréhension des transferts de chaleur et de matière lors de la cuisson du pain. Industrie des Céréales 147:5–11

Sommier A, Chiron H, Colonna P, Della Valle G, Rouille J (2005) An instrument pilot scale oven for the study of French bread baking. J Food Eng 69(1):97–106

Wagner MJ, Lucas T, Le Ray D, Trystram G (2007) Water transport in bread during baking. J Food Eng 78(4):1167–1173

Yin Y, Walker CE (1995) A quality comparison of breads baked by conventional versus nonconventional ovens: a review. J Sci Food Agric 67(3):283–291

Zanoni B, Peri C, Pierucci S (1993) A study of the bread-baking process. I: a phenomenological model. J Food Eng 19:389–398

Chapter 12
Traditional Rye Sourdough Bread in the Baltic Region

Grazina Juodeikiene

12.1 Introduction

Many European supermarkets offer bread loaves from around the continent—from the French baguette through to the Italian ciabatta and Germany's dark pumpernickel. But ironically, despite the variety, many consumers are turning back to local bakeries or even rolling up their sleeves to make their own bread at home like some of their grandparents. On top of this, for home and kitchen cooking and baking equipment, one can buy easily in every electric home appliances shop. This new trend does not surprise food scientists who think that consumers are searching for the kind of flavour and texture often sacrificed during industrial bread making. Humans ate rye sourdough bread in ancient times, and it remains a traditional part of the diet in the Baltic region (Sahlstrom and Knutsen 2010). Rye sourdough is a mixture of rye flour and water and contains metabolically active lactic acid bacteria (LAB), which contribute to the bread's special flavour and taste (Rosenquist and Hansen 2000; Hammes et al. 1996; Venskaityte et al. 2005). Sourdough is used as an essential ingredient for acidification, leavening and production of flavour compounds and biopreservation of bread (Katina et al. 2005; De Vuyst and Leroy 2007; Sadeghi 2008). In addition, several researchers have reported on how sourdough bread can resist microbiological spoilage by moulds and rope-forming bacilli due to the production of inhibitory substances (Katina et al. 2002; Hassan and Bullerman 2008; Sadeghi 2008; Valerio et al. 2009). As in the other Baltic countries, Lithuanians still eat sourdough bread—a pleasant-tasting bread that uses natural leavening, which traces back to the ancient Egyptians (Imbrasiene 2008).

G. Juodeikiene (✉)
Kaunas University of Technology, K. Donelaičio g. 73, Kaunas, Lithuania
e-mail: grazina.juodeikiene@ktu.lt

© Springer Science+Business Media New York 2016
K. Kristbergsson, J. Oliveira (eds.), *Traditional Foods*, Integrating Food
Science and Engineering Knowledge Into the Food Chain 10,
DOI 10.1007/978-1-4899-7648-2_12

12.2 Ancient Art in Modern Bread Making

The leavening technique of modern bread making was replaced in many countries by industrially processed yeast and food additives, but sourdough bread has continued to be the main staple in the Baltics and some other regions of the world. In comparison with the rye bread processes used in the EU, the major difference in the Baltic region is the scalding step in the process which gives the product a pleasant sweet tone (Fig. 12.1).

In sourdough, the secret ingredient is a "starter" or "mother" of flour and water that ferments when a lactobacillus bacteria culture is added. That starter gives the lightness to the dough. This living culture is fed and preserved for use in successive loaves and often passed down through generations. Malted rye flour, often enzymatically non-active, is added to scald when traditional sourdough production is used in order to give the bread a dark brown colour, which Baltic consumers like. Besides, caraway seeds or sometimes linseeds are an essential part of rye bread. The seeds are mostly used whole and often they are added to the scald process. Some breads may be even decorated with caraway seeds. Traditionally, country rye bread loaves are often very big, e.g. today in Latvia one can buy a rye loaf weighing up to 5 kg. These breads are still produced with traditional methods using much hand work, and they may be baked in big stone ovens. For the up-to-date consumer's convenience, these huge loaves are available sliced and packed. The most common weight of rye bread loafs is 0.7–1.0 kg. In Lithuania, it is still common to buy an unsliced 2 kg loaf, packed in a colourful plastic bag. Traditional rye breads in the Baltic countries are oval in shape. The crust of the traditional loaf is very thick, often more than 1 cm. Due to the flour cooking and gelatinization processes, bread crumb is very moist and sometimes sticky and has a long shelf life.

Rye is rich in dietary fibre and this type of cereal is an excellent raw material for sourdough preparation as well as for a healthy and tasty bread product. In a time where people prefer a healthy diet, the option of sourdough bread making is attractive. The method is ideally suitable for making rye bread—which has lower calorie content than many other types of bread as well as containing more dietary fibre. Other particularity is that baker's yeast does not work well as a leavening agent with rye flour. The leavening action is taken over partly by LAB so that one can decrease the amount of baker's yeast.

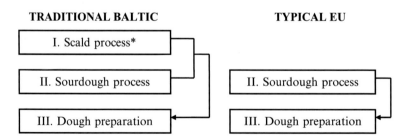

Fig. 12.1 Comparison of rye sourdough processes used in the Baltic countries and elsewhere in the EU. *The scald process is actually a heat treatment of a flour matrix with hot water or steam. This matrix is brought in a certain amount of time and by a certain temperature to gelatinization

12.3 Microbial Population of Lithuanian Spontaneous Rye Sourdoughs

12.3.1 General Aspects

LAB are an important group of industrial starter cultures, which has a long tradition in bread making. The LAB that developed in the dough may originate from selected natural occurrence in the flour or from a starter culture containing one or more known species of LAB. In Lithuania the oldest and most popular rye bread technology is with spontaneous sourdough. The organoleptic and structural properties of bread, as well as their keeping quality and the changes in the properties associated with staling, are remarkably affected by the fermentation process and the microbial ecology of the sourdough (Röcken and Voysey 1995; Rosenquist and Hansen 1998; Gänzle and Vogel 2003). High-quality sourdough bread is dependent on a consistent and microbial stable sourdough. LAB and yeast cohabit in the sourdough and interact for their carbohydrate, amino acid and vitamin requirements. During rye dough fermentation, LAB play the main role; therefore, their amount in the dough is ten times higher than those of yeast. Both homo-fermentative and hetero-fermentative LAB lower the pH by the production of lactic acid, thus allowing the swelling of pentosans and proteins in rye, improving the sensory properties and shelf life (Stolz and Böcker 1996). Moreover, hetero-fermentative LAB also contribute to the leavening and affect flavour.

The microflora responsible for sourdough fermentation in different countries has been characterised, for example, in Germany (Müller et al. 2001; Meroth et al. 2003), Finland (Salovaara and Katunpää 1984), Russia (Kazanskaya et al. 1983), Sweden (Spicher and Lönner 1985), Denmark (Rosenquist and Hansen 2000) and other countries. About 43 different sourdough species of LAB have been reported until now. The majority are members of the LAB core genera *Lactobacillus* and to a lesser extent of *Lactococcus*, *Pediococcus*, *Weissella* and *Leuconostoc*. The most frequently isolated species are *Lactobacillus sanfranciscensis*. This microorganism is effective in the fermentation of maltose and is able to hydrolyse raffinose, galactose and ribose. Therefore, sourdough with this strain is characterised by very good microbiological, biochemical and baking properties. During spontaneous fermentation, both lactobacilli (homo-fermentative *Lactobacillus plantarum*, *L. delbrueckii*, *L. farciminis*, *L. casei*, *L. acidophilus* and *L. johnsonii* and hetero-fermentative *L. brevis*, *L. fermentum*, *L. buchneri*, *L. fructivorans*, *L. pontis* and *L. panis*) and pediococci (*Pediococcus pentosaceus*, *P. acidilactici*) develop. The homo-fermentative organisms are dominating, and no significant differences exist between rye and wheat. *Leuconostoc* and *Weissella* may play a role during the first phase of fermentation, and they can be important for the growth of lactobacilli. Pediococci usually exist at the end of the fermentation process of plant material (De Vuyst and Neysens 2005).

Several factors account for the dominance of sourdough lactobacilli during traditional dough preparation. First, their carbohydrate metabolism is highly adapted to the main energy sources in dough, maltose and fructose. Utilisation of maltose via maltose phosphorylase and the pentose phosphate shunt with fructose as co-substrate results in a higher energy yield than homo-fermentative maltose degradation. Second, the growth requirements of *L. sanfranciscensis* with respect to temperature and pH match the conditions encountered during sourdough fermentation. Similarly growth of *L. amylovorus* occurs optimally under rye sourdough fermentation conditions. Third, antimicrobial compounds may contribute to the stable persistence of lactobacilli in sourdough fermentations.

Many studies have been carried out on the lactic microflora of sourdough and the microflora may vary significantly with types, thus contributing to the peculiar characteristics of many traditional breads. Because of scarce data in literature of microbial population and of antimicrobial activity of Lithuanian sourdoughs, the traditional spontaneous rye sourdoughs have been taken in consideration (Digaitiene et al. 2005).

12.3.2 Physicochemical and Microbiological Characteristics of Sourdoughs in Lithuania

The physicochemical and microbiological characteristics of the sourdoughs in Lithuania have been studied in different sourdoughs: three rye sourdoughs (samples G, H, K) were received from different bakeries in Lithuania, and sourdoughs A to F were prepared from Lithuanian whole meal rye flour according to the most common traditional procedure used locally by changing the rates of flour and water (1:1 or 1:2), fermentation temperatures (25, 30 or 35 °C) and time (24 or 48 h). Conditions of sourdough preparations are shown in Table 12.1.

Table 12.1 shows the fermentation time (FT), pH, titratable acidity (TTA) and population of tested sourdoughs (LAB, CFU g^{-1}). In all investigated samples (a total of nine sourdoughs), the pH values ranged from 4.21 to 3.72 and the TTA from

Table 12.1 Physicochemical and microbiological characteristics of some sourdoughs in Lithuania

Sample	Fermentation temperature (°C)	FT[a] (h)	pH	TTA[b]	LAB (CFU g^{-1})	Consistency
A	30	48	4.09	20.8 ± 0.11	3.1×10^8	Thick
B	30	48	3.94	18.1 ± 0.33	3.4×10^8	Fluid
C	25	72	3.94	22.2 ± 0.23	2.8×10^8	Thick
D	25	72	3.72	19.5 ± 0.06	2.1×10^8	Fluid
E	35	72	3.90	21.9 ± 0.19	4.0×10^8	Thick
F	35	72	3.78	18.9 ± 0.09	2.0×10^8	Fluid
G	–	–	4.21	16.0 ± 0.14	4.5×10^6	Thick
H	–	–	4.05	27.9 ± 0.20	2.2×10^6	Thick
K	–	–	3.85	29.9 ± 0.10	1.2×10^6	Fluid

[a]Fermentation time
[b]Total titratable acidity—determined on 10 g of sample homogenised with 90 mL of distilled water and expressed as the amount (mL) of 0.1 M NaOH to get pH of 8.2

16.0 ± 0.14 to 29.9 ± 0.10. The lowest pH of 3.72 and 3.78 was determined for the sourdough samples D and F, respectively, and the highest—for sample G. The highest TTA was noticed for samples K (29.9) and H (27.9), while the lowest TTA was noticed for sample G (16). The rather high pH of 4.21 of sample G could be attributed to the dominance of low-acidifying species of LAB due to the different technology applied in sourdough preparation as well as the buffering capacity of the flour used. The total number of LAB of the samples ranged from 2.2×10^6 to 4.0×10^8 CFU g^{-1}. The highest count of LAB was observed in the sourdough sample E. These results of the number of LAB and acidity in the sourdoughs were similar to other published results on rye sourdoughs (Kazanskaya et al. 1983; Salovaara and Katunpää 1984; Spicher and Lönner 1985; Spicher and Stephan 1993; Müller et al. 2001).

12.3.3   Phenotypic Characteristics of Isolates

To obtain a first overview of the fermentation flora, a total of 56 LAB were randomly chosen for phenotypic and genotypic identification. All isolates were Gram positive, catalase is negative, and 24 strains were recognised as rods and the others as *cocci* (Table 12.2).

Based on their morphological character and carbohydrate fermentation patterns, the isolates were initially divided into six groups. Group No. 1, containing 42 % of the isolates, was rods hydrolysing sucrose, maltose, trehalose and glucose. Only one isolate (group No. 2) fermented all the five tested sugars. Group No. 3 (38 %) and No. 4 (17 %) were differentiates due to their ability to ferment trehalose. The *cocci* were divided into two groups No. 5 (69 %) and No. 6 (31 %) based on their ability to ferment sucrose and maltose. All of them could ferment trehalose and glucose. Accordingly, the rods showed greater variation than the *cocci* as they were divided into four groups.

12.3.4   Genetic Characterisation of the LAB

In order to determine to which species the isolate belonged to, representatives for each group were analysed genetically. The internally transcribed spacer PCR method was applied on isolates K6, K3, K28, H5, K4, K5, H8, H15, C5, C11, C27, D6, E4, E8 and E16. The isolates could be divided in five clusters (Fig. 12.2) based on mobility and their number of amplified ITS fragments separated after agarose gel electrophoresis. Isolates belonging to the same group gave identical band pattern of ITS fragments. Thus, group No. 1 had pattern X1, group No. 3, X3; group No. 4, *X*2; group No. 5, X4; and group No. 6, X5.

One or a few representatives from each group (No. 1, K6; No. 2, K1; No. 3, K3; No. 4, H15; No. 5, C2, C11, D12, F27; No. 6, E16, F15) were further analysed by amplification and sequencing of 16S rDNA. The identification of *Lactobacillus*

Table 12.2 Phenotypic characteristics of isolates

Group	Strain	Morphology	Carbohydrate fermentation					Growth at 50 °C
			Arabinose	Sucrose	Maltose	Trehalose	Glucose	
1	G3, G5, G11, G20, K6, K7, H9, H24, H27, H29	Rods	–	+	+	+	+	n.d.
2	K1		+	+	+	+	+	n.d.
3	K3, K9, K11, K26, K28, K29, K30, H1, H5		–	–	+	+	+	n.d.
4	K4, K5, H8, H15		–	–	+	±	+	n.d.
5	A11, A26, A27, B2, B9, B11, B27, C1, C2, C5, C11, C21, C27, D5, D6, D7, D12, D23, E10, E30, F4, F27	Cocci	–	±	+	+	+	–
6	D8, D25, E2, E4, E8, E16, E22, E27, F3, F15		–	–	–	+	+	+

"+" ferment, "–" not ferment, ± weak fermentation, n.d. not determined

Ż X3 X3 X2 X2 X1 X3 X2 X2 X4 X4 X4 X4 X5   X5 X5 Ż

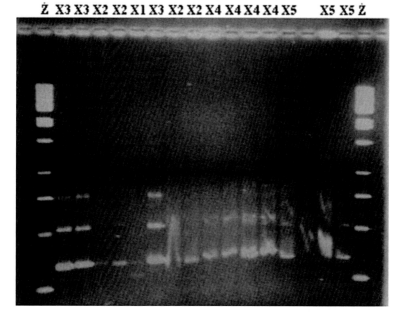

Fig. 12.2 Clustering of ITS patterns generated by PCR with 16S-1500F and 23S-32R primers. *M* marker 50 bp DNR GeneRuler, *X1* group No. 1 (K6), *X2* group No. 4 (K4, K5, H8, H15), *X3* group No. 3 (K3, K28, H5), *X4* group No. 5 (C5, C11, C27, D6) and *X5* group No. 6 (E4, E8, E16)

strains belonging to cluster X3 was not possible, because the tested strain (K3) did not generate any product after PCR amplification. The obtained sequences (Table 12.3) were compared to strains in the NCBI Blast Library (Table 12.4).

The homology search showed that strains K1 and H15 are either *L. sakei* or *L. curvatus* and K6 either *L. farciminis* or *L. paralimentarius*, while the other strains showed high similarity to both *P. pentosaceus* and *P. acidilactici.*

In order to differentiate between the two possibilities, the strains were grown at different temperatures and on selected carbon sources. *L. sakei* differs from *L. curvatus* in the fermentation of arabinose, sucrose, maltose and trehalose. It was determined that the strain K1 was able to ferment these sugars and so it was designated as belonging to *L. sakei.* The LAB H15 fermented trehalose very weakly and did not ferment arabinose and sucrose, showing that it belongs to *L. curvatus. L. farciminis* and *L. paralimentarius* differ in the fermentation of maltose; LAB K6 fermented maltose very well so it belongs to *L. farciminis.*

P. pentosaceus differs from *P. acidilactici* in the fermentation of maltose and growth at 50 °C. As isolates C2, C11, D12 and F27 fermented maltose and did not grow at 50 °C, they identify themselves as isolates belonging to *P. pentosaceus.* Strains E16 and F15 did not ferment maltose but was grown at 50 °C showing that they belong to *P. acidilactici.*

Table 12.3 Species identification based on 16S rDNA gene sequence comparison

LAB strain	16S rDNR gene sequence
K6	GGTGGCTTCCCCCCTGCTTCCCCGGGTCTNCTGCGTGCGAGCCTCTTTCGCTCGATCTGNG
	AAGTNTTGTTTGCGGTTTGCCCTGTTTCCGATGTTATCCCCCNCTTTTGGGCCGGTTACCCA
	CGTGTTACTCACCCGTCCGCCACTCACCANGGTCTGAATCNTGAAGCGAGCTTCNTTCTTC
	NGGATGGTTCGTTCGTACTTGCNNTGTATTAGTGCATGCCGCCAGACGTTCGTCCTGAGCC
	ATGATC
K1	GGGTCCTCCTAAAGTGATAGCCGAAACCATCTTTCAACCCTACACCATGCGGTGTTAGGTT
	TTATGCGGTATTAGCATCTGTTTCCANNATGTTATCCCCCACTTTAGGGCAGGTTACCCACG
	TGTTACTCACCCGTCCGCCACTCACTCAAATGTTTATCAATCAGGAGCAAGCTCCTTCAAT
	CTAAACGAGAGTGCGTTCGACTTGCATGTAT
H15	CGCCTGCTGCCGTNGGTGGCNTNNCCCCNCTNCTGTGCCCGCGGGTCCTCCTNGTGTGCCG
	NCCTCTTTCTCCCTCCCTGCGGTGTTGGTTTTTGCGGTTTGCTCTGTTTCCATGTTTCCCCCC
	TTTGGGCGGTTACCCCGTGTTACTCCCCGTCCGCCCTCCTCNAATGTNTATCATCNGGAGC
	AAGCTCCTTCAATCTAAACGAGAGTGCGTNCCGACTTGCATGTATTAGGCACGCCGCCNG
	CGTTCGTCCTGAGCC
C2	GGTCCATCCAGAAGTGATAGCAGAGCCATCTTTTAAAAGAAAACCATGCGGTTTTCTCTGT
	TATACGGTATTAGCATCTGTTTCCAGGTGTTATCCCCTACTTCTGGGCAGGTTACCCACGTG
	TTACTCACCCGTTCGCCACTCACTTCGTGTTAAAATCTCAATCAGTACAAGTACGTCATAAT
	CAATTAACGGAAGTTCGTTCGACTTGCATGTATTAGGCACGCCGCCAGCG
C11	AGTCCTTCCGTGGAGGGNCGCTANCNCCNCCAACTACTAATCNCCGCGGGGTCNTNCGAA
	AANTGNATAGCAGAGCCATCTTTTAAAAGAAAACCATGCGGTTTTCTCTGTTATACGGTAT
	TAGCATCTGTTTCCAGGNGTTATCCCCTACTTCTGGGCAGGTTACCCACGTGTTACTCACCC
	GTTCGCCACTCACTTCGTGTTAAAATCTCAATCAGTACAAGTACGTCATAATCAATTAACG
	GAAGTTCGTTCGACTTGGCATGTATNAGGGCACGCCGCCAGCGTTCATCCT
D12	TCGGCTACGTATCACTGCCTTGGTGAGCCTTTACCTCACCAACTAGCTAATACGCCGCGGG
	TCCATCCAGAAGTGATAGCAGAGCCATCTTTCAAAAGAAAACCATGCGGTTTTCTCTGTTA
	TACGGTATTAGCATCTGTTTCCAGGTGTTATCCCCTACTTCTGGGCAGGTTACCCACGTGTT
	ACTCACCCGTTCGCCACTCACTTCGTGTTAAAATCTCAATCAGTACAAGTACGTCATAATC
	AATTAACGGAAGTNCGTTCGACTTGCATGTATTAGGCACGCCGCCAGCGTTCAT
F27	CGCCTTGAGTGAGCCGTTACCTCACCAACTAGCTAATGCGCCGCGGGTCCATCCAGAAGTG
	ATAGCAGAGCCATCTTTTAAAAGAAAACCAGGCGGTTTTCTCTGTTATACGGTATTAGCAT
	CTGTTTCCAGGTGTTATCCCCTGCTTCTGGGCAGGTTACCCACGTGTTACTCACCCGTCCGC
	CACTCACTTCGTGTTAAAATCTCATTCAGTGCAAGCACCTCATNGATCAATTAACGGAAGT
	TC
E16	CCGATTACCCTCTCATGTCGGCTACGTATCACTGCCTTGGTGAGCCTTTACCTCACCAACTA
	GCTAATACGCCGCGGGTCCATCCAGAAGTGATAGCAGAGCCATCTTTTAAAAGAAAACCA
	TGCGGTTTTCTCTGTTATACGGTATTAGCATCTGTTTCCAGGTGTTATCCCCTACTTCTGGG
	CAGGTTACCCACGTGTTACTCACCCGTTCGCCACTCACTTCGTGTTAAAATCTCAATCAGTA
	CAAGTACGTCATAATCAATTAACGGAAGTTCGTTCGACTTGCATGTATTAGGCACGCCGCC
	AGCGTTCATCCTGAGCCATGA
F15	GTCGGCTACGCATCATCGCCTTGGTGAGCCGTTACCTCACCAACTAGCTAATGCGCCGCGG
	GTCCATCCAGAAGTGATAGCAGAGCCATCTTTTAAAAGAAAACCAGGCGGTTTTCTCTGTT
	ATACGGTATTAGCATCTGTTTCCAGGTGTTATCCCCTGCTTCTGGGCAGGTTACCCACGTGT
	TACTCACCCGTCCGCCACTCACTTCGTGTTAAAATCTCATTCAGTGCAAGCACGTCCTGATC
	AATTAACGGAAGTTCGTTCGACTTGCATGTATTAGGCACGCCGCCAGCGT

Table 12.4 Species identification based on 16S rDNA gene sequence comparison

Isolates	NCBI lactic acid bacteria strain	Similarity (%)	Accession number
K6	*L. farciminis* DSM2018	90	AJ417499.1
	L. paralimentarius DSM13238	88	AJ417500.1
K1	*L. sakei* HNSL5a	99	AY204896.1
	L. curvatus subsp. melibiosus CCUG 34545	99	AY204889.1
H15	*L. sakei subsp. carnosus* CCUG31331	91	AY204892.1
	L. curvatus subsp. melibiosus CCUG 34545	91	AY204889.1
C2	*P. pentosaceus* LM 2632	100	AY675245.1
	P. acidilactici RO 17	97	AF515229.1
C11	*P. pentosaceus* LM 2632	98	AY675245.1
	P. acidilactici RO 17	95	AF515229.1
D12	*P. pentosaceus* LM 2632	99	AY675245.1
	P. acidilactici RO 3	94	AY375299.1
F27	*P. pentosaceus* LM 2632	96	AY675245.1
	P. acidilactici YDW 17	100	AF375935.1
E16	*P. pentosaceus* LM 2632	99	AY675245.1
	P. acidilactici RO 17	96	AF515229.1
F15	*P. pentosaceus* LM 2632	95	AY675245.1
	P. acidilactici DSMZ 202	100	AJ249539.1

12.3.5 Genetic Characterisation of the Species

According to the species genetic characterisation, *L. farciminis* belongs to group No. 1, *L. sakei* to group No. 2, *Lactobacillus ssp.* to group No. 3, *L. curvatus* to group No. 4, *P. pentosaceus* to group No. 5 and *P. acidilactici* to group No. 6. The composition of lactic microflora of sourdoughs is shown in Table 12.5.

For each sample, the percentage of isolates (referred to the total number of isolates for a given sample) belonging to different groups is shown. The bacterial flora of the commercial sourdoughs (G, H, K) consisted of four species belonging to genus *Lactobacillus*. A high percentage of isolates identified as *L. farciminis* (from 50 % for sample H to 100 % for sample G) was found in most samples. Moreover, sample H had the same quantity (25 %) of *L. curvatus* and not identified *Lactobacillus* spp. In sample K a more heterogeneous composition of the microflora was found as it consisted of *Lactobacillus* spp. (58 %). Also *L. farciminis* (17 %), *L. curvatus* (17 %) and *L. sakei* (8 %) were isolated. In contrast to the commercial sourdoughs, the samples prepared from Lithuanian rye flour (A–F) also harbour strains belonging to predominant species of *P. pentosaceus* (from 60 % for sample F and 71 % for sample D to 100 % for sample A, B and C); however, in the sample E, *P. acidilactici* (86 %) dominated.

Based on the investigations, one can conclude that the lactic microflora of spontaneous rye sourdoughs used for the production of the traditional bread in Lithuania was dominated by *Pediococcus* and *Lactobacillus*: overall, 39 % of the isolates were identified as *P. Pentosaceus*, 18 % as *P. acidilactici* and *L. Farciminis*, 7 % as *L. Curvatus*, 2 % as *L. sakei* and 16 % as *Lactobacillus ssp.*

Table 12.5 Percentage distribution of LAB belonging to different groups in sourdough samples

Samples	Isolates	Groups					
		Nr. 1 *L. farciminis*	Nr. 2 *L. sakei*	Nr. 3 *Lactobacillus* spp.	Nr. 4 *L. curvatus*	Nr. 5 *P. pentosaceus*	Nr. 6 *P. acidilactici*
A	3	–	–	–	–	100	–
B	4	–	–	–	–	100	–
C	6	–	–	–	–	100	–
D	7	–	–	–	–	71	29
E	8	–	–	–	–	25	75
F	4	–	–	–	–	50	50
G	4	100	–	–	–	–	–
H	8	50	–	25	25	–	–
K	12	17	8	58	17	–	–

12.4 Technological Aspects of Traditional Rye Sourdough in the Commercial Bread Process

Today most bakeries use the two-stage sourdough process or multiple-stage rye sourdough process, characterised by continuous propagation. Depending on the bakery, the sourdough is either liquid (palpable; dough yield 300) or stiff (dough yield 170).

In the two-stage sourdough process, sourdough (fermented scald) is made as in the one-stage process, but it is refreshed before use. In refreshing, a certain portion of saccharafied scald is added to the sourdough (fermented scald), and it is fermented till pH 3.2–3.8 (depending on used LAB). Today the multiple-stage process is used if special aroma preferable for consumers is desired.

Multistage processes for the production of rye bread are outlined in Fig. 12.3.

The recipe of rye bread (kg) in different steps of the process and some technological parameters are presented in the Table 12.6.

The stages and conditions for rye dough preparation with a multiple-stage process are the following:

Scald preparation and saccharification. Scalds are prepared in the scalding machine. According to the recipe water, sifted rye flour and caraway seeds are added into the scalding machine. Temperature of mixture is adjusted to $(66 \pm 2)\,°C$ by steam

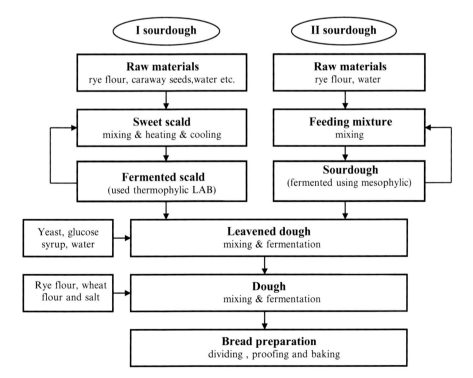

Fig. 12.3 Production of sourdough rye bread using a multistage process

Table 12.6 Recipe and technological parameters of rye bread (kg) using multistage sourdough production (8 %, R700, and 20 %, W10505B)

Raw materials and semimanufactured products, parameters of process	Steps of technological process					
	Sweet scald	Fermented scald	Feeding mixture	Fermented sourdough	Liquid leavened dough	Dough
Sifted rye flour 700	12.0	–	12.0	–	–	56.0
Wheat flour (type 1050B)	–	–	–	–	–	20.0
Water	25.07	–	6.65	–	27.10	–
Caraway seeds (kg)	0.2	–	–	–	–	–
Pressed yeast (kg)	–	–	–	–	0.5	–
Glucose syrup					3	
Salt (kg)	–	–	–	–	–	1.5
Sweet scald (kg)	–	37.27[a]	–	–		–
Fermented scald (kg)	–	74.54	–		37.27	–
Feeding mixture (kg)	–	–	–	18.65[b]		–
Fermented sourdough (kg)	–	–	–	37.27	18.65	
Liquid leavened dough (kg)	–	–	–	–	–	86.52
Saccharification time (min)	60±5	–	–	–	–	–
Fermentation time (min)	–	150±20	–	120±20	90±20	180±20
Temperature (°C)	66±2	49±2	30±2	29±2	29±2	30±2
Moisture (%)	72±2	73±2	45±2	45±2	73±2	47±1
Titrable acidity (°)	–	6±1	–	6.5±1	6.5±1	6.5±1

Remark: Semi-products from previous production: [a]fermented scald or [b]sourdough

in the jacket of the scalding machine. The duration of heating is $30±5$ min, and then the scald is mixed throughout and left to cool down to $(49±2)$ °C temperature, which is optimal for scald saccharification (duration $60±5$ min).

Fermented scald preparation. The sweet scald is cooled till optimal temperature for thermophylic LAB and added to the tank of the scalding machine, in which 1/2 of fermented scald (TTA of $6.0±1.0$) remains. Everything is mixed well. Scald is left to ferment for $(150±20)$ min at $(49±1)$ °C, while TTA of $6.0±1.0$ is achieved.

Feeding mixture for sourdough preparation. Sifted rye flour and water are mixed in the scalding machine until a homogenous mass develops. Mixture temperature should be $(30±2)$ °C and moisture $(45±2)$ %. The prepared feeding mixture is pumped to the tank for liquid sourdough fermentation.

Sourdough preparation. One half of the feeding mixture is added to the tank in which 1/2 of the produced sourdough remains (from previous production). Everything is mixed well and then left to ferment for $(120±20)$ min., while TTA of $(6.5±1.0)$ is achieved. The temperature of the liquid sourdough is $(29±2)$ °C.

Leavened dough preparation. The necessary amount (1/2 of the tank) of the fermented scald (TTA $6±1$) is added into a tank and cooled down to $(29±2)$ °C temperature. Then the required amount (1/2 of tank) of the fermented sourdough (TTA $6.5±1.0$) is added and mixed. The liquid leavened dough is fermented for $(90±20)$ min., while TTA of $(6.5±1)$ is achieved. Yeast is added to the fermented scald and sourdough mixture right before dough preparation and is mixed well.

Dough preparation, dividing, proofing and baking. Dough is mixed in a continuous stirrer or periodical stirrer. After mixing the dough is fermented for (180 ± 20) min. The dough is divided in 0.9 kg pieces using an appropriate equipment, e.g. a BTW II type machine. The mass of the dough pieces depends on the final product (bread) mass, as well as on the accuracy of the dividing machines, mass losses during baking and cooling. Dough pieces are proofed in a conveyer proofing board. Proofing time is (45 ± 5) min (depends on proofing conditions). The optimum proofing conditions for rye dough pieces are temperature (34 ± 2) °C and relative air humidity (66–70 %). Before baking the dough pieces are marked (cutting, pricking) and sprayed with water. Bread is baked in a continuous baking oven and the baking time is approximately (47 ± 2) min.

12.5   Advantages of the Traditional Rye Sourdough Bread Processes

The incorporation of sourdoughs using scald fermentation in rye bread formula is recommended for a more aromatic bread flavour, and this type of sourdough bread has a higher content of volatile compounds and a higher score in sensory tests (Venskaityte et al. 2005). From market studies, it is well known that 85 % of all consumers say that the smell and taste of this type of bread is the most pleasant. The homo-fermentative LAB mainly used for scald fermentation are mostly responsible for acidification of the dough. Sourdough fermented with hetero-fermentative LAB have, aside from a much higher content of acetic acid and ethanol, a higher content of ethyl acetate and ethyl hexanoate compared to sourdoughs fermented with homo-fermentative LAB, which have a higher content of diacetyl and some other carbonyls. When sourdough yeast is added in the preparation of the sourdough (in case of traditional process in leavened dough), the production of ethanol, iso-alcohols, esters and diacetyl increases considerably.

Sourdough addition is the most promising procedure to preserve bread from microbial spoilage (ropiness and mould). Several sourdough LAB produce inhibitory substances in varying degrees such as organic acids (lactic acid, acetic acid), ethanol, diacetyl, hydrogen peroxide and carbon dioxide (Rosenquist and Hansen 1998). Inhibition, however, can also be caused by bacteriocins or bacteriocin-like inhibitory substances (BLIS) (Narbutaite et al. 2008; Digaitiene et al. 2012). Bacteriocins are extracellularly released and ribosomally synthesized low molecule mass peptides or proteins, with a bactericidal or bacteriostatic mode of action, in particular against a wide range of mostly closely related Gram-positive bacteria (Klaenhammer 1993; Savadogo et al. 2006) and even against food-borne pathogens (Garneau et al. 2002), but the producer cells are immune to their own bacteriocins (De Vuyst and Leroy 2007). During recent years, health-conscious consumers are looking for natural food which can fit into their healthy lifestyles. This includes food without additives as chemical preservatives, and so the use of LAB and their antibacterial substances for biopreservation has become more attractive to the food industry (Gálvez et al. 2007; Zotta et al. 2009). Besides, the preparation of bread with scald results in the reduction of the staling process and retrogradation of the product (Juodeikiene et al. 2011).

The addition of sourdough to the bread recipe has a positive influence on the nutritive value of the bread, as the minerals become bioavailable, and the blood glucose and insulin responses are lowered after eating sourdough bread compared to wheat bread. This effect is believed to be due to the presence of lactic acid formed during fermentation, which will promote interactions between starch and gluten (Östman et al. 2002) during heating, or increases the hardness and decreases the porosity of the bread (Autio et al. 2003). Also, the perceived taste of salt is enhanced in sourdough rye bread compared to wheat bread, so less salt can be added in sourdough rye bread and thus improves the nutritional properties of wholemeal cereals.

References

Autio K, Liukkonen K-L, Juntunen K, Katina K, Laaksonen DE, Mykkänen H, Niskanen L, Poutanen K (2003) Food structure and its relation to starch digestibility, and glycaemic response. In: Proceedings of the 3rd international symposium food rheology and structure (ISFRS 2003). Institute of Food Science and Nutrition, Zürich, pp 7–11

De Vuyst L, Leroy F (2007) Bacteriocins from lactic acid bacteria: production, purification, and food applications. J Mol Microbiol Biotechnol 13:194–199

De Vuyst L, Neysens P (2005) The sourdough microflora: biodiversity and metabolic interactions. Trends Food Sci Technol 16:43–56

Digaitiene A, Hansen A, Juodeikiene G, Josephsen J (2005) Microbial population in Lithuanian spontaneous rye sourdoughs. Ekologia i Technika 13(5):193–198

Digaitiene A, Hansen A, Juodeikiene G, Eidukonyte D, Josephsen J (2012) Lactic acid bacteria isolated from rye sourdoughs produce bacteriocin-like inhibitory substances active against Bacillus subtilis and fungi. J Appl Microbiol 112(5):732–742

Gálvez A, Abriouel H, López RL, Omar NB (2007) Bacteriocin-based strategies for food biopreservation. Int J Food Microbiol 120:51–70

Gänzle MG, Vogel RF (2003) Contribution of reutericyclin production to the stable persistence of Lactobacillus reuteri in an industrial fermentation. Int J Food Microbiol 80:31–45

Garneau S, Martin NI, Vederas JC (2002) Two-peptide bacteriocins produced by lactic acid bacteria. Biochimie 84:577–592

Hammes WP, Stolz P, Gänzle MG (1996) Metabolism of lactobacilli in traditional sourdoughs. Adv Food Sci 18:176–184

Hassan YI, Bullerman LB (2008) Antifungal activity of Lactobacillus paracasei ssp. tolerans isolated from a sourdough bread culture. Int J Food Microbiol 121:112–115

Imbrasiene B (2008) Lithuanian traditional meals. Baltos lankos, Vilnius, 38 p

Juodeikiene G, Salomskiene J, Eidukonyte D, Vidmantiene D, Narbutaitė V, Vaiciulyte-Funk L (2011) The impact of novel fermented products containing extruded wheat material on the quality of wheat bread. Food Technol Biotechnol 49(4):502–510

Katina K, Sauri M, Alakom L, Sandholm MT (2002) Potential of lactic acid bacteria to inhibit rope spoilage in wheat sourdough bread. Lebensm Wiss Technol 35:38–45

Katina K, Arendt E, Liukkonen KH, Autio K, Flander L, Poutanen K (2005) Potential of sourdough for healthier cereal products. Trends Food Sci Technol 16:104–112

Kazanskaya LN, Afanasyeva OV, Patt VA (1983) Microflora of rye sours and some specific features of its accumulation in bread baking plants of the USSR. In: Holas J, Kratochvil F (eds) Developments in food science. Progress in cereal chemistry and technology, vol 5B. Elsevier, London

Klaenhammer TR (1993) Genetics of bacteriocins produced by lactic acid bacteria. FEMS Microbiol Rev 12:39–86

Meroth CB, Walter J, Hertel C, Brandt M, Hames WP (2003) Monitoring the bacterial population dynamics in sourdough fermentation processes by using PCR-denaturing gradient gel electrophoresis. Appl Environ Microbiol 69:475–482

Müller MRA, Wolfru G, Stolz P, Ehrmann MA, Vogel RF (2001) Monitoring the growth of Lactobacillus species during a rye flour fermentation. Food Microbiol 18:217–227

Narbutaite V, Fernandez A, Horn N, Juodeikiene G, Narbad A (2008) Influence of baking enzymes on antimicrobial activity of five bacteriocin-like inhibitory substances produced by lactic acid bacteria isolated from Lithuanian sourdoughs. Lett Appl Microbiol 47(6):555–560

Östman EM, Nilsson M, Liljeberg-Elmstahl HGM, Molin G, Björk IME (2002) On the effect of lactic acid on blood glucose and insulin responses to cereal products: mechanistic studies in healthy subjects and in vitro. J Cereal Sci 36:339–346

Röcken W, Voysey PA (1995) Sourdough fermentation in bread making. J Appl Bacteriol 79:38–48

Rosenquist H, Hansen A (1998) The antimicrobial effect of organic acids, sour dough and nisin against Bacillus subtilis and B. licheniformis isolated from wheat bread. J Appl Microbiol 85:621–631

Rosenquist H, Hansen A (2000) The microbial stability of two bakery sourdoughs made from conventionally and organically grown rye. Food Microbiol 17:241–250

Sadeghi A (2008) The secrets of sourdough: a review of miraculous potentials of sourdough in bread shelf life. Biotechnology 7:413–417

Sahlstrom S, Knutsen SH (2010) Oats and rye: production and usage in Nordic and Baltic countries. Cereal Food World 55(1):12–14

Salovaara H, Katunpää H (1984) An approach to the classification of Lactobacilli isolated from Finnish sour rye dough ferments. Acta Aliment Pol 10:231–239

Savadogo A, Outtara Cheik AT, Bassole Imael HN, Traore SA (2006) Bacteriocins and lactic acid bacteria: a mini review. Afr J Biotechnol 5:678–683

Spicher G, Lönner C (1985) Die Mikroflora des Sauerteiges. XXI. Mitteilung: Die Sauerteigen schwedischer Bäckereien vorkommenden Lactobacillen. Z Lebensm Unters Forsch 181:9–13

Spicher G, Stephan H (1993) Handbuch Sauerteig, 4th edn. Behr's Verlag GmbH, Hamburg

Stolz B, Böcker G (1996) Technology, properties and applications of sourdough products. Adv Food Sci 18:234–236

Valerio F, Favilla M, De Bellis P, Sisto A, de Candia S, Lavermicocca P (2009) Antifungal activity of strains of lactic acid bacteria isolated from a semolina ecosystem against Penicillium roqueforti, Aspergillus niger and Endomyces fibuliger contaminating bakery products. Syst Appl Microbiol 32:438–448

Venskaityte A, Juodeikiene G, Hansen A, Petersen MA (2005) Application of dynamic headspace technique and gas chromatography: mass spectrometry methods for bread. Ekologia i Technika 13(5):182–186

Zotta T, Parente E, Ricciardi A (2009) Viability staining and detection of metabolic activity of sourdough lactic acid bacteria under stress conditions. World J Microbiol Biotechnol 25:1119–1124

Chapter 13
The Legume Grains: When Tradition Goes Hand in Hand with Nutrition

Marta Wilton Vasconcelos and Ana Maria Gomes

13.1 Legume Grains and Tradition

13.1.1 Nomenclature and Classification

The word "legume" is derived from the Latin word *legumen,* which means seeds harvested in pods. Their seeds usually fall into one of two main classes: one in which the principal storage material is a polysaccharide, usually starch, and another in which the principal storage material is fat, often described as oilseeds (Kingman 1991). In many parts of the world, legume grains are also called "pulses", a word derived from the Latin term *puls*, meaning pottage or pulp. The latter is the word most commonly used in most English-speaking countries to describe legume seeds which have a low content of fat, such as kidney beans, broad beans, peas or lentils (Aykroyd et al. 1982).

All members of the *Leguminosae* share the characteristic of producing pods, but their growth habits can vary from being small, herbaceous plants to full-grown trees. They comprise 18,000–19,000 species belonging to about 670 genera and are distributed worldwide, with woody genera mostly in the southern hemisphere and the tropics; herbaceous genera are mostly present in temperate regions and are very numerous in Mediterranean-climate areas (Langran et al. 2010).

The majority of the most commonly consumed legume grains originated from Asia, Africa and, to a lesser extent, the Mediterranean region and India. Bean, chickpea and lentil, crops commonly consumed in the Mediterranean region, originated in this area but are now widespread in temperate regions throughout the world

M.W. Vasconcelos (✉) • A.M. Gomes
Centro de Biotecnologia e Química Fina – Laboratório Associado, Escola Superior de Biotecnologia, Universidade Católica Portuguesa,
Rua Arquiteto Lobão Vital, Apartado 2511, Porto 4202-401, Portugal
e-mail: mvasconcelos@porto.ucp.pt

© Springer Science+Business Media New York 2016
K. Kristbergsson, J. Oliveira (eds.), *Traditional Foods*, Integrating Food Science and Engineering Knowledge Into the Food Chain 10,
DOI 10.1007/978-1-4899-7648-2_13

(Aykroyd et al. 1982). Soybean, which is now mostly produced in the United States, originated as a wild crop in East Asia, was domesticated in China and was later brought to other countries (Guo et al. 2010).

13.1.2 Origins and Production

Legumes have been consumed by humans since the earliest practice of agriculture and have been used both for their medicinal, cultural as well as nutritional properties (Phillips 1993). Other parts of the plant besides the grain, such as leaves, shoots, flowers, immature pods and tubers, in addition to sprouted seeds (Aykroyd et al. 1982) are commonly consumed.

The species grown include important grain, pasture and agroforestry species. Their wide agricultural distribution makes them important crops in numerous parts of the world, such as Africa, Latin America, Asia and Mediterranean regions. Legumes are one of the world's most important sources of food supply, especially in developing countries, in terms of food energy as well as nutrients (Reyes-Moreno and Paredes-López 1993; Wang et al. 2003). In Africa and Latin America, they are especially valuable as a source of dietary protein to complement cereals, starchy roots and tubers (Phillips 1993). Their domestication for human food and animal feed have been reported back to as early as 9500 years before present (BP) in the Fertile Crescent of the Near East (Phillips 1993). Jatobá, also known as guapinol, or Brazilian cherry, is a member of the *Fabaceae*, genus *Hymenaea*, which was used as a food source in Amazonian prehistory (Roosevelt et al. 1996). Beans of several species were domesticated in tropical America thousands of years ago, and *Phaseolus* identified in archaeological sites in Mexico and Peru indicated the presence of domesticated beans as early as 10,000 years ago (Kaplan and Lynch 1999). There is evidence for the cultivation in Mexico of common beans, *P. vulgaris*, and teparies, *P. acutifolius*, before about 2500 BP in the Tehuacán Valley (Kaplan and Lynch 1999). In the Peruvian Andes, there are reports of domesticated common beans and lima beans by about 4400 BP and by about 3500 BP, respectively, and lima beans by about 5600 BP in the coastal valleys of Peru. Reports on chickpea indicate as possible origin the fertile crescent of the Mediterranean, though some ethnobotanists report early findings in the Himalayas (Kaplan and Lynch 1999).

The use of legumes in pastures and for soil improvement dates back to the Romans, with Varro (37 BC), cited by Graham and Vance (2003) noting "Legumes should be planted in light soils, not so much for their own crops as for the good they do to subsequent crops". This was intuitive information from farmers that later on was explained by the fact that legumes can fix atmospheric nitrogen through a symbiotic interaction with *Rhizobia* and, by doing so, are able to reduce atmospheric nitrogen to ammonia, resulting in high protein content of the soil. Because of this ability to "fix" nitrogen, legumes help stabilize soil and prevent erosion and are essential for healthy ecosystems and agriculture.

Legumes species can be adapted to climates ranging from temperate to tropical and humid to arid. The dry mature seeds, which have high food value and store well for long periods of time, play an important role in the diets of the peoples of the world. In rank order, dry bean (*Phaseolus vulgaris*), pea (*Pisum sativum*), chickpea (*Cicer arietinum*), fava bean (*Vicia faba*), pigeon pea (*Cajanus cajan*), cowpea (*Vigna unguiculata*) and lentil constitute the primary dietary legumes. Peanut (*Arachis hypogaea*) and soybean (*Glycine max*) are dominant sources of cooking oil and protein. They are also major food sources in many regions (National Academy of Science 1994).

Legumes are second only to the *Graminiae* in their importance to humans (Graham and Vance 2003), with soybean being the biggest produced legume crop. In fact, the Food and Agriculture Organization statistics for 2012 (http://faostat.fao.org/site/339/default.aspx) show that about 276 million metric tons of soybean were produced across the world, ranking it seventh on the world's top food and agriculture commodities (the largest being milk). Soybean came right after rice and wheat, which ranked second and sixth (http://faostat.fao.org/site/339/default.aspx), respectively. Figure 13.1 shows that the world's top producer of soybean in 2013 was the United States, with a production of 89 million metric tons, followed by Brazil and Argentina. It is noticeable that Ukraine, which is listed as the ninth biggest soybean producer in the world, is the only European country in the top 10 list, followed by Italy, in 15th place.

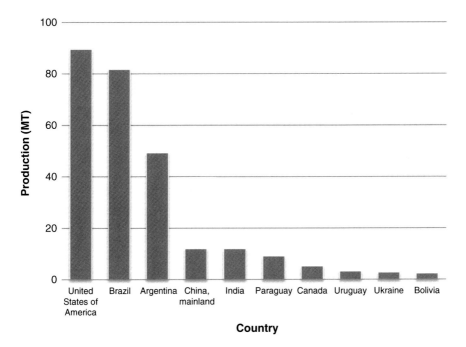

Fig. 13.1 World's top 10 soybean-producing countries (million metric tons). *Source*: FAOSTAT (2013)

Common bean (*Phaseolus vulgaris*, $2n = 22$) is the world's most important grain legume for direct food consumption, especially in Latin America and Africa (Pedrosa-Harand et al. 2006). The world's top bean producers are Brazil and India, with a production of 3.4 and 1.1 million metric tons in 2008, respectively, according to FAO statistics.

Grain quality of the legumes is determined by factors such as acceptability by the consumer, soaking characteristics, cooking quality and nutritive value. Acceptability characteristics include a wide variety of attributes, such as grain size, shape, colour, appearance, stability under storage conditions, cooking properties, quality of the product obtained and flavour (Reyes-Moreno and Paredes-López 1993). In addition to traditional food and forage uses, legume grains can be milled into flour (Loayza Jibaja and Bressani 1988); used to make bread, doughnuts, tortillas, chips, spreads and extruded snacks; or used in liquid form to produce milks, yoghurt and infant formula. Legume grains can also be used as a source of food additives. Gums for thickeners (e.g. arabic gum, guar gum and tragacanth gum) are derived from legumes, and soybean derivatives (e.g. soybean lecithin) are used extensively in processed foods.

Many more nonedible uses of legume grains exist, such as in the production of plastics, biodiesel, cosmetics, dyes, insecticides, etc., but these would fall out of the scope of this chapter.

13.1.3 Legume Grains in the Mediterranean Diet

The Mediterranean diet can be described as the dietary pattern found in the olive-growing areas of the Mediterranean region, in the late 1950s and early 1960s, when the consequences of World War II were overcome, but the fast-food culture had not reached the area yet (Trichopoulou 2001).

Recent studies, capturing the evidence accumulated over the last three decades, have documented that the traditional Mediterranean diet meets several important criteria for a healthy lifestyle (Trichopoulou 2001). Although it is difficult to describe the "typical" traditional Mediterranean diet, one can define eight aspects that tend to characterize it, of which two of the most important ones are a high ratio of monounsaturated/saturated fat and a high intake of legumes (including beans, lentils, peas and chickpeas) (Trichopoulou et al. 2003b, 2009). Although variable, it is estimated that the average person following a Mediterranean diet ingests about 102.5 g of legumes per day (Martinez-Gonzalez et al. 2002).

Total lipid consumption may be moderately high in the populations consuming Mediterranean diets, reaching around 40 % of total energy intake in Greece and around 30 % of total energy intake in Italy (Trichopoulou 2001), but the ratio of monounsaturated to saturated dietary lipids is much higher than in other places of the world, including northern Europe and North America (Trichopoulou 2001).

The legumes in the Mediterranean diet can have a significant impact in plasma carotenoid levels. It was found that amongst different components of the

Mediterranean diet, fruits and vegetables tended to increase levels of some carotenoids, whereas olive oil had no apparent effect (Trichopoulou et al. 2003a).

Also, although soybean is not a traditional component in the Mediterranean diet, researchers have found that a daily intake of 40 mg of soy isoflavones together with a Mediterranean diet and exercise can reduce insulin resistance in postmenopausal women (Llaneza et al. 2010).

Some of the best described health benefits of the Mediterranean diet are the lower incidence of osteoporosis (Puel et al. 2007) and different types of cancer such as gastric adenocarcinoma (Buckland et al. 2009), cancers of the upper aerodigestive tract (Bosetti et al. 2009), colorectal cancer (Bingham et al. 2003; Stamatiou et al. 2007), prostate cancer (Itsiopoulos et al. 2009) and other types of cancer (Materljan et al. 2009). The prevention of these diseases through dietary means is linked to specific food items that contain a complex array of naturally occurring bioactive molecules with antioxidant, anti-inflammatory and alkalinising properties (Puel et al. 2007). As legume grains are high in compounds such as proteins, mono- and di-unsaturated fatty acids, phenolic compounds, trypsin inhibitors, phytates and oligosaccharides components known for having important antioxidant and anti-inflammatory activities, it is probable that their consumption is linked to lower prevalence of these ailments.

As an example, certain legume grains, such as lentils, fava beans and split beans, are known to be very fibre rich (Table 13.2); high fibre intakes are related to lower rates of colon cancer. In fact, in populations with low average intake of dietary fibre, an approximate doubling of total fibre intake from foods could reduce the risk of colorectal cancer by 40 % (Bingham et al. 2003).

13.2 Legume Grains and Nutrition

Legumes are important sources of adequate proportions of different nutrients, such as protein, minerals, lipids, vitamins, starch, sugars and other non-starch polysaccharides, which have made them targets of justified nutritional interest (Table 13.2). Bearing in mind that diet and fat ingestion are the factors most directly related to high cholesterol levels and the increase of arteriosclerosis and fatal coronary diseases in various countries, the quantity and type of fat in pulses is another factor to be considered when evaluating the nutritional importance of these substances. Some of these nutrients and their impact on human and animal health will be discussed next.

Table 13.1 shows the typical nutritional content of a selected number of legume grains, in terms of water, proteins, fat, energy, sugars, fibre and carbohydrates. One can observe that there is a substantial variability in terms of protein content, with soybean having the highest amount per 100 g FW basis. Of these grain legumes, chickpea has the lowest amount of protein but, on the other hand, shows the highest amount of total sugars in the grain. The fattiest legume grain, by far, is the peanut, with almost half its fresh weight corresponding to total lipids. The legume showing the highest fibre content is lentil, followed by split pea and fava bean (Table 13.1).

Table 13.1 Nutrient content of selected legume grains, mature seeds, raw, per 100 g fresh weight (FW) basis

Legume	Water (g)	Protein (g)	Total lipid (fat) (g)	Fibre (g)	Energy (kcal)	Sugars, total (g)	Carbohydrate by difference (g)
Soybean (*Glycine max*)	8.54	36.49	19.94	9.3	446	7.33	30.16
Chickpea (*Cicer arietinum*)	11.53	19.30	6.04	17.4	364	10.70	60.65
Lentils (*Lens culinaris*)	10.40	25.8	1.06	30.5	353	2.03	60.08
Split pea (*Pisum sativum*)	11.27	24.55	1.16	25.5	341	8.00	60.37
Groundnut (*Arachis hypogaea*)	6.50	25.80	49.24	8.5	567	3.97	16.13
Fava bean (*Vicia faba*)	10.98	26.12	1.53	25.0	341	5.70	58.29
Black beans (*Phaseolus vulgaris*)	11.02	21.60	1.42	15.2	341	2.12	62.36

Source: USDA (2010)

More and more new crops are being identified which seem to have a good combination in terms of nutrient composition, being very well balanced not just in one nutrient category but for several. The morama bean (*Tylosema esculentum*), for example, is a leguminous oilseed native to the Kalahari Desert and neighbouring sandy regions of Botswana, Namibia and South Africa that has recently been reported to be an excellent source of good quality protein (29–39 %), rich in mono- and di-unsaturated fatty acids and a good source of micronutrients (calcium, iron, zinc, phosphate, magnesium and B vitamins including folate) (Jackson et al. 2010).

Moreover, it is also reported to be a potential source of phytonutrients (phenolic compounds, trypsin inhibitors, phytates and oligosaccharides), components which have been shown to be important in the prevention of noncommunicable diseases such as cardiovascular diseases (CVDs), diabetes and some cancers (Jackson et al. 2010).

Winged bean (*Psophocarpus tetragonolobus*), also known as the Goa bean, is an important underexploited legume of the tropics. All the plant parts, viz., seeds, immature pods, leaves, flowers and tubers are edible. Mature seeds contain 29–37 % protein, 15–18 % oil and fairly good amounts of phosphorus, iron and vitamin B. Its essential amino acid composition is very similar to that of soybean, and the fatty acid composition is very much comparable to groundnut (Kadam and Salunkhe 1984). The trypsin inhibitor in winged bean has been shown to be heat resistant.

These less commonly used crops, given their desirable nutritional aspects, should be considered more carefully, and their utilization should be pondered in non-native environments.

13.2.1 Carbohydrates

Carbohydrates in the legumes can be divided into water-soluble components such as sugars and certain pectins and insoluble ones such as starch and cellulose (Aykroyd et al. 1982). Both of these groups include carbohydrates that are very useful in energy production, but also the more indigestible ones, that cannot be so effortlessly utilized because they are not easily "breakable" by the digestive enzymes. In general, legumes are richer in fibre than most cereals and include cellulose, hemicellulose, lignin and galactomannans (Aykroyd et al. 1982). The primary action of fibres in the human organism occurs in the gastrointestinal tract, presenting different physiological effects. It has been demonstrated that diets rich in these nondigestible fermentable carbohydrates favour the development of beneficial species in detriment of pathogens in the gut (Grizard and Barthomeuf 1999). According to various investigations, metabolites from the fermentation of complex carbohydrates can be beneficial to health because they show a protective effect against colon or rectal cancer; decrease infectious intestinal diseases by inhibiting putrefactive and pathogenic bacteria (*Clostridium perfringens*, *Escherichia coli*, *Salmonella*); increase mineral bioavailability such as that of calcium; aid in the decreases of hypercholesterolemia, hyperlipoproteinemia and hyperglycaemia; and stimulate the immune system (Grizard and Barthomeuf 1999).

The carbohydrate content of legumes varies roughly between 16 and 60 g/100 FW basis (Table 13.2). Legume starches are particularly rich in the unbranched moiety amylose. Thus, they are particularly prone to retrogradation and the devel-

Table 13.2 Lipid content of selected legume grains, mature seeds, raw, per 100 g FW basis

| Legume | Lipids (total) | | | |
	Saturated (g)	Monounsaturated (g)	Polyunsaturated (g)	Phytosterol (mg)
Soybean (*Glycine max*)	2.884	4.404	11.255	161
Split peas (*Pisum sativum*)	0.161	0.242	0.495	135
Chickpea (*Cicer arietinum*)	0.626	1.358	2.694	35
Lentil (*Lens culinaris*)	0.156	0.189	0.516	nd[a]
Peanut (*Arachis hypogaea*)	6.834	24.429	15.559	220
Fava bean (*Vicia faba*)	0.254	0.303	0.629	124
Black beans (*Phaseolus vulgaris*)	0.366	0.343	0.610	nd[a]

Source: USDA (2010)

[a]*nd* no available data

opment of resistant starch after heat treatment. Legume starch may also resist digestion due to entrapment within thick-walled cells, leading to the commonly reported
issues of flatulence associated with the ingestion of legume grains (Phillips 1993).
Legume starch is more slowly digested than starch from cereals and tubers and produces less abrupt changes in plasma glucose and insulin upon ingestion.

13.2.2 Proteins

Protein calorie malnutrition is prevalent in many developing countries of the tropics
and subtropics (Kadam and Salunkhe 1984). In many countries, human nutrition is
highly dependent on grain legumes for protein. It is estimated that about 20 % of
food protein worldwide is derived from legumes (National Academy of Science
1994). The highest consumption occurs in the former Soviet Union, South America,
Central America, Mexico, India, Turkey and Greece.

Different legumes, such as common bean (Reyes-Moreno and Paredes-López
1993), pea (Casey et al. 1984) or soybean (Xu and Chang 2009a; Young et al. 1979),
are very high in protein content (Table 13.2). Besides the grain, the nutritive value
of sprouts as an important source of protein and other nutrients has also been recognized (Ribeiro et al. 2011; von Hofsten 1979). Soy milk is also an excellent dietary
protein source for common consumers, vegetarians and people with lactose intolerance and milk allergy (Xu and Chang 2009a).

Protein contents in legume grains range from 19 to 40 % (Table 13.2), contrast-
ing with 7–13 % of cereals and being equal to the protein contents of meats (18–25
%) (Genovese and Lajolo 2001). However, although quantitative profiles are similar, the same does not apply from a qualitative profile point of view. It is estimated
that the protein in legume grains is generally rich in lysine (Phillips 1993) but,
compared with meat, which is the main source of protein, it is deficient in sulphur-
containing amino acids (Wang et al. 2003), such as cystine (Bertagnolli and Wedding
1977) and methionine (National Academy of Science 1994). On the other hand,
cereal-grain proteins are low in lysine but have adequate amounts of sulphur amino
acids. Therefore, the supplementation of cereals with legumes has been advocated
as a way of combating protein-calorie malnutrition problems in developing countries, and in some countries such combination is part of the food cultural heritage.

Asparagine accounts for 50–70 % of the nitrogen in translocatory channels serving fruit and seed of white lupin (*Lupinus albus* L.) (Atkins et al. 1975). The nutritional quality of legume protein is limited by the presence of heat labile and heat
stable antinutrients as well as an inherent resistance to digestion of the major globulins. In addition to its nutritional impact, legume protein has been shown to reduce
plasma low density lipoprotein (LDL cholesterol) when consumed.

On the down side, many legume seeds have been proven to contain high amounts
of lectin. Lectin is a sugar-binding protein present in plants and animals. These
proteins have the ability to agglutinate erythrocytes, acting as haemagglutinins
(Grant et al. 1983). As described before in this chapter, soybean is the most impor-

tant grain legume crop, the seeds of which contain high activity of soybean lectins. These are able to disrupt small intestinal metabolism and damage small intestinal villi via the ability of lectins to bind with brush border surfaces in the distal part of the small intestine (Imberty and Perez 1994). Heat processing can reduce the toxicity of lectins, but low temperature or insufficient cooking may not completely eliminate their toxicity, as some plant lectins are resistant to heat. In addition, lectins can result in irritation and oversecretion of mucus in the intestines, causing impaired absorptive capacity of the intestinal wall. Other protein non-nutritional compounds include protease inhibitors—Kunitz trypsin inhibitor (KTI) and Bowman-Birk inhibitor. Protease inhibitors represent 6 % of the protein present in soybean seed. Approximately, 80 % of the trypsin inhibition is caused by KTI, which strongly inhibits trypsin and therefore reduces food intake by diminishing their digestion and absorption. Another effect of KTI is the induction of pancreatic enzyme, hyper secretion and the fast stimulation of pancreas growth, hypertrophy and hyperplasia.

However, reports have shown that germinating or soaking in 0.5 % sodium bicarbonate can decrease most antinutritional factors such as phytic acid, tannin, trypsin inhibitor and haemagglutinin activity, at least in soybean, lupin and bean seeds (el-Adawy et al. 2000). Furthermore, efforts have been made to use selection and breeding to improve antinutritional factors content in legumes, and low antinutritional factor lines with acceptable seed size have been developed (Mikić et al. 2009). However, environment and agricultural management influences such quality of legumes to a considerable extent, and this should be noted while breeding for antinutritional factor lowering potential.

13.2.3   Lipids

The legume grains, in particular soybean and groundnut (peanut), are often used to extract edible oils. Of all edible oils, that from soybean is the second most important, with an estimate world production of 36 million metric tonnes in 2009 (FAO stats), followed by palm oil, which produced 41 million metric tonnes in the same year. Groundnut oil is the sixth most important edible oil of vegetable origin and had an annual production of about 5.2 million metric tonnes in 2009 (FAO stats). In contrast to the leguminous oilseeds, most pulses contain only a small amount of fats (usually less than 2 %, Table 13.2). Chickpeas are unusual amongst the pulses as they have a relatively high fat percentage (around 6 %, Table 13.2). Whereas soybean and groundnut have around 20 and 50 g of total lipids per 100 g FW basis, respectively, split peas and lentils have only about 1 g of total lipids per 100 g FW basis. However, even in the high fat legumes, the lipid component is highly unsaturated. Table 13.3 gives more detailed information on the types of lipids present in the legume grains. Of the total lipid content in soybean and groundnut, over 85 % are comprised of unsaturated fat, with peanut having more polyunsaturated fat and soybean more monounsaturated fat. Even though peanuts are high in fat, their

Table 13.3 Vitamin content of selected legume grains, mature seeds, raw, per 100 g FW basis

Legume	Vitamin					
	Vitamin A (IU)	Vitamin E (alpha-tocopherol) (mg)	Vitamin C (mg)	Thiamine (mg)	Riboflavin (mg)	Vitamin B-6 (mg)
Soybean (*Glycine max*)	22	0.85	6	0.874	0.870	0.377
Split peas (*Pisum sativum*)	149	0.09	1.8	0.726	0.215	0.174
Chickpea (*Cicer arietinum*)	67	0.82	4	0.477	1.541	0.535
Lentil (*Lens culinaris*)	39	0.49	4.4	0.873	0.211	0.540
Groundnut (*Arachis hypogaea*)	0	8.33	0	0.640	0.135	0.348
Fava bean (*Vicia faba*)	53	0.05	1.4	0.555	0.333	0.366
Black beans (*Phaseolus vulgaris*)	0	0.21	0	0.900	0.193	0.286

Source: USDA (2010)

consumption is viewed as a significant contribution to a healthy diet. They contain important nutrients and bioactive constituents that can provide a wide range of health benefits, such as phytosterols (Shin et al. 2010). More recently, peanuts have also been studied in satiety control and body weight management.

Phytosterols are unsaponifiable triterpenes found in the lipid fraction, possessing a cyclopentanoperhydrophenanthrene ring structure and are sometimes referred to as 4-desmethyl sterols. Several other legume grains besides peanuts are rich in phytosterols, in particular soybean, split peas and fava beans (Table 13.3). The health benefits of phytosterol consumption as natural components of a regular diet include anticarcinogenic, anti-osteoarthritic (Gabay et al. 2009), anti-coronary heart disease (Escurriol et al. 2009; Teupser et al. 2010) and hypocholesterolemic (Demonty et al. 2009; Hernandez-Mijares et al. 2010; Lin et al. 2009) actions. Due to the importance of phytosterols in human health, genetic engineering approaches have been utilized and have been successful in generating transgenic plants with increased phytosterol content in the seeds (Harker et al. 2003; Hey et al. 2006). However, these tests were conducted in tobacco plants, which are highly amenable for transformation and should be repeated for these crops.

Also, modulating endogenous levels of novel fatty acids of oils has gained significant attention in recent years, due to the increasing awareness of consumers of the impact that dietary lipids have on health (Clemente and Cahoon 2009).

Commodity soybean oil is composed of five fatty acids (palmitic acid, stearic acid, oleic acid, linoleic acid and linolenic acid) with average percentages of 10 %, 4 %, 18 %, 55 % and 13 %, respectively (Clemente and Cahoon 2009). This fatty acid profile sometimes results in low oxidative stability that limits the uses of soybean oil in food products and industrial applications. Therefore, there is an interest in utilizing genetic transformation techniques to modulate the fatty acid composition of soybean oil.

13.2.4   Vitamins

The vitamin content of different legume grains is quite variable (Table 13.4). Vitamin A concentration is highest in split pea, whereas groundnut has no trace of this particular vitamin. However, groundnut is very rich in alpha-tocopherol, having 8.33 g per 100 g FW basis. Alpha-tocopherol is quantitatively the major form of vitamin E in humans and animals and has been extensively studied (Jiang et al. 2001). It is a class of compounds that may have vitamin E activity and function as an antioxidant protector of the organism. Oxidative damage is a major contributor to the development of cancer, CVD and neurodegenerative disorders. Antioxidant vitamins defend against oxidative injury and are therefore believed to provide

Table 13.4   Mineral content of selected legume grains, mature seeds, raw, per 100 g FW basis

Legume	Minerals (mg)						
	Calcium (Ca)	Iron (Fe)	Magnesium (Mg)	Phosphorous (P)	Potassium (K)	Copper (Cu)	Zinc (Zn)
Soybean (*Glycine max*)	277	15.7	280	704	1797	1.66	4.89
Split peas (*Pisum sativum*)	55	4.4	115	366	981	0.866	3.01
Chickpea (*Cicer arietinum*)	105	6.2	115	366	875	0.847	3.43
Lentil (*Lens culinaris*)	56	7.5	122	451	955	0.519	4.78
Groundnut (*Arachis hypogaea*)	92	4.6	168	376	705	1.144	3.27
Fava bean (*Vicia faba*)	103	6.7	192	421	1062	0.824	3.14
Black beans (*Phaseolus vulgaris*)	123	5.0	171	352	1483	0.841	3.65

Source: USDA (2010)

protection against various diseases. In the form of food supplement, alpha-tocopherol can be found under the label E307.

Another key vitamin also found in the legume grains but in smaller amounts is vitamin C (also called L-ascorbic acid or L-ascorbate). Vitamin C is also an important antioxidant, protecting the body against oxidative stress. In the legumes, the germinated seeds are richer in this vitamin than the grains, as the vitamin degrades rapidly after long storage (Aykroyd et al. 1982).

Starchy legumes are valuable sources of dietary fibre as well as thiamine and riboflavin. Starchy legumes are a valuable component of a prudent diet, but their consumption is constrained by low yields, the lack of convenient food applications and flatulence (Phillips 1993). In a study where chicks were fed test diets containing 35 % soybean meal (control) or raw (40 %) or dehulled (35 %) lupine seed meal lupin diet, all broilers fed with lupine-based diets had significantly higher ($P < 0.001$) riboflavin in the plasma (Olkowski et al. 2005).

13.2.5 Minerals

Legume seeds are also an important source of dietary minerals, with the potential to provide all of the essential minerals required by humans (Grusak 2002). The plants contain 14 mineral elements defined as essential for plant growth and reproductive success, which are N, S, P, K, Ca, Mg, Cl, Fe, Zn, Mn, Cu, B, Mo and Ni. However, iron bioavailability is not always favourable amongst the different legume seeds; it has been shown to be quite low for lupin, soybean (Macfarlane et al. 1988, 1990) and urd bean (Chitra et al. 1997), whereas in chickpea it is reported to be the highest (Chitra et al. 1997).

It seems that the amount of available nutrients in the legumes is influenced by processing methods (Chau and Cheung 1997; Xu and Chang 2009b). Cooking, as well as autoclaving the seeds of, e.g. *Phaseolus* brings about a more significant improvement in the digestibility of both protein and starch as compared with soaking and germination (Chau and Cheung 1997). This is an important factor for these particular nutrients but not as much for the minerals, because mineral bioavailability seems to be relatively undisturbed with different processing methods (Trugo et al. 2000). Therefore, using natural variation can be one of the possible options to help reduce the problems associated with lower bioavailability of minerals. Phytic acid is one of the chief antinutrients present mainly in seeds of grain crops such as legumes and cereals and has the potential to bind mineral micronutrients in food and reduce their bioavailability. Germination can reduce or eliminate appreciable amounts of phytic acid in legumes and hence improves mineral bioavailability (Sutardi and Buckle 1985). Perlas and Gibson (2002) demonstrated that soaking can be used to reduce the inositol penta- and hexaphosphate (phytic acid) content of complementary foods based on mung bean (legume) flour and thus enhance the bioavailability of iron and zinc. Low phytic acid lentils have been identified in Saskatchewan that have low phytic acid concentrations, making them a promising material to be used in regions with widespread micronutrient malnutrition (Thavarajah et al. 2009).

Legume grains, given their high micronutrient status, should be considered priority picks in food dietary choices, both for human food and animal feed.

13.2.6 Nutraceuticals

Some research works have reported bioactive components in plant foods with nutraceutical properties related to the maintenance of health or prevention of chronic disease. Selection of the right legume varieties combined with a suitable germination process could provide good sources of bioactive compounds from legumes and their germinated products for nutraceutical applications (Lin and Lai 2006; Xu and Chang 2009b). Amongst these substances are proteins, carbohydrates, antioxidant molecules, polyunsaturated fatty acids and inositol derivates (Ribeiro et al. 2011).

Legume seeds contain two major classes of soluble α-galactosyl: the raffinose family oligosaccharides and the galactosyl cyclitols (Ribeiro et al. 2011). Legumes are also a unique source of natural products such as flavonoids, isoflavonoids, alkaloids and saponins, many of which have documented antimicrobial, pharmaceutical and/or nutraceutical properties (Leia et al. 2011). Soybeans and soy foods contain significant amounts of health-promoting components of these isoflavones, flavon-3-ols and phenolic compounds (Xu and Chang 2009a) with important antioxidant activities. The dark-coat seeds, such as azuki beans, black soybeans (Lin and Lai 2006) and the hulls of *Phaseolus vulgaris* (Oomah et al. 2010), contain high amounts of phenolic compounds with high antioxidative ability. Flavonoids are widely present in plants, usually in the form of water-soluble glycosides (Wollenweber and Dietz 1981). Isoflavones are polyphenolic compounds found mainly in legumes, the benefits of which have been widely studied and attributed in particular to their phytoestrogenic activity (Romani et al. 2011).

Besides the grain and grain derivatives, seed sprouts have also been used as a good source of nutraceutical compounds (Ribeiro et al. 2011). The high nutritional value of seeds derives from the deposition of compounds during development. Bean sprout, for example, is an important source of proteins, sucrose, glucose and myo-inositol (Ribeiro et al. 2011), antioxidants and phytoestrogens (Diaz-Batalla et al. 2006). Additionally, bean sprouts have low levels of raffinose, an antinutritional compound for flatulence symptoms when ingested in high quantities (Ribeiro et al. 2011).

Legume grains have also been used traditionally for their antifungal activity in traditional medicine. Extracts obtained from seeds of *Psoralea corylifolia* (Fabaceae), a commonly occurring medicinal plant in South India where it is known as *babchi*, showed several degrees of antifungal activity, having important anti-dermatophytic properties (Rajendra Prasad et al. 2004). The seeds of this legume plant are also used as stomachic, deobstruent and diaphoretic in febrile conditions and also in leucoderma and other skin diseases (Rajendra Prasad et al. 2004).

Other examples of the medicinal uses of species from the *Fabaceae* come from the Mayan populations, which used them routinely for the treatment of diarrhoea and eye infections. These properties seem to be associated to the presence of flavonoids and quinones that showed strong activity against both bacteria and fungi (Rosado-Vallado et al. 2000). The tropical forage legume *Clitoria ternatea* (L.) also has shown to have insecticidal and antimicrobial activities (Kelemu et al. 2004).

Finally, early studies of the effects of dietary carbohydrates on plasma lipid levels identified leguminous seeds as potential hypocholesterolaemic agents.

13.3   Legume Grains for Animal Feed

Forage legumes are also widely used in many grassland farming areas of the world; their importance arose principally because of their ability to fix atmospheric nitrogen biologically and secondly because of their nutritional value. They can reduce nitrate leaching and reduce global warming potential of livestock farming systems. In animal feeding, they can be used as green forage, forage dry matter, forage meal, silage, haylage, immature grain, mature grain and straw, while some species may be used for grazing too (Mikić et al. 2009). Forage legumes can sustain high animal performances and improve the nutritional quality of ruminant products. This, by itself, can indirectly improve human health, as ruminant products are part of the diet of large segments of the population. These benefits have been shown in research conducted in a wide range of situations in Europe (Rochon et al. 2003). Amongst the most commonly used legume grains for animal feed are soybean (Solanas et al. 2005), pea (Yanez-Ruiz et al. 2009), chickpea (Yanez-Ruiz et al. 2009), alfalfa (Hoffman et al. 1999; McAllister et al. 2005; Tongel and Ayan 2010), crown vetch (McAllister et al. 2005), white clover (*Trifolium repens*, also known as Dutch White, New Zealand White or Ladino) (Lacefield and Ball 2000; Rochon et al. 2003; Smith and Valenzuela 2002) and several *Lotus* species (McAllister et al. 2005). White clover (*Trifolium repens*) is the main legume to be found in pastures and meadows of temperate regions and is adapted to survive in a range of grassland environments (Rochon et al. 2003).

The plants of the genus *Medicago* L. (*Fabaceae*) belong to 50 species of annual or perennial herbs, rarely shrubs, of which *Medicago sativa* L. (alfalfa) is one of the best known, as it is used as a forage crop widely (Alessandra et al. 2010). Lucerne, or alfalfa, is sometimes called "queen of forage crops" in Turkey, as it is the most promising forage crop for Turkish farmers, producing high-quality feed during summer on lowland fields (Tongel and Ayan 2010).

Feed legumes have a great significance in animal feeding as one of the most quality and least expensive solutions for a long-term demand for plant protein in animal husbandry. Forages help meet the protein requirements of ruminants by providing degraded substrate for microbial protein synthesis plus protein that escapes ruminal degradation (Broderick 1995). The potential of different legume seed species, including recently new developed varieties as protein supplements to low-quality forages, has been evaluated with positive outcomes (Yanez-Ruiz et al. 2009). Due to a different amino acid composition between species, each of the feed legumes may easily find its useful place in animal feeding, supplementing and replacing each other (Mikić et al. 2009). Evidence from numerous feeding studies with lactating dairy cows indicates that excessive ruminal protein degradation may be the most limiting nutritional factor in higher-quality temperature legume forages (Broderick 1995). Condensed tannins found in legumes are known to decrease protein degradation, either by altering the forage proteins or by inhibiting microbial proteases (Broderick 1995). Compared to soybean meal, beans and peas showed similar suitability as protein supplements for sustaining in vitro fermentation of low-quality forages (Yanez-Ruiz et al. 2009). In order to increase protein digestibility, one common strategy is to mix grain legumes with cereals (Chauhan et al. 1988; Geervani et al. 1996; Mupangwa et al. 2000; Solanas et al. 2007) and grasses (Hoffman et al. 1999; Roffler and Thacker 1983).

Legumes are also important forages because of their ability to convert atmospheric nitrogen through a symbiotic interaction with *Rhizobia*. During this interaction, atmospheric nitrogen is reduced to ammonia which is incorporated in amino acid biosynthesis and ultimately results in high protein content. It was estimated that $7–10 billion worth of nitrogen is fixed by legumes annually and a proportion returned to the soil for subsequent crops. Thus, even modest use of alfalfa (*Medicago sativa*) in rotation with corn could save farmers $200–300 million in nitrogen fertilizer costs annually (Leia et al. 2011).

13.4 Conclusions

Legume grains have been grown for thousands of years and have a long tradition of cultivation in many parts of the world, including the Mediterranean region, where they are part of the staple diet. The grain legumes are cultivated primarily for their grains and are used either for human consumption or for animal feed. However, plant sources rich in proteins, used for feeding, are still highly deficient in Europe. Covering this deficit requires massive imports of soya bean meal that is mainly produced in the United States. These imports demand transport over long distances, resulting in adverse environmental impacts. Legumes have the potential to positively contribute to sustainable agriculture, as they can lower the needs for mineral nitrogen, which is an expensive, unavailable resource in many countries. It allows producing human food and animal feed with very limited ecologically harmful inputs, reducing greenhouse gas emissions. Grain legumes are more adapted than grasses to global changing environments, some being less susceptible to drought and others more productive under high temperatures. In spite of the importance of legumes in agriculture, increases in yield over the past few decades have lagged behind those of cereals, mainly because of numerous stresses such as salinity, acidity, nutrient limitations and various diseases and pests. However, on the whole, introducing grain legumes into European crop rotations offers interesting options for reducing environmental burdens, especially in a context of depleted fossil energy resources and climate change. Moreover, the heavily substantiated and proven health benefits associated to the legume grains make them ideal choices when faced with the increased available dietary items in the market.

Acknowledgments This work received financial support from Fundação para a Ciência e a Tecnologia (FCT, Portugal), through projects PEst-OE/EQB/ LA0016/2013, PTDC/AGRPRO/3972/2014 and PTDC/AGRPRO/3515/2014

References

Alessandra B, Ciccarelli D, Garbari F, Pistelli L (2010) Flavonoids isolated from Medicago littoralis Rhode (Fabaceae): their ecological and chemosystematic significance. Caryologia 63:106–114

Atkins CA, Pate JS, Sharkey PJ (1975) Asparagine metabolism-key to the nitrogen nutrition of developing legume seeds. Plant Physiol 56:807–812

Aykroyd WR, Doughty J, Walker A (1982) Legumes in human nutrition. In: FAO food and nutrition paper. Food and Agriculture Organization of the United Nations, Rome, p 152

Bertagnolli BL, Wedding RT (1977) Cystine content of legume seed proteins: estimation by determination of cysteine with 2-vinylquinoline, and relation to protein content and activity of cysteine synthase. J Nutr 107:2122–2127

Bingham SA, Day NE, Luben R, Ferrari P, Slimani N, Norat T, Clavel-Chapelon F, Kesse E, Nieters A, Boeing H, Tjonneland A, Overvad K, Martinez C, Dorronsoro M, Gonzalez CA, Key TJ, Trichopoulou A, Naska A, Vineis P, Tumino R, Krogh V, Bueno-de-Mesquita HB, Peeters PH, Berglund G, Hallmans G, Lund E, Skeie G, Kaaks R, Riboli E (2003) Dietary fibre in food and protection against colorectal cancer in the European Prospective Investigation into Cancer and Nutrition (EPIC): an observational study. Lancet 361:1496–1501

Bosetti C, Pelucchi C, La Vecchia C (2009) Diet and cancer in Mediterranean countries: carbohydrates and fats. Public Health Nutr 12:1595–1600

Broderick GA (1995) Desirable characteristics of forage legumes for improving protein utilization in ruminants. J Anim Sci 73:2760–2773

Buckland G, Agudo A, Lujan L, Jakszyn P, Bueno-de-Mesquita HB, Palli D, Boeing H, Carneiro F, Krogh V, Sacerdote C, Tumino R, Panico S, Nesi G, Manjer J, Regner S, Johansson I, Stenling R, Sanchez MJ, Dorronsoro M, Barricarte A, Navarro C, Quiros JR, Allen NE, Key TJ, Bingham S, Kaaks R, Overvad K, Jensen M, Olsen A, Tjonneland A, Peeters PH, Numans ME, Ocke MC, Clavel-Chapelon F, Morois S, Boutron-Ruault MC, Trichopoulou A, Lagiou P, Trichopoulos D, Lund E, Couto E, Boffeta P, Jenab M, Riboli E, Romaguera D, Mouw T, Gonzalez CA (2009) Adherence to a Mediterranean diet and risk of gastric adenocarcinoma within the European Prospective Investigation into Cancer and Nutrition (EPIC) cohort study. Am J Clin Nutr 91:381–390

Casey R, Domoney C, Stanley J (1984) Convicilin mRNA from pea (Pisum sativum L.) has sequence homology with other legume 7S storage protein mRNA species. Biochem J 224:661–666

Chau C-F, Cheung PC-K (1997) Effect of various processing methods on antinutrients and in vitro digestibility of protein and starch of two Chinese indigenous legume seeds. J Agric Food Chem 45:4773–4776

Chauhan GS, Verma NS, Bains GS (1988) Effect of extrusion processing on the nutritional quality of protein in rice-legume blends. Nahrung 32:43–47

Chitra U, Singh U, Rao PV (1997) Effect of varieties and processing methods on the total and ionizable iron contents of grain legumes. J Agric Food Chem 45:3859–3862

Clemente TE, Cahoon EB (2009) Soybean oil: genetic approaches for modification of functionality and total content. Plant Physiol 151:1030–1040

Demonty I, Ras RT, van der Knaap HC, Duchateau GS, Meijer L, Zock PL, Geleijnse JM, Trautwein EA (2009) Continuous dose-response relationship of the LDL-cholesterol-lowering effect of phytosterol intake. J Nutr 139:271–284

Diaz-Batalla L, Widholm JM, Fahey GC, Castano-Tostado E, Paredes-Lopez O (2006) Chemical components with health implications in wild and cultivated Mexican common bean seeds (Phaseolus vulgaris L.). J Agric Food Chem 54:2045–2052

el-Adawy TA, Rahma EH, el-Bedawy AA, Sobihah TY (2000) Effect of soaking process on nutritional quality and protein solubility of some legume seeds. Nahrung 44:339–343

Escurriol V, Cofan M, Moreno-Iribas C, Larranaga N, Martinez C, Navarro C, Rodriguez L, Gonzalez CA, Corella D, Ros E (2009) Phytosterol plasma concentrations and coronary heart disease in the prospective Spanish EPIC cohort. J Lipid Res 51:618–624

Food and Agriculture Organization of the United Nations (FAOSTAT) (2013) FAOSTAT database. Latest update 07 Mar 2014. http://data.fao.org/ref/262b79ca-279c-4517-93de-ee3b7c7cb553. html?version=1.0. Accessed 8 July 2015

Gabay O, Sanchez C, Salvat C, Chevy F, Breton M, Nourissat G, Wolf C, Jacques C, Berenbaum F (2009) Stigmasterol: a phytosterol with potential anti-osteoarthritic properties. Osteoarthritis Cartilage 18:106–116

Geervani P, Vimala V, Pradeep KU, Devi MR (1996) Effect of processing on protein digestibility, biological value and net protein utilization of millet and legume based infant mixes and biscuits. Plant Foods Hum Nutr 49:221–227

Genovese MI, Lajolo FM (2001) Atividade inibitória de tripsina do feijão (Phaseolus vulgaris L.): avaliação crítica dos métodos de determinação. Archivos Latinoamericanos de Nutrição 51:386–394

Graham PH, Vance CP (2003) Legumes: importance and constraints to greater use. Plant Physiol 131:872–877

Grant G, More LJ, McKenzie NH, Stewart JC, Pusztai A (1983) A survey of the nutritional and haemagglutination properties of legume seeds generally available in the UK. Br J Nutr 50:207–214

Grizard D, Barthomeuf C (1999) Non-digestible oligosaccharides used as prebiotic agents: mode of production and beneficial effects on animal and human health. Reprod Nutr Dev 39:563–568

Grusak MA (2002) Enhancing mineral content in plant food products. J Am Coll Nutr 21:178S–183S

Guo J, Wang Y, Song C, Zhou J, Qiu L, Huang H, Wang Y (2010) A single origin and moderate bottleneck during domestication of soybean (Glycine max): implications from microsatellites and nucleotide sequences. Ann Bot 106:505–514

Harker M, Holmberg N, Clayton JC, Gibbard CL, Wallace AD, Rawlins S, Hellyer SA, Lanot A, Safford R (2003) Enhancement of seed phytosterol levels by expression of an N-terminal truncated Hevea brasiliensis (rubber tree) 3-hydroxy-3-methylglutaryl-CoA reductase. Plant Biotechnol J 1:113–121

Hernandez-Mijares A, Banuls C, Rocha M, Morillas C, Martinez-Triguero ML, Victor VM, Lacomba R, Alegria A, Barbera R, Farre R, Lagarda MJ (2010) Effects of phytosterol ester-enriched low-fat milk on serum lipoprotein profile in mildly hypercholesterolaemic patients are not related to dietary cholesterol or saturated fat intake. Br J Nutr 104:1018–1025

Hey SJ, Powers SJ, Beale MH, Hawkins ND, Ward JL, Halford NG (2006) Enhanced seed phytosterol accumulation through expression of a modified HMG-CoA reductase. Plant Biotechnol J 4:219–229

Hoffman PC, Brehm NM, Hasler JJ, Bauman LM, Peters JB, Combs DK, Shaver RD, Undersander DJ (1999) Development of a novel system to estimate protein degradability in legume and grass silages. J Dairy Sci 82:771–779

Imberty A, Perez S (1994) Molecular modelling of protein-carbohydrate interactions. Understanding the specificities of two legume lectins towards oligosaccharides. Glycobiology 4:351–366

Itsiopoulos C, Hodge A, Kaimakamis M (2009) Can the Mediterranean diet prevent prostate cancer? Mol Nutr Food Res 53:227–239

Jackson JC, Duodu KG, Holse M, de Faria MD, Jordaan D, Chingwaru W, Hansen A, Cencic A, Kandawa-Schultz M, Mpotokwane SM, Chimwamurombe P, de Kock HL, Minnaar A (2010) The morama bean (Tylosema esculentum): a potential crop for southern Africa. Adv Food Nutr Res 61:187–246

Jiang Q, Christen S, Shigenaga MK, Ames BN (2001) gamma-Tocopherol, the major form of vitamin E in the US diet, deserves more attention. Am J Clin Nutr 74:714–722

Kadam SS, Salunkhe DK (1984) Winged bean in human nutrition. Crit Rev Food Sci Nutr 21:1–40

Kaplan L, Lynch TF (1999) Phaseolus (Fabaceae) in archaeology: AMS radiocarbon dates and their significance for pre-Colombian agriculture. Econ Bot 53:261–272

Kelemu S, Cardona C, Segura G (2004) Antimicrobial and insecticidal protein isolated from seeds of Clitoria ternatea, a tropical forage legume. Plant Physiol Biochem 42:867–873

Kingman SM (1991) The influence of legume seeds on human plasma lipid concentrations. Nutr Res Rev 4:97–123

Lacefield G, Ball D (2000) White clover. Circular. University of Kentucky and Auburn University, Kentucky, pp 1–2

Langran X, Dezhao C, Xiangyun Z, Puhua H, Zhi W, Ren S, Dianxiang Z, Bojian B, Delin W, Hang S, Xinfen G, Yingxin L, Yingxin L, Zhaoyang C, Jianqiang L, Mingli Z, Podlech D, Ohashi H, Larsen K, Welsh SL, Vincent MA, Gilbert MG, Pedley L, Schrire BD, Yakovlev GP, Thulin M, Nielsen IV, Choi B, Turland NJ, Polhill RM, Larsen SS, Hou D, Iokawa Y, Wilmot-Dear M, Kenicer G, Nemoto T, Lock JM, Salinas AD, Kramina TE, Brach AR, Bartholomew B, Sokoloff DD (2010) Fabaceae. In: Wu ZY, Raven PH, Hong DY (eds) Flora of China. Science Press/Missouri Botanical Garden Press, Beijing, pp 1–577

Leia Z, Daia X, Watsona BS, Zhao PX, Sumner LW (2011) A legume specific protein database (LegProt) improves the number of identified peptides, confidence scores and overall protein identification success rates for legume proteomics. Phytochemistry 72(10):1020–1027

Lin P-Y, Lai H-M (2006) Bioactive compounds in legumes and their germinated products. J Agric Food Chem 54:3807–3814

Lin X, Ma L, Racette SB, Anderson Spearie CL, Ostlund RE Jr (2009) Phytosterol glycosides reduce cholesterol absorption in humans. Am J Physiol Gastrointest Liver Physiol 296:G931–G935

Llaneza P, Gonzalez C, Fernandez-Inarrea J, Alonso A, Diaz-Fernandez MJ, Arnott I, Ferrer-Barriendos J (2010) Soy isoflavones, Mediterranean diet, and physical exercise in postmenopausal women with insulin resistance. Menopause 17:372–378

Loayza Jibaja C, Bressani R (1988) Evaluation of the protein quality of legume flours obtained by roasting in heated beds. Arch Latinoam Nutr 38:152–161

Macfarlane BJ, Baynes RD, Bothwell TH, Schmidt U, Mayet F, Friedman BM (1988) Effect of lupines, a protein-rich legume, on iron absorption. Eur J Clin Nutr 42:683–687

Macfarlane BJ, van der Riet WB, Bothwell TH, Baynes RD, Siegenberg D, Schmidt U, Tal A, Taylor JR, Mayet F (1990) Effect of traditional oriental soy products on iron absorption. Am J Clin Nutr 51:873–880

Martinez-Gonzalez MA, Sanchez-Villegas A, De Irala J, Marti A, Martinez JA (2002) Mediterranean diet and stroke: objectives and design of the SUN project. Seguimiento Universidad de Navarra. Nutr Neurosci 5:65–73

Materljan E, Materljan M, Materljan B, Vlacic H, Baricev-Novakovic Z, Sepcic J (2009) Multiple sclerosis and cancers in Croatia—a possible protective role of the "Mediterranean diet". Coll Antropol 33:539–545

McAllister TA, Martinez T, Bae HD, Muir AD, Yanke LJ, Jones GA (2005) Characterization of condensed tannins purified from legume forages: chromophore production, protein precipitation, and inhibitory effects on cellulose digestion. J Chem Ecol 31:2049–2068

Mikić A, Perić V, Đorđević V, Srebrić M, Mihailović V (2009) Anti-nutritional factors in some grain legumes. Biotechnol Anim Husb 25:1181–1188

Mupangwa JF, Ngongoni NT, Topps JH, Hamudikuwanda H (2000) Effects of supplementing a basal diet of Chloris gayana hay with one of three protein-rich legume hays of Cassia rotundi-folia, Lablab purpureus and Macroptilium atropurpureum forage on some nutritional parameters in goats. Trop Anim Health Prod 32:245–256

National Academy of Science (1994) Biological nitrogen fixation: research challenges—a review of research grants funded by the U.S. Agency for International Development. National Academy Press, Washington, p 51

Olkowski BI, Classen HL, Wojnarowicz C, Olkowski AA (2005) Feeding high levels of lupine seeds to broiler chickens: plasma micronutrient status in the context of digesta viscosity and morphometric and ultrastructural changes in the gastrointestinal tract. Poult Sci 84:1707–1715

Oomah B, Corb D, Balasubramanian P (2010) Antioxidant and anti-inflammatory activities of bean (Phaseolus vulgaris L.) hulls. J Agric Food Chem 58:8225–8230

Pedrosa-Harand A, de Almeida CC, Mosiolek M, Blair MW, Schweizer D, Guerra M (2006) Extensive ribosomal DNA amplification during Andean common bean (Phaseolus vulgaris L.). Theor Appl Genet 6:1–10

Perlas LA, Gibson RS (2002) Use of soaking to enhance the bioavailability of iron and zinc from rice-based complementary foods used in the Philippines. J Sci Food Agri 82:1115–1121

Phillips RD (1993) Starchy legumes in human nutrition, health and culture. Plant Foods Hum Nutr 44:195–211

Puel C, Coxam V, Davicco MJ (2007) Mediterranean diet and osteoporosis prevention. Med Sci (Paris) 23:756–760

Rajendra Prasad N, Anandi C, Balasubramanian S, Pugalendi KV (2004) Antidermatophytic activity of extracts from Psoralea corylifolia (Fabaceae) correlated with the presence of a flavonoid compound. J Ethnopharmacol 91:21–24

Reyes-Moreno C, Paredes-López O (1993) Hard-to-cook phenomenon in common beans—a review. Crit Rev Food Sci Nutr 33:227–286

Ribeiro ED, Centeno DD, Figueiredo-Ribeiro RD, Fernandes KV, Xavier-Filho J, Oliveira AE (2011) Free cyclitol, soluble carbohydrate and protein contents in Vigna unguiculata and Phaseolus vulgaris bean sprouts. J Agric Food Chem 59(8):4273–4278

Rochon JJ, Doyle CJ, Greef JM, Hopkins A, Molle G, Sitzia M, Scholefield D, Smith CJ (2003) Grazing legumes in Europe: a review of their status, management, benefits, research needs and future prospects. Grass Forage Sci 63:197–214

Roffler RE, Thacker DL (1983) Early lactational response to supplemental protein by dairy cows fed grass-legume forage. J Dairy Sci 66:2100–2108

Romani A, Vignolini P, Tanini A, Pampaloni B, Heimler D (2011) HPLC/DAD/MS and antioxidant activity of isoflavone-based food supplements. Nat Prod Commun 5:1775–1780

Roosevelt AC, Costa ML, Machado CL, Michab M, Mercier N, Valladas H, Feathers J, Barnett W, Silveira MI, Henderson A, Sliva J, Chernoff B, Reese DS, Holman JA, Toth N, Schick K (1996) Paleoindian cave dwellers in the Amazon: the peopling of the Americas. Science 272:373–384

Rosado-Vallado M, Brito-Loeza W, Mena-Rejon GJ, Quintero-Marmol E, Flores-Guido JS (2000) Antimicrobial activity of Fabaceae species used in Yucatan traditional medicine. Fitoterapia 71:570–573

Shin EC, Pegg RB, Phillips RD, Eitenmiller RR (2010) Commercial peanut (Arachis hypogaea L.) cultivars in the United States: phytosterol composition. J Agric Food Chem 58(16):9137–9146

Smith J, Valenzuela H (2002) Cover crops: white clover. In: Sustainable agriculture cover crops. College of Tropical Agriculture and Human Resources (CTAHR), Manoa

Solanas E, Castrillo C, Balcells J, Guada JA (2005) In situ ruminal degradability and intestinal digestion of raw and extruded legume seeds and soya bean meal protein. J Anim Physiol Anim Nutr (Berl) 89:166–171

Solanas E, Castrillo C, Calsamiglia S (2007) Effect of extruding the cereal and/or the legume protein supplement of a compound feed on in vitro ruminal nutrient digestion and nitrogen metabolism. J Anim Physiol Anim Nutr (Berl) 91:269–277

Stamatiou K, Delakas D, Sofras F (2007) Mediterranean diet, monounsaturated: saturated fat ratio and low prostate cancer risk. A myth or a reality? Minerva Urol Nefrol 59:59–66

Sutardi, Buckle KA (1985) Reduction in phytic acid levels in soybeans during tempeh production, storage and frying. J Food Sci 50: 260–263

Teupser D, Baber R, Ceglarek U, Scholz M, Illig T, Gieger C, Holdt LM, Leichtle A, Greiser KH, Huster D, Linsel-Nitschke P, Schafer A, Braund PS, Tiret L, Stark K, Raaz-Schrauder D, Fiedler GM, Wilfert W, Beutner F, Gielen S, Grosshennig A, Konig IR, Lichtner P, Heid IM, Kluttig A, El Mokhtari NE, Rubin D, Ekici AB, Reis A, Garlichs CD, Hall AS, Matthes G, Wittekind C, Hengstenberg C, Cambien F, Schreiber S, Werdan K, Meitinger T, Loeffler M, Samani NJ, Erdmann J, Wichmann HE, Schunkert H, Thiery J (2010) Genetic regulation of serum phytosterol levels and risk of coronary artery disease. Circ Cardiovasc Genet 3:331–339

Thavarajah P, Thavarajah D, Vandenberg A (2009) Low phytic acid lentils (Lens culinaris L.): a potential solution for increased micronutrient bioavailability. J Agric Food Chem 57:9044–9049

Tongel OM, Ayan I (2010) Nutritional contents and yield performances of Lucerne (Medicago sativa L.) cultivars in southern Black Sea shores. J Anim Vet Adv 9:2067–2073

Trichopoulou A (2001) Mediterranean diet: the past and the present. Nutr Metab Cardiovasc Dis 11:1–4

Trichopoulou A, Benetou V, Lagiou P, Gnardellis C, Stacewicz-Sapunzakis M, Papas A (2003a) Plasma carotenoid levels in relation to the Mediterranean diet in Greece. Int J Vitam Nutr Res 73:221–225

Trichopoulou A, Costacou T, Bamia C, Trichopoulos D (2003b) Adherence to a Mediterranean diet and survival in a Greek population. N Engl J Med 348:2599–2608

Trichopoulou A, Bamia C, Trichopoulos D (2009) Anatomy of health effects of Mediterranean diet: Greek EPIC prospective cohort study. BMJ 338:b2337

Trugo LC, Donangelo CM, Trugo NMF, Bach Knudsen KE (2000) Effect of heat treatment on nutritional quality of germinated legume seeds. J Agric Food Chem 48:2082–2086

U.S. Department of Agriculture, Agricultural Research Service (2010) USDA National Nutrient Database for Standard Reference, Release 23. Nutrient Data Laboratory Home Page. http://www.ars.usda.gov/ba/bhnrc/ndl

von Hofsten B (1979) Legume sprouts as a source of protein and other nutrients. J Am Oil Chem Soc 56:382

Wang TL, Domoney C, Hedley CL, Casey R, Grusak MA (2003) Can we improve the nutritional quality of legume seeds? Plant Physiol 131:886–891

Wollenweber E, Dietz VH (1981) Occurrence and distribution of free flavonoid aglycones in plants. Phytochemistry 20:869–932

Xu B, Chang SK (2009a) Isoflavones, flavan-3-ols, phenolic acids, total phenolic profiles, and antioxidant capacities of soy milk as affected by ultrahigh-temperature and traditional processing methods. J Agric Food Chem 57:4706–4717

Xu B, Chang SKC (2009b) Total phenolic, phenolic acid, anthocyanin, flavan-3-ol, and flavonol profiles and antioxidant properties of pinto and black beans (Phaseolus vulgaris L.) as affected by thermal processing. J Agric Food Chem 57:4754–4764

Yanez-Ruiz DR, Martin-Garcia AI, Weisbjerg MR, Hvelplund T, Molina-Alcaide E (2009) A comparison of different legume seeds as protein supplement to optimise the use of low quality forages by ruminants. Arch Anim Nutr 63:39–55

Young VR, Scrimshaw NS, Torun B, Viteri F (1979) Soybean protein in human nutrition: an overview. J Am Oil Chem Soc 56:110–120

Chapter 14
Traditional Food Products from *Prosopis* sp. Flour

Leonardo Pablo Sciammaro, Daniel Pablo Ribotta, and Maria Cecilia Puppo

14.1 Cultivation and Potential Uses of *Prosopis* Species

The *Prosopis* genus belongs to the leguminous family and the Mimosoideae subfamily. It includes 44 species (Burkart 1976a) with great importance in the tree and shrub flora of arid and semiarid regions. *Prosopis* distribution includes southwest Asia (three native species), tropical Africa (one native species), and America (40 native species). In the latter *Prosopis* is extended through southwest of USA toArgentinean Patagonia and Chile, being Argentina the major center of diversification, with the greatest proportion of endemic species.

Several species from the Americas with useful attributes may have been introduced to neighboring regions (Pasiecznik et al. 2001). *Prosopis* trees have been introduced widely by man during the last 100–150 years. The first documented introduction of *Prosopis* in Kenya was in 1973, since then it has spread widely (Choge et al. 2007). *P. juliflora* is the most common naturalized species in Kenya, but *P. pallida* also occurs. This expansion was due to the perceived value of the trees' products, the multiple products obtained, and the high yields under even the poorest conditions. Many *Prosopis* species are valuable multipurpose resources in their native extension, providing timber, firewood, livestock feed, human food, shade, shelter, and soil improvement. Tolerance of drought and poor soils, and their value in agroforestry systems, including tolerance of repeated cutting and the amelioration of soils would have been the cause. The value of *Prosopis* as producing quality timber, excellent fuel, and sweet pods was well known by those who had

L.P. Sciammaro • M.C. Puppo (✉)
Centro de Investigación y Desarrollo en Criotecnología de Alimentos (CONICET, UNLP),
Calle 47 y 116, CC 553, La Plata 1900, Argentina
e-mail: mcpuppo@quimica.unlp.edu.ar

D.P. Ribotta
CONICET-Universidad Nacional de Córdoba, CC 509, Córdoba 5000, Argentina

© Springer Science+Business Media New York 2016
K. Kristbergsson, J. Oliveira (eds.), *Traditional Foods*, Integrating Food
Science and Engineering Knowledge Into the Food Chain 10,
DOI 10.1007/978-1-4899-7648-2_14

seen these trees in their native areas, and there was a desire for selected species to fulfill similar roles in other countries (Choge et al. 2007). The pods, which are high in sugars, carbohydrates, and protein, have been a historic source of food for human populations in North and South America providing flour and other edible products. However, this indigenous knowledge has not followed the *Prosopis* trees, and the fruits are unused or provide only fodder for livestock in most of Africa and Asia (Choge et al. 2007). *Prosopis juliflora* and *P. pallida*, prevalent in tropical zones, and *P. glandulosa* and *P. velutina*, found in subtropical regions, were the major species introduced. *P. alba* and *P. chilensis* have proved to be well adapted and are locally common in some South American regions.

In Argentina, *Prosopis flexuosa* and *P. chilensis* are native from an arid-temperate zone, the phytogeographical woodland province ("Provincia Fitogeográfica del Monte"). This zone includes a corridor parallel to the Andes from Salta to La Pampa provinces (Villagra et al. 2004). The white "*algarrobo*" (*Prosopis alba*) is a tree that inhabits the center of Argentina (Córdoba), the ecology region of the "Gran Chaco" (Chaco, Formosa, and Santiago del Estero), and part of Mesopotamia (Corrientes and Misiones).

Prosopis species are commonly trees and shrubs of varied appearance with bipinnate foils and in some cases without foils, with flowers in raceme and fleshy, iridescent, and sweet fruits, with high amount of seeds (Dimitri 1977). Due to its abundance and ecologic behavior, *Prosopis* species are important elements in arid regions of several countries. They offer shade, wood for constructions, firewood, and food for wild and domesticated animals and also for humans.

The indigenous people conferred different names to the distinct "*algarrobo*" species according to their quality, i.e., the Guaraní called the white *Prosopis* (*algarrobo blanco*) as "igopé-para" that means "tree put in the pathway for eating." Quechuas named it "*yaná-tacú*" which means "the tree" (Roig 1993). These two cultures, among others, prove the great importance of the "*algarrobo*" for Argentinean and South American original communities.

The great majority of *Prosopis* species present elevated resistance to drought (López Lauenstein et al. 2005), extreme temperatures (Verga 2005), and high salinity content (Felker et al. 1981), counting on a high capacity of fixing nitrogen and being easily adaptable to a great weather diversity, mainly subhumid and dry climates (Villagra et al. 2004).

In Argentina, it has been estimated that the natural coverage of *Prosopis* forests was reduced to between one quarter and one half of its original area between 1500 and 1975, due to the activities of man (D'Antoni and Solbrig 1977; Villagra et al. 2004).

Reforest dessert lands with *Prosopis* species would allow marginalize rural population of the northwest of Argentina (NOA) to guarantee their food sovereignty with autochthonous products (Abraham de Vázquez and Prieto 1981). In this way, they would cease depending on seeds and inputs that do not ensure a good production or are not adapted to arid zones (Gobierno de Formosa 2005; Verga et al. 2005). Another advantage of reforestation with "*algarrobos*" is environmental benefits. The "*algarrobo*" increases biodiversity, favors major infiltration and soil water retention, and decreases atmosphere contamination and erosion processes (Villagra et al. 2004).

Research on the food properties of pods of different native Argentinean *Prosopis* species would enable revalorization of the fruit. Native communities associate "*algarrobo*" pods and their products to their origins and identity. In addition, using this crop as food ingredient would contribute to reforestation of the vast desert zones of Argentina.

14.2   Pods and Flours

Prosopis fruit is an indehiscent drupaceous legume classified by as "lomento drupáceo" (lomentum drupaceum) (Burkart 1952, 1976a, b). Latin American inhabitants called it "*algarrobo*." This fruit varies in size, color, and chemical composition, depending on the species (Fagg and Stewart 1994) (Fig. 14.1).

Fig. 14.1 Pods of different *Prosopis* species

Prosopis alba fruits are rectus and semicircular with yellowish color of 12–25 cm of length (Fig. 14.1) (Burkart 1976a, b). Their seeds are brown reddish and oblong with an asymmetric endosperm and vitreous and yellowish cotyledons (Galera 2000). These fruits contain 25–28 % of glucose, 11–17 % of starch, iron and calcium, low lipid content, and acceptable digestibility. Seeds are in the fruit in a 10 % (w/w) proportion and contain 32 % protein and 2–7 % lipids with high content of linoleic (42–48 %) and oleic (25–27 %) acids.

Prosopis pallida and *Prosopis juliflora* are varieties with especially large and sweet fruits (Fig. 14.1). Pods are yellow with a size up to 30 cm long and average weight of approximately 12 g. The sugar content of its pulp is 46 % (Grados and Cruz 1996). *Prosopis chilensis* is a straw-colored fruit with an "S" form of 12–18 cm long with little brown ovoid seeds (Burkart 1976a, b). *P. chilensis* var. Catamarcana has pulpy yellowish fruits, while *P. chilensis* var. Riojana pods are next to black. *P. chilensis* pod composition is about 25–28 % glucose, 11–17 % starch, and 7–11 % proteins. *Prosopis nigra* produces cylindrical and purple pods of 7–18 cm length. These fruits are pulpy and sweet, and between seeds there is a narrowing of the pod giving an aspect of a string of beads (Fig. 14.1). This species shows the least damage by bruchids due to its hard endocarp. *P. nigra* seeds contain 30–32 % proteins and 12–14 % lipids.

It is customary for local people to harvest *algarrobo* pods collectively, forming convoys of men, women, and children that "awaken" the summer season. They collect the best pods and leave them 1 or 2 days under the sun and the pods become dry. This practice is a party called "*La Algarrobeada*" with the contribution of aborigines and also "criollos" (Creole people). In times of harvest, the *algarrobo* pods are directly consumed raw or are used in the elaboration of different products like "*añapa*" or "aloja." Pods are roasted and ground and then incorporated to milk. For preserving, native people place pods in "*trojas*" (saddlebag or slammer) or convert them into different products: flour, "*patay*," "*arrope*," and "*bolanchao*" (INCUPO 1991).

Prosopis flour is prepared in a very traditional way. After harvest, pods are dried under the sun, placed on the roof of the houses or in "*trojas*." The dried material is kept in the "*troja*" and must be protected from Bruchidae, an insect that attacks the seeds. Local people place some plants, like "*atamisqui*," "*paico*," and "*ancoche*," that are used as repellents, in the saddlebag with the pods. Milling of dried pods is traditionally performed with a homemade mortar but currently a hammer mill is used. The flour then is sieved and conserved in hermetic recipients (Juarez et al. 2003). Tobas people in the community of Chaco province detect when pods are in optimum conditions for making flour: if the pods produce a noise when they are shaken, it is time to prepare the flour. In general, white "*algarrobo*" is preferred by native families to the black one to make flour because it has a softer taste. On the other hand, aborigines prefer black "*algarrobo*" because it is sweeter but has a stronger taste.

These flours are highly energetic because of its elevated content of sugars. *Algarroba* flour (*P. alba*) analyzed by Prokopiuk et al. (2000) presented 2.5 % water, 3.1 % ashes, 59.1 % sugars (2.8 reducing sugars), 2.4 % crude fiber, 7.2 % proteins, and 2.2 % lipids. These percentages vary according to species, year, and region of harvest. This flour also presented a high amount of minerals, mainly calcium (1274 ppm) and iron (450 ppm) (Prokopiuk et al. 2000). Flour from *P. pallida* contains higher quantities of carbohydrates, iron, and phosphorous although lesser amount of fiber and proteins than *P. alba* (Prokopiuk et al. 2000).

Comparing flours from different *Prosopis* species, *P. nigra* showed the highest levels of crude protein (11.33 g/100 g DM) and ashes (4.12 g/100 g DM) (González Galán et al. 2008). *P. juliflora* presented the lowest levels of lipids (0.79 g/100 g DM), crude protein (8.84 g/100 g DM), and dietary fiber (40.15 g/100 g DM) and the highest levels of nonreducing sugars (52.51 g/100 g DM) and in vitro protein digestibility (66.45 %). Trypsin inhibitors concentration (0.29–9.32 UTI/mg DM) was inferior to that of raw soy, presenting *P. juliflora* the highest values. Regarding saponin, hemagglutinin, and polyphenol values, the levels found are considerably low. The levels of phytate vary from 1.31 to 1.53 g/100 g between species (González Galán et al. 2008).

The high content of sugar, fiber, and minerals and acceptable amount of proteins make *algarroba* flour a suitable ingredient for several food products.

14.3   Products and Regional Activities

Prosopis pods of highest quality are employed for human food but only in the native region of the crop (GESER 2006). *Prosopis* fruits constitute an ancient human food source, becoming an excellent source of carbohydrates and proteins. The products prepared by NOA people are the following:

Patay. It is a sweet (17.2 % glucose, 26.9 % sucrose) product made with fine and dried flour. Flour is mixed with water up to forming dough. This dough is introduced in circular molds and put in the oven. They are sold in regional markets and constitute an important regional food in Santiago del Estero and other provinces of Argentina (Fig. 14.2).

Añapa. It is a refreshing beverage made with the mature algarroba fruits. People prefer *Prosopis alba* to *P. nigra*. Pods are grinded in a mortar and mixed with an excess of cold water. Dispersion is agitated and the filtered. The extract is then reposed in special vessels (Fig. 14.2).

Aloja. It is an ancient autochthonous alcoholic fermented beverage (Fig. 14.2) of sweet taste. It is drunk very cold on the streets. *Prosopis alba* is grinded in a mortar and put in a vessel with water to favor the fermentation process. The mix is stored in a fresh, dry, and dark place. After 2 days, remnants of the pods are manually discarded and a fresh amount of ground *algarroba* must be incorporated, so as to favor fermentation. The optimum point of the beverage is detected by tasting. *Aloja* must be consumed in a few days because it has short time preservation. It is utilized by regional people as a diuretic beverage.

| Añapa | Aloja | Arrope |

| Patay | Snack | Alfajor |

Fig. 14.2 Traditional food products of *Prosopis* flour

Arrope. The "*arrope*," "*algarrobina*," or "*miel de algarrobo*" is a dark and dense liquid that is obtained by cooking grinded algarroba pods with water during 8 h. In this process, sugars are concentrated. During the middle of cooking, seeds are removed. A dark syrupy product (Fig. 14.2) that looks like honey is obtained.

Bolanchao. The "*bolanchao*" is a traditional dessert of the north of Argentina, with a spherical form. It is prepared by grinding fruits (*Ziziphus mistol* Griseb) in a mortar until forming a granulose paste. This recipe does not need water because these fruits provide all the water needed. Ball-shaped products are obtained from the dough. Balls are powdered with roasted *algarroba* flour and cooked in an oven.

At homemade level, the variety of recipes available including *algarroba* flour has the amplitude of the creativity and customs of local people. The Institute of Popular Culture of Argentina (Instituto de Cultura Popular, INCUPO) assayed different products: candies with honey and algarroba flour, hot beverages with roasted flour like chocolate that replaces coffee, and bread-making products such as bread and cakes. Communities of Santiago del Estero, Argentina (Juarez et al. 2003), experimented with *algarroba* flour in several products: cakes, snacks (Fig. 14.2), bonbons (with honey), honey, liquor, "*aloja*," biscuits, pizza, spaghetti, coffee, and "*bolanchao*," among others.

A commercial product with *algarroba* named "*alfajor*" (Fig. 14.2), a typical Argentinean sweet commonly made with wheat flour, is being commercialized all around the country.

References

Abraham de Vázquez EM, Prieto MR (1981) Enfoque diacrónico de los cambios ecológicos y de las adaptaciones humanas en el NE. árido mendocino. In: Ruiz JC (ed) Cuaderno N° 8. CEIFAR-CONICET-UNC, Mendoza, pp 108–139

Burkart A (1952) Las leguminosas Argentinas silvestres y cultivadas, 2nd edn. Acme Agency, Buenos Aires, pp 138–139

Burkart A (1976a) A monograph of the genus Prosopis (Leguminosae subfamily Mimosoideae). J Arnold Arbor 57(3–4):219–249

Burkart A (1976b) A monograph of the genus Prosopis (Leguminosae, subfamily Mimosoideae), continuation. J Arnold Arbor 57(3–4):450–523

Choge SK, Pasiecznik NM, Harvey M, Wright J, Awan SZ, Harris PJC (2007) Prosopis pods as human food, with special reference to Kenya. Water SA 33:419–424

D'Antoni HL, Solbrig OT (1977) Algarrobos in South American cultures: past and present. In: Simpson B (ed) Mesquite: its biology in two desert ecosystems. Dowden Hutchinson and Ross, Stroudsburg, pp 189–199

Dimitri MJ (1977) Enciclopedia Argentina de Agricultura y Jardinería. Tomo I. ACME S.A.C.I, Buenos Aires, pp 483–484

Fagg C, Stewart J (1994) The value of Acacia and Prosopis in arid and semi-arid environments. J Arid Environ 27:3–25

Felker P, Clark PR, Laag AE, Pratt PF (1981) Salinity tolerance of the tree legumes: mesquite (*Prosopis glandulosa* var. torreyana, *P. Velutina* and *P. Articulata*) algarrobo (*P. Chilensis*), kiawe (*P. Pallida*) and tamarugo (*P. Tamarugo*) grown in sand culture on nitrogen-free media. Plant and Soil 61:311–317

Galera FM (2000) Las especies del género prosopis (algarrobos) de américa latina con especial énfasis en aquellas de interés económico. http://www.fao.org/DOCREP/006/AD314S/AD314S00.HTM. Accessed July 2005

GESER (2006) Misiones: La Algarroba, cultivo tradicional, Latitud 2000 Revista digital de Turismo y Ambiente. http://www.latitud2000.com.ar. Accessed 2015

Gobierno de Formosa (2005) Fomento de la forestación con algarrobo. Plan Provincia Formosa, Argentina. http://www.formosa.gov.ar/obras.html. Accessed 2015

González Galán A, Duarte Correa A, Patto de Abreu CM, Piccolo Barcelos MF (2008) Caracterización química de la harina del fruto de *Prosopis* spp. procedente de Bolivia y Brasil. Arch Latinoam Nutr 58(3):309–315

Grados N, Cruz G (1996) New approaches to industrialization of algarrobo (Prosopis pallida) pods in Peru. In: Felker P, Moss J (eds) Prosopis: semiarid fuelwood and forage tree. Building consensus for the disenfranchised. Center for Semi-Arid Forest Resources, Kingsville, pp 3.25–3.42

INCUPO (1991) El Monte nos da comida. En: Salud y Alimentación. Tomo I. Instituto de Cultura Popular, Argentina, pp 3–7. http://www.incupo.org.ar/. Accessed 2015

Juarez E, Campos N, Poggi J, Hoyos G, Maldonado S, Feuillade D (2003) Aprovechamiento familiar de la algarroba. Informe Equipo Pro Huerta. Dpto Capital y Silípica. INTA 2003. http://www.inta.gov.ar/santiago/actividad/extension.htm. Accessed 2015

López Lauenstein D, Melchiorre M, Verga A (2005) Respuestas de los algarrobos al estrés hídrico. IDIA XXI 8:210–214

Pasiecznik NM, Felker P, Harris PJC, Harsh LN, Cruz G, Tewari JC, Cadoret K, Maldonado LJ (2001) The Prosopis juliflora—Prosopis pallida complex: a monograph. HDRA, Coventry, p 172

Prokopiuk D, Cruz G, Grados N, Garro O, Chiralt A (2000) Comparative study among the fruits of Prosopis alba and Prosopis pallid. Multequina 9:35–45

Roig F (1993) Aportes a la Etnobotánica del Género Prosopis. In: Instituto Argentino de Investigaciones de las Zonas Áridas. Conservación y Mejoramiento de Especies del Género

Prosopis. Quinta Reunión Regional para América Latina y el Caribe de la Red de Forestación del Centro Internacional de Investigaciones para el Desarrollo, Mendoza, Argentina, pp 99–119

Verga A (2005) Recursos Genéticos, Mejoramiento y Conservación de Especies del Género Prosopis. In: Norberto CA (ed) Mejores árboles para más forestadores: El Programa de Producción de Material de Propagación Mejorado y el Mejoramiento Genético en el Proyecto Forestal de Desarrollo. SAGPyA-BIRF, pp 205–221

Verga A, Mottura M, López Lauenstein D, Melchiorre M, Joseau J, Carranza C, Ledesma M, Recalde D, Tomalino L, Mendoza S, Vega R (2005) El Proyecto Algarrobo del INTA. Para el Chaco árido argentine. IDIA XXI 8:195–200

Villagra PE, Cony MA, Mantován NG, Rossi BE, González Loyarte MM, Villalba R, Marone L (2004) Ecología y manejo de los algarrobales de la Provincia Fitogeográfica del Monte. In: Arturi MF, Frangi JL, Goya JF (eds) Ecología y manejo de bosques nativos de Argentina. Universidad Nacional de La Plata, La Plata, pp 1–32

Chapter 15
Amaranth: An Andean Crop with History, Its Feeding Reassessment in America

Myriam Villarreal and Laura Beatriz Iturriaga

15.1 Amaranth History

The term amaranth comes from the Greek language *Amarantón* which is *a* (without) and *marainein* (wilt, pale) and stands for unwitherable.

Amaranth is referred to depending on the country and the city which it is native to as "kiwicha," "coimi," "grano inca," "sangorache," "ataco," "huautli," "bledo," or "chaclion."

Amaranth grains have been grown in America for more than 7000 years. It was first consumed and utilized by the Mayas and spread among the Aztecs and the Incas afterwards. It was also used ancestrally by the Quechua peoples in Bolivia. Within the last 50 years, the growing of amaranth expanded noticeably in China, Korea, India, and the United States.

By the time of colonization, an intensive exchange of crops with the Old World occurred where some crops became more important than others that eventually disappeared. The reasons for this profound change are said to be based on aspects connected to adaptation, production, and harvesting concerns of the new crops brought from Europe and to the relationship between the autochthonous crops and social, religious, and cultural conditions of that time. Amaranth consumption was deeply rooted among Aztecs which in addition to using its grains and leaves for food they used the flour for "tzoalli," figures of deities with which the aborigines took communion in religious meetings. Spaniards associated this with the Catholic Eucharist and prohibited its use like they did to other native costumes and prosecuted all who continued the practice. These events reduced amaranth production dramatically. Only small farmers continued growing amaranth at small scale for

M. Villarreal • L.B. Iturriaga (✉)
Facultad de Agronomía, y Agroindustrias, Instituto de Ciencia y Tecnología de Alimentos,
Universidad Nacional de Santiago del Estero, Santiago del Estero, Argentina
e-mail: laura.iturriaga@gmail.com; litur@unse.edu.ar

© Springer Science+Business Media New York 2016   217
K. Kristbergsson, J. Oliveira (eds.), *Traditional Foods*, Integrating Food
Science and Engineering Knowledge Into the Food Chain 10,
DOI 10.1007/978-1-4899-7648-2_15

domestic consumption which, thanks to its deep insertion in the traditions of these native people, kept the practice through the centuries (Hernández Garciadiego and Herrerías Guerra 1998). It was not until after several centuries that the nutritional importance of this crop was taken into consideration again.

15.2   Description

Amaranth is a dicotyledonous, annual, herbaceous, or shrubby plant species that belongs to the Amaranthaceae family which includes more than 60 species and 50 varieties out of which three are domesticated species from the pre-Columbian America and major grain producers: *Amaranthus hypochondriacus*, *Amaranthus cruentus* y *Amaranthus caudatus*. It develops rapidly, demands little care while growing, is water stress and heat and pesticide resistant, adapts to different altitudes and climates—though it is quite sensitive to very low temperatures—and produces good grain yields. It provides grain and tasty leaves with high nutritional quality suitable for human and animal consumption and shows high versatility for its transformation and industrialization. Additionally, it is an ornamental plant due to the appealing color of its head.

15.3   The Seed

15.3.1   Morphological Structure

The amaranth seed is spherical, golden brown in color (CIELab color parameters scale: $L^*=63.6$; $a^*=6.54$; $b^*=22.5$), and between 1.0 and 1.7 mm in diameter approximately. Abalone et al. (2004) studied *Amaranthus cruentus* grains and reported their average dimensions (length, 1.42 mm; width, 1.20 mm; thickness, 0.8 mm). The average pycnometric volume of a grain is 0.65 mm³.

It has a monolayered skin strongly associated with the perisperm excepting the area of the embryo where it is linked to the bigger and denser endosperm cells. The embryo has the cotyledons, the radicle or primary root, and the procambium (Fig. 15.1). The bran (seed coat) of amaranth is thinner and softer than that of wheat (Irving et al. 1981; Zapotoczny et al. 2006).

The particular amaranth grain structure and morphology along with the compartmentalized location of its major components makes it possible to apply the methods for milling and separating its anatomical constituents that result in the yield of fractions with different composition, rich in carbohydrates or proteins, minerals, and/or fibers (Bestchart et al. 1981; Tosi et al. 2000).

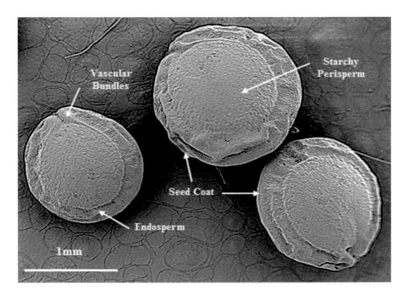

Fig. 15.1 Scanning electron micrograph of *Amaranthus cruentus* seeds (×150)

15.3.2   Main Components

Their broad genetic basis of different amaranth species and varieties impede the provision of a single compositional profile; consequently, value ranges are given in accordance with the data provided by several authors (Becker et al. 1981). The main grain components are starch, proteins, lipids, and fibers.

Starch is the main constituent of the amaranth seed. The content varies with species and ranges from 48 to 72 % of the grain total weight. Starch is located in the perisperm as a tightly packed matrix (Willet 2001).

Amaranth starch granules (Fig. 15.2) are polyhedral, having faceted smooth-surfaced faces, with no fissures or pores that would be the indication of low-amylase activity within the seed (Calzetta Resio et al. 1999; Marcone 2001). Their quite small size with a diameter estimated between 0.4 and 2 μm has been given special attention because of its potential industrial use (Hoover et al. 1998; Marcone 2001; Villarreal et al. 2009). Both granules' shape and surface are critical factors when the starch is to be used as color and flavor carrier.

Amaranth starches can be considered as waxy- or low-amylase content in accordance with the amylose content ranging between 2.7 and 12.5 %. Some authors, however, reported amylose contents of up to 35 % in *Amaranthus retroflexus* (Wuand Corke 1999). Amylopectin, the amaranth starch granule main component, shows a typical A and B chains bimodal distribution; its average molecular weight is 2.7×10^8 g/mol, which is similar to that in several waxy cereals (e.g., corn, sorghum, and wheat) but with a more compact structure (Kong et al. 2009a).

Fig. 15.2 Scanning electron micrograph of *Amaranthus cruentus* starch (×24,200)

Several authors (Radosavljevic et al. 1998; Marcone 2001; Konishi et al. 2006; Gunaratne and Corke 2007; Villarreal et al. 2009) have confirmed variations among the thermal properties of the various amaranth species and varieties. The gelatinization temperatures of the native amaranth starches are similar to those of both waxy rice and sorghum starches and higher than for waxy cornstarch. The gelatinization temperature range for water–starch mixtures (75:25) informed in different articles lies between 69 and 78.5 °C, while their enthalpies range from 14.1 to 18.4 J/g (Gunaratne and Corke 2007).

Likewise cereal starches, amaranths show an "A" type X-ray diffraction pattern with peaks at 3.8 and 5.8 and an overlapped double peak at 4.9 and 5.2 Å and a high degree of crystallinity (about 40 %) caused by the granule high amylopectin content (Villarreal et al. 2009).

The amaranth grains from various species and varieties show a protein content that ranges between 14 and 18 %, higher than that of cereals.

The main protein fractions present in amaranth are globulins, albumins, and glutelins. They contain little or none prolamin storage proteins, which are the main storage proteins and toxic proteins for the celiac disease as well (Barba de la Rosa et al. 2009; Alvarez Jubete et al. 2010). In globulins, the 7S and 11S fractions have been identified (the first one in lowest quantities) together with the P globulins. These proteins that are distinctive to amaranth possess a high polymerizing capacity likely due to the presence of a specific 56 kD polypeptide (Martínez et al. 1997).

Silva-Sánchez et al. (2008), Tironi and Añon (2010), and Fritz et al. (2011) demonstrated the presence of different biopeptides, in several storage proteins of amaranth grains. These active molecules possess free radical scavenging activity and antihypertensive and anticarcinogenic properties. Mendonça et al. (2009) also provided evidence for the hypocholesterolaemic effect of protein isolates from

Table 15.1 Comparison of essential amino acids in *Amaranthus* grains and FAO/WHO/UNU amino acid requirement pattern

Essential amino acid	*Amaranthus* [mg/g crude protein][a]	Requirement FAO/WHO/UNU[b]
Isoleucine	35–38.5	28
Leucine	55–63	66
Lysine	54–69.5	58
Sulfurs amino acids	44–46	25
Total aromatic amino acids	67.2–94	63
Threonine	33–42	34
Tryptophan	10.6	11
Valine	32–37	35

[a]Values reported by Dowton (1973), Bestchart et al. (1981), Becker et al. (1981), and Irving et al. (1981). The values were divided by 5.85 to put on mg/g crude protein
[b]Based on amino acid requirements of preschool-age child (1–2 years)

Amaranthus cruentus in hamsters that when ingested cause an important reduction in non-HDL cholesterol which might be produced by a decrease in its synthesis.

Table 15.1 includes the essential amino acids composition of amaranth grains as reported by several authors and compares them with the FAO/WHO/UNU amino acids requirement pattern. It is seen that lysine content in amaranth grains is higher as to that in most cereals. The limiting amino acids in amaranth grains are in decreasing importance leucine, valine, and threonine. The ßamaranth amino acidic profile appears as more balanced than that of the most commonly eaten (used) cereal and bean proteins.

The in vitro amaranth protein digestibility ranges between 78 and 85 % but is always lower to that of casein (91.2 %) used as reference.

The total dietary fiber is divided into two fractions: the one soluble in water and the other non-soluble. The latter relates particularly to intestinal regulation (fecal bulk increase, lowered fecal material passage time), while the former is involved in the reduction of both in-blood cholesterol and glucose intestinal absorption. The beneficial effects on health from food with good dietary fiber content have widely been studied. Total dietary fiber contents between 9 and 16.5 % have been reported for various species and varieties of amaranth being in some cases higher to those in corn, wheat, oat, rice, and sorghum but always lower to those in barley and rye (Picolli da Silva and Santorio Ciocca 2005). The non-soluble dietary fiber represents 75–88 % of the total dietary fiber informed (Tosi et al. 2001; Repo-Carrasco-Valencia et al. 2009). Such high dietary fiber content in amaranth grains becomes interesting for the formulation of food diets.

Amaranth grains contain between 5 and 10 % of lipids; among them tocopherols and tocotrienols and phytosterols that regulate cholesterol metabolism (Lehmann 1996) are present in high percentages. Squalene is also present in a significant percentage (4.5–8 %) of the total seed oil. The squalene is a hydrocarbon (triterpene) with high demand during the last years due to its apparent hypocholesterolemic effects (Chatuverdi et al. 1993; Plate and Arêas 2002); additionally it shows physicochemical features that are of interest for different industries: as lubricant, photoprotector in cosmetics, and chemical preventive in colon cancer treatment among others. Today the only squalene sources are shark and whale oils that are protected sea species; this explains the importance its potential extraction from amaranth grains acquires.

Plate and Arêas (2002) showed that rabbits fed with extruded amaranth for 21 days reduced their total cholesterol in about 50 %, whereas the control diet and the diet using amaranth oil reduced it in about 14 % and 18 %, respectively. These results suggest that consuming extruded amaranth and amaranth oil may become an alternative in the prevention of the coronary heart disease.

15.3.3 Minor Components

The mineral contents of several species and varieties of amaranth grains are given in Table 15.2. The contents vary according to species, genetic background, and crop conditions. Many authors have emphasized the exceptional mineral content of the seed, especially Mg, Ca, and K (Bestchart et al. 1981; Becker et al. 1981; Alvarez Jubete et al. 2009).

According to studies on mineral distribution in *A. hypochondriacus* and *A. cruentus* grains carried out using energy dispersive X-ray microanalyzer (EDX) (Konishi et al. 1998), K, P, and, Mg are only found in the embryo (cotyledons and radicles), but P may be as phytic acid (inositol 6-P) in the procambium and K and Mg were complexed to phytic acid. S is present in embryonic tissues including the

Table 15.2 Mineral contents of amaranth grains

Mineral	ppm[a]
Calcium	1300–2850
Copper	7.9–13.2
Iron	72.2–174
Manganese	15.9–45.9
Magnesium	2300–3320
Nickel	0.8–2.4
Phosphorus	540
Potassium	2900–5200
Sodium	160–480
Zinc	36.2–40.0

[a]Values on dry basis. Values reported by Dowton (1973), Bestchart et al. (1981), Becker et al. (1981), Irving et al. (1981), and Alvarez Jubete et al. (2009)

procambium though not complexed to the phytic acid like K and Mg. Ca is found as part of the grain skin and between the perisperm and the embryo.

Fe, Cu, and Zn could not be detected by this method since they are present at lower levels than the other elements.

Amaranth seeds are an excellent source of vitamin E (5.7 mg/100 g) and niacin (4.6 mg/100 g) with reasonable content of riboflavin (0.70 mg/100 g seeds), but they are not very rich in vitamin C (Bestchart et al. 1981).

The non-starchy polysaccharides, lignin, and resistant starches can be considered as the second group of minority components. A study on two varieties of *A. caudatus* showed a cellulose content ranging from 2.5 to 3.8 %, 0.6 % of beta-glucans, and between 0.72 and 0.98 % of pentosans—relatively low values as to medium for lignin at 4 % and approximately 0.1 % of resistant starch (Repo-Carrasco-Valencia et al. 2009).

Other constituents of interest in amaranth seeds are the phenolic compounds (i.e., phenolic acid and flavonoids) that are the major contributors to the antioxidant capacity of amaranth due to their free radical scavenging activity. In the presence of excess free radicals, the body must neutralize and defend in order to prevent tissue damage. In this case, biomolecules are damaged by oxidative stress generating several pathologies with high incidence in humans such as atherosclerosis, cancer, cardiovascular disease, and degenerative neurological and even natural aging processes. One of the natural ways to stop this process is the supply of antioxidants that protect DNA, protein, and membrane lipids from oxidative damage in biological systems and may provide health benefits by disease prevention and health promotion. The content of total phenolic compounds, reported as gallic acid equivalent in different species of de amaranth, ranged between 0.1 and 0.4 g gallic acid/100 g (Barba de la Rosa et al. 2009). The authors identified and found traces of three flavonoids (rutin, isoquercitrin, and nicotiflorin), while Repo-Carrasco-Valencia et al. (2010) did not find quantifiable content of these components in *A. caudatus* or *A. cruentus*.

15.4 The Leaves

Aletor et al. (2002) reported that the leaves of *Amaranthus hybridus* have the following proximal composition: 32.3 % crude protein, 9.1 % ether extract, 7.4 crude fiber, 19.5 % ash, 23.0 % nitrogen-free extract, and 8.7 % moisture. Major components are sodium (848 mg/kg), calcium (699 mg/kg), magnesium (694 mg/kg), potassium (689 mg/Kg), and phosphorus (130 mg/kg), while there are traces of iron, zinc, and manganese. Their high mineral ratio compared to other foods such as legumes and tubercles confirm their importance as a rich source of these components. However, their consumption should be controlled when they are part of sodium- and potassium-restricted diets for humans.

Amaranthus leaves have a total antioxidant activity ranging between 44 and 61 % and contribute to such an activity with carotenoids, ascorbic acid, flavonoids, and phenolic acids. The plant tissues' color diversity (red, burgundy, yellow, and green) of the various amaranth species and varieties have strong influence on the total phenol content that can be from 70 to 110 g/kg (Amin et al. 2006).

Amaranth leaf amino acid profiles, if exception is made for the relatively low methionine content, compare positively with soybean grain, beef, fish, and eggs and exceed the FAO/WHO essential amino acid patterns which make them a cheap and accessible source of protein for the people in underdeveloped countries. The solubility in alkaline media of their proteins suggests their potential use in alkaline foods. On the other hand, ingesting such leaves can supply high proportion of the daily mineral requirements, especially of Ca, P, and Mg that are involved in bone and teeth strength.

Toxic factors or antinutrient components present in vegetable leaves affect their nutritional value as food and feed. Even though most of them are counted as traces, they have profound effect of their nutritional quality (Aletor and Adeogum 1995).

Anti-nutritional constituents present in amaranth leaves are as follows: phytic acid, oxalate, tannin, and hydrocyanic acid. The former is considered an antinutrient since it functions as a strong mineral captor of Ca, Fe, Mg, and Zn and reduces niacin absorption that prevents phosphor bioavailability for monogastric animals including humans. Values in the order of 600 mg/100 g of phytic acid, 150 mg/100 g of phytin phosphorus, and 14 % phytin phosphorous as percentage of total P were found in fresh amaranth leaves. Oxalates on their part are considered toxic because in the presence of Ca, ions form calcium oxalate that lowers calcium content in the body and can promote renal illnesses by crystallization. The oxalate content informed was about 550 mg/100 g. Although some specific tannins may be healthy for human beings, they are toxic in general since they react with proteins and can cause cross-linking. Values for tannins are 70 mg/100 g and in the order of 45 mg/100 g for hydrocyanic acid (Aletor and Adeogum 1995; Fasuyi 2006, 2007).

15.5 Functional Properties

Food functionality is defined as "every non-nutritional property influencing the usefulness of an ingredient in a food;" it can also be referred to as "every physico-chemical property that contributes to foods for a specific desirable characteristic." Functional properties affect directly or indirectly food processing and storage and its quality and acceptability and have influence on its sensoric characteristics, especially texture.

15.5.1 Hydration Properties

Amaranth hydration properties are essentially linked to the interaction between proteins and water and have influence over issues related to food formulation, processing, and storage.

The water absorption capacity for amaranth flours was in the order of 1.8 g/g flour, higher than that in soybean, peas, and sunflower flours. Adding NaCl and NaHCO$_3$ enhanced water absorption significantly up to a maximum of 0.04 M NaCl (2.58 g/g flour) and to 0.6 M NaHCO$_3$ (2.076 g/g flour) (Mahajan and Dua 2002).

Proteins solubility in amaranth flours is in the order of 24.6. This solubility increased when NaCl was added, due to the salting-in effect of the salt over the hydrophobic interactions which might decrease their aggregating tendency giving rise to an increased solubility. The least amaranth proteins solubility occurred at pH 4 and increased linearly above and below this value and reached its peak at pH 12. High solubility, i.e., 80–90 %, at extreme acid and alkaline pH are also shown by colza and guar flours (Mahajan and Dua 2002).

Amaranth flours can be used in soups, meat-based products, and beverages due to their preferable solubility, viscosity, and water absorption capacity. However, even though they show good hydration properties at pH 10, at alkaline values the electrostatic repulsion increases causing scission of disulfide bonds to an extent that limits the use of these flours in leavened baked products since the dough elasticity is lowered.

In the case of protein concentrates, their water solubility ranges widely from 20 to 60 %. The lowest solubility (i.e., between 19 and 50 %) occurs at pH 5 and increases at alkaline pH up to 80 % in some cases. Concentrate solubility may be affected by the thermal treatment (drying) applied during processing unless the potential protein denaturation is kept under careful control (Bejosano and Corke 1999).

15.5.2 Foaming Properties

The maximum foaming capacity of amaranth flour proteins is 13.6 % at pH 2 and decreases at higher pH values, whereas maximum foam stability measured as the percentage of the initial value after 60 min occurs at pH 6. Similar behaviors were reported for linseed and poppy seed flours. Maximum foaming capacity at pH 2 would be associated with high solubility of proteins at this pH level, something that is not verified at pH 12 probably because proteins are not adsorbed easily in the interphase at this pH. The foaming capacity decreased when NaCl and NaHCO$_3$ were added though its stability increased. These aspects might restrict the use of amaranth flours in cakes and shake creams in the presence of such salts (Mahajan and Dua 2002).

The foaming capacity of the protein concentrate obtained as by-product of the *A. cruentus* y *A. hybridus* starch extraction process, taken as the foam volume after the mixture stage, varies between 35 and 90 mL, while its foaming stability, measured as volume after 30 min at rest at pH 7, shows values between 18 and 56 % (Bejosano and Corke 1999). Similar to solubility, stability was higher than for soybean protein concentrates.

15.5.3   Emulsifying Properties

Emulsification is a functional property widely utilized in the food industry; many foods are totally or partially emulsified like mayonnaise, cream, sauces, desserts, comminuted meat products, and some beverages. The most important emulsifiers in such foods are proteins, lipids, and phospholipids (Lizarraga et al. 2008).

The emulsifying activity of amaranth flours in oil shows its peak (60 %) at pH 10 while its minimum (47 %) at pH 4. This activity does not seem to be affected by the presence of salts (e.g., NaCl and NaHCO$_3$). Likewise foam stability, the emulsifying stability measured as the percentage of the remaining emulsifying activity after heating at 80 °C for 30 min reached its peak (62 %) at pH 6 and its minimum (49 %) at pH 12 (Mahajan and Dua 2002).

Advantages from the emulsifying properties of amaranth flours were used in the design and construction of biodegradable edible films with good mechanical properties, low solubility, and excellent barrier properties, due to their low water vapor and oxygen permeability, compared to other commercial films. Such features of the amaranth films result from both the interaction among its components where lipids are tightly associated to proteins and homogeneously distributed in the starchy phase and the natural composition of components (Tapia Blácido et al. 2005, 2007; Colla et al. 2006).

Konishi and Yoshimoto (1989) were among the first researchers studying the emulsifying properties of amaranth globulins and characterized them as having excellent characteristics as an emulsifying agent. Tomoskozi et al. (2008) studied the functional properties of amaranth protein fractions and of protein isolates and found that their emulsifying activity was relatively poor compared to legumes. Ventureira et al. (2010) reported that *Amaranthus hypochondriacus* protein isolates contain protein species capable of forming and stabilizing emulsions, mainly at acid pH (2.0) and to a lesser extent at pH 8. These emulsions are sensitive to creaming and flocculation; they do not destabilize by coalescence. Their emulsifying stability can be enhanced by enzymatic hydrolysis.

15.5.4   Gelling Properties

15.5.4.1   In Proteins of Amaranth

Gelation is one of the most important functional properties of proteins. Protein gels are made up of a protein matrix where the aqueous phase is trapped. Both protein gelation capacity and gelling properties are related to rheological properties such as viscoelasticity and texture that in turn are strongly associated to the gel matrix microstructure.

Regarding amaranth proteins, Avanza et al. (2005) report that the required concentration and critical temperature assuring appropriate gelation are 7 % w/v and 70 °C, respectively. Gels obtained at high protein concentration and heating temperatures (above critical conditions) showed viscoelastic behavior with elastic predominance similar to that of strong-like gels. These gels show high hardness, fracturability, cohesion, and low adhesiveness especially at 20 % of solid content. These results indicate that amaranth proteins would be able to form self-supporting gels with different rheological properties that broaden their potential use in foods.

15.5.4.2 In Amaranth Starches

The amylose/amylopectin ratioin starches, polymers showing structures that differentiate clearly (linear for amylose and quite branched for amylopectin), and the amylopectin chains distribution and length have strong influence on the functional properties of the native starch and its derivatives like viscosity, shear strength, gelatinization, texture, solubility, stickiness, gel stability, swelling by cooling, and retrogradation.

Amaranth starches pasting properties (pasting temperature, viscosity peak) are lower than for regular cornstarch and quite similar to waxy cornstarch (Radosavljevic et al. 1998; Kong et al. 2009b). This makes amaranth starch an optimal ingredient for the making of instant soups.

The nature of gels can be determined empirically by analyzing their mechanical spectra (G' and G″ vs. frequency. Figure 15.3 shows how the elastic (G') and viscous (G″) modules evolve with frequency for *A. cruentus* and *A. caudatus* at different concentrations. The predominance of the elastic component is observed in all the dispersions. *A. cruentus*, with lower amylose content, proved that both modules depend on frequency while in the caudatus species, G' was independent on frequency at 20 % of solids. This particular behavior might be caused by the fact that when the dispersions of *Amaranthus cruentus* at 7 and 20 % of solids and of *Amaranthus caudatus* at 7 % solids are heated, they form a macromolecular suspension with hydrogen bonds between the chains of amylopectin. On its part,

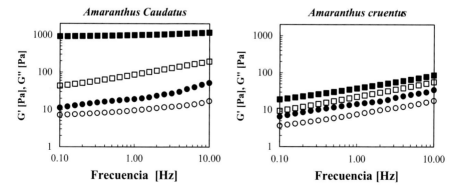

Fig. 15.3 Frequency sweeps of amaranth starch pastes and gels prepared at different concentrations: (*filled circle, open circle*) 7 % and (*filled square, open square*) 20 %. G' *filled symbols*, G″ *open symbols*

Amaranthus caudatus at 20 % solids form a strong gel which is more elastic probably due to enhanced molecular entanglement (Villarreal et al. 2010).

During storage at room temperature or in cold, starch pastes and gels tend to retrograde. This brings about, on the one hand, that the molecules of the gelatinized starch undergo a progressive aggregation process by setting hydrogen bonds that increase their structural order and thus their crystallinity and, on the other hand, that the water adsorbed in the amylose molecules may be expelled by syneresis. This transformations lead to significant changes in starch-based food textural and rheological properties. The amaranth starches that, in general, are of low amylose content retrograde slower than normal corn, barley, and oat starches. Konishi et al. (2006) report that the *Amaranth caudatus* starches required 14-day storage at 5 °C in order for the retrogradation process to start. This suggests that the amaranth starch gels might be used as stabilizing agent in food systems and as a thickening agent in pastry, sauces, ice creams, instant soup, canned foods, and frozen products.

Gels of *Amaranthus hybridus* and *Amaranth hypochondriacus w*axy and low amylose content starches show higher stability for up to four freezing–thawing cycles before a noticeable syneresis occurs in comparison with gels of regular corn, wheat, and rice starches (Yanez and Walker 1986; Baker and Rayas Duarte 1998). Such a difference in stability during the freezing–thawing process shown by starches may be due to difference in their amylose content even though multiple factors may intervene during such a behavior. Increased stability during freezing–thawing for a higher number of cycles was also observed by adding salts or sugars. This is explained by competition for available water exserted by the additatives.

15.6 Potential Industrial Applications

Amaranth grains and leaves are an excellent nutritional alternative due to their protein quality, the unique properties of its proteins, the high mineral and dietary fiber content, and the presence of lipids that are favorable to health and high oxidative stability.

Table 15.3 shows some of the potential industrial uses of the different amaranth leaves and grain components.

15.7 Current Challenges

However, increasing amaranth cultivation in tropical regions including Africa, India, Bangladesh, Sri Lanka, the United States, China, Caribbean, Southeastern Asia, and several countries in Latin America (Peru, Ecuador, Mexico, Bolivia, and Argentina) presents a great challenge to most of the countries in America today. On one hand, the crop areas should be broadened in order to increase production volume and assure industrialization. This involves (a) availability of seed suppliers and

Table 15.3 Potential industrial applications of amaranth

Components	Industrial applications
Grains	Puffed amaranth, crispy bars, flakes for breakfast (Berger et al. 2003), snacks
	Extraction of flavonoids and phenolic acids to produce nutraceutical compounds
Meal	Whole meal and refined for use in bakery (Marcilio et al. 2003)
	Gluten-free flours for the manufacture products for celiac (Tosi et al. 1996)
	Extruded products
Grain protein	Baby food rich in protein
	Protein isolates with different properties
	Functional protein concentrates
	Self-supporting gels with different rheological properties
Lipids	Extraction of squalene
	Source of health, promoting bioactive compounds
Starches	Stabilizer and thickener in formulations of custards, sauces, jellies, ice cream, soups, desserts, and canned and frozen food
	Bakery products
	Laundry starch, paper coating (Choi et al. 2004)
	Low-fat fried formulation (Ahamed et al. 1997)
	Biodegradable films and fillers
	Products for the cosmetic industry
	Carrier of colors, flavors, and seasonings
	Synthesis of nontoxic, biodegradable, and superabsorbent materials (Ahamed et al. 1996; Teli and Waghmare 2010)
Grain fiber	Inclusion in rich-in-fiber diets
Leaf protein	Production leaf protein concentrates for their subsequent use as additives to stabilize emulsions in soups and cakes and to provide viscosity in soups and sauces
Leaf pigments	Extraction of the coloring sheets for industrial uses
Leaf	Extraction of ascorbic acid, carotenoids, flavonoids, and phenolic acids to produce nutraceutical compounds

germ banks in sufficient quantities to supply farm producers; (b) improvement of agricultural management practices (fertilization, soil management, etc.) that increases crop yields; and (c) availability of specific amaranth sowing, harvesting, and storage machinery that is suitable for the extremely small size of its grain (UIA, SECyT, ANPCyT, and PROFECyT Reports 2008). On the other hand, research on amaranth functional properties should be deepened in order to include all its different components in feeding formulations existing today and the development of new formulations that modify nutritional and functional properties, for the design of new products.

All the above suggests that the search for industrial applications of the various amaranth grain and leave components does not pose a problem by itself, but the most important drawback has to do with finding a profitable production method, in appropriate volumes and with high yields for eventual commercialization.

References

Abalone R, Cassinera A, Gaston A, Lara MA (2004) Some physical properties of amaranth seeds. Biosyst Eng 89(1):109–117

Ahamed NT, Singhal RS, Kulkarni PR, Pal M (1996) Studies on Chenopodium quinoa and Amaranthus paniculatas starch as biodegradable fillers in LDPE films. Carbohydr Polym 31:157–160

Ahamed NT, Singhal RS, Kulkarni PR, Pal M (1997) Deep fat-fried snacks from blends of soya flour and corn, amaranth and chenopodium starches. Food Chem 58(4):313–317

Aletor VA, Adeogum OA (1995) Nutrient and anti-nutrient component of some tropical leafy vegetables. Food Chem 53:375–379

Aletor VA, Oshodi AA, Ipinmoroti K (2002) Chemical composition of common leafy vegetables and functional properties of their leaf protein concentrates. Food Chem 78:63–68

Alvarez Jubete L, Arendt EK, Gallagher E (2009) Nutritive value and chemical composition of pseudocereals as gluten-free ingredients. Int J Food Sci Nutr 60(1):240–257

Alvarez Jubete L, Arendt EK, Gallagher E (2010) Nutritive value of pseudocereals and their increasing use as functional gluten-free ingredients. Trends Food Sci Technol 21:106–113

Amin I, Norazaidah Y, Emmy Hainida KI (2006) Antioxidant activity and phenolic content of raw and blanched *Amaranthus* species. Food Chem 94:47–52

Avanza MV, Puppo MC, Añon MC (2005) Rheological characterization of amaranth protein gels. Food Hydrocoll 19:889–898

Baker LA, Rayas Duarte P (1998) Retrogradation of amaranth starch at different storage temperature and the effects of salts and sugars. Cereal Chem 75:308–314

Barba de la Rosa AP, Fomsgaard IS, Laursen B, Mortensen AG, Olevera Martínez L, Silva Sánchez C, Mendoza Herrera A, González Cañeda J, De León Rodríguez A (2009) Amaranth (*Amaranthus hypochondriacus*) as an alternative crop for sustainable food production: phenolic acids and flavonoids with potential impact on its nutraceutical quality. J Cereal Sci 49:117–121

Becker R, Wheeler EL, Lorenz K, Stafford AE, Grosjean OK, Betschart AA, Saunders RM (1981) A compositional study of amaranth grain. J Food Sci 46:1175–1180

Bejosano FP, Corke H (1999) Properties of protein concentrates and hydrolysates from Amaranthus and buckwheat. Ind Crop Prod 10:175–183

Berger A, Monnard I, Dionisi F, Gumy D, Lambelet P, Hayes KC (2003) Preparation of amaranth flakes, crude oils, and refined oils for evaluation of cholesterol-lowering properties in hamster. Food Chem 81:119–124

Bestchart AA, Irving DW, Shepherd AD, Saunders RM (1981) *Amaranthus cruentus*: milling characteristics, distribution of nutrient within seed components, and the effect temperature on nutritional quality. J Food Sci 46:1181–1186

Calzetta Resio A, Aguerre RJ, Suárez C (1999) Analysis of the sorptional characteristics of amaranth starch. J Food Eng 42:51–57

Chatuverdi A, Sarojini G, Devi NL (1993) Hypocholesterolemic effects of amaranth seed (*Amaranthus esculantus*). Plant Foods Hum Nutr 44:63–70

Choi H, Kim W, Shin M (2004) Properties of Korean amaranth starch compared to waxy miller and waxy sorghum starches. Starch 56:460–477

Colla E, Sobral PJA, Menegalli FC (2006) *Amaranthus cruentus* flour edible films: influence stearic acid addition, plasticizer concentration, and emulsion stirring speed on water vapor permeability an mechanical properties. J Agric Food Chem 54:6645–6653

Dowton WJS (1973) Amaranthus edulis: a high lysine grain amaranth. World Crop 25(1):20–26

Fasuyi AO (2006) Nutritional potentials of some tropical vegetable leaf meals: chemical characterization and functional properties. J Biotechnol 5(1):49–53

Fasuyi AO (2007) Bio-nutritional evaluation of three tropical leaf vegetable (*Telfairia occidentalis*, *Amaranthus cruentus* and *Talinum triangulare*) as sole dietary protein sources in rat assay. Food Chem 103:757–765

Fritz M, Vecchi B, Rinaldi M, Añon MC (2011) Amaranth seed protein hydrolysates have in vivo and in vitro antihypertensive activity. Food Chem 126:878–884

Gunaratne A, Corke H (2007) Gelatinizing, pasting and gelling properties of potato and amaranth starch mixtures. Cereal Chem 84(1):22–29

Hernández Garciadiego R, Herrerías Guerra G (1998) Amaranto: Historia y Promesa. Horizonte del Tiempo, Tehuacán, pp 64–83. www.alternativas.org.mx

Hoover R, Sinnot AW, Perera C (1998) Physicochemical characterization of starches from Amaranthus cruentus grains. Starch 56:460–477

Irving DW, Betschart AA, Saunders RM (1981) Morphological studies on Amaranthus cruentus. J Food Sci 46:1170–1175

Kong X, Corke H, Bertoft E (2009a) Fine structure characterization of amylopectins from grain amaranth starch. Carbohydr Res 344:1701–1708

Kong X, Bao J, Corke H (2009b) Physical properties of Amaranthus starch. Food Chem 113:371–376

Konishi Y, Yoshimoto N (1989) Amaranth globulin as a heat stable emulsifying agent. Agric Biol Chem 53:3327–3328

Konishi Y, Takezoe R, Murase J (1998) Energy dispersive X-ray microanalysis of element distribution in amaranth seed. Biosci Biotechnol Biochem 62(11):2288–2290

Konishi Y, Arnao Salas I, Calixto Cotos MR (2006) Caracterización del almidón de Amaranthus caudatus por barrido calorimétrico diferencial. Rev Sociedad Química del Perú 72(1):12–18

Lehmann JW (1996) Case history of grain amaranth as an alternative crop. Cereal Food World 41:399–404

Lizarraga MS, Pan LG, Añon MC (2008) Stability of concentrated emulsions measured by optical and rheological methods. Effect of processing conditions—I. Whey protein concentrate. Food Hydrocoll 22(5):868–878

Mahajan A, Dua S (2002) Salts and pH induced changes in functional properties of amaranth (Amaranthus tricolors L.) seed meal. Cereal Chem 79(6):834–837

Marcilio R, Amaya Farfan F, Ciacco CF, Spehar CR (2003) Fraccionamento do grao de Amaranthus cruentus brasileiro por moagem e suas características composicionais. Cienc Tecnol Aliment 23(3):511–516

Marcone MF (2001) Starch properties of Amaranthus pumilus (seabeach amaranth): a threatened plant species with potential benefits for the breeding/amelioration of present Amaranthus cultivars. Food Chem 71:61–66

Martínez EN, Castellani OF, Añon MC (1997) Common molecular features among amaranth storage protein. J Agric Food Chem 45:3832–3839

Mendonça S, Saldiva PH, Cruz RJ, Arêas JAG (2009) Amaranth protein present cholesterol-lowering effect. Food Chem 116:738–742

Picolli da Silva L, Santorio Ciocca M (2005) Total, insoluble and soluble dietary fiber measured by enzymatic-gravimetric method in cereal grains. J Food Compost Anal 18:113–120

Plate AYA, Arêas JAG (2002) Cholesterol-lowering effect of extruded amaranth (Amaranthus caudatus L.) in hypercholesterolemic rabbits. Food Chem 76:1–6

Radosavljevic M, Jane J, Johnson LA (1998) Isolation of amaranth starch by diluted alkaline-protease treatment. Cereal Chem 75(2):212–216

Repo-Carrasco-Valencia R, Peña J, Kallio H, Salminen S (2009) Dietary fiber and other functional components in two varieties of crude and extruded kiwicha (Amaranthus caudatus). J Cereal Sci 49:219–224

Repo-Carrasco-Valencia R, Hellstrom JH, Pihlava JM, Mattila PH (2010) Flavonoids and other phenolic compounds in Andean indigenous grains: quinoa (Chenopodium quinoa), kaniwa (Chenopodium pallidicaule) and kiwicha (Amaranthus caudatus). Food Chem 120:128–133

Silva-Sánchez C, Barba de la Rosa AP, León-Galván MF, de Lumen BO, de León Rodríguez A, González de Mejia E (2008) Bioactive peptides in amaranth (Amaranthus hypochondriacus) seed. J Agric Food Chem 56:1233–1240

Tapia Blácido D, Sobral PJA, Menegalli FC (2005) Development and characterization of biofilms based on Amaranthus flour (*Amaranthus caudatus*). J Food Eng 67:215–223

Tapia Blácido D, Mauri AN, Menegalli FC, Sobral PJA, Añon MC (2007) Contribution of the starch, protein, and lipids fractions of the physical, thermal, and structural properties of amaranth (*Amaranthus caudatus*) flour films. J Food Sci 72(5):293–300

Teli MD, Waghmare NG (2010) Synthesis of superabsorbents from Amaranthus starch. Carbohydr Polym 81:695–699

Tironi VA, Añon MC (2010) Amaranth proteins as a source of antioxidant peptides: effect of proteolysis. Food Res Int 43:315–322

Tomoskozi S, Gyenge L, Pelceoder A, Varga J, Abonyl T, Lasztity R (2008) Functional properties of protein preparations from amaranth seeds in model system. Eur Food Res Technol 226(6):1343–1348

Tosi EA, Ciappini MC, Masciarelli R (1996) Utilización de la harina de amaranto (Amaranthus cruentus) en la fabricación de galletas para celíacos. Alimentaria 33:49–55

Tosi EA, Ré E, Lucero H, Masciarelli R (2000) Acondicionamiento del grano de amaranto (Amaranto spp.) para la obtención de una harina hiperproteica mediante molienda diferencial. Food Sci Technol Int 5(6):60–63

Tosi EA, Ré E, Lucero H, Masciarelli R (2001) Dietary fiber obtained from amaranth (*Amaranth cruentus*) grain by differential milling. Food Chem 73:441–443

UIA, SECyT, ANPCyT y PROFECyT (2008) Quinua y Amaranto. Desafíos y debilidades tecnológicas del sector productivo. http://www.cofecyt.mincyt.gov.ar

Ventureira J, Martínez EN, Añon MC (2010) Stability of oil: water emulsion of amaranth proteins. Effect of hydrolysis and pH. Food Hydrocoll 24:551–559

Villarreal ME, Lescano NE, Iturriaga LB (2009) Comportamiento físico y térmico de harinas y almidones de amaranto (*Amaranthus cruentus*), III Congreso Internacional de Ciencia y Tecnología de los Alimentos. Tomo 4, pp 390–398

Villarreal ME, Lescano NE, Iturriaga LB (2010) Propiedades viscoelásticas de geles de almidón de *Amaranthus cruentus y caudatus*. Investigaciones en Facultades de Ingeniería del NOA. pp 539–544

Willet JL (2001) Packing characteristic of starch granules. Cereal Chem 78(1):64–68

Wu H, Corke H (1999) Genetic diversity in physical properties of starch from a world collection of *Amaranthus*. Cereal Chem 76(6):877–883

Yanez GA, Walker CE (1986) Effect of tempering parameters on extract and ash of prosomillet flours and partial characterization. Cereal Chem 63:164–167

Zapotoczny P, Markowski M, Majewska K, Ratajski A, Konopko H (2006) Effect of temperature on the physical, functional, and mechanical characteristic of hot-air-puffed amaranth seeds. J Food Eng 76:469–476

Part IV
Traditional Meat and Fish Products

Chapter 16
Yunnan Fermented Meat: Xuanwei Ham, Huotui

Huang Aixiang, Sarote Sirisansaneeyakul, and Yusuf Chisti

16.1 Introduction

16.1.1 History

Xuanwei dry-cured ham is reputed as one of the three famed hams produced in China. (The other well-known Chinese hams are *Jinhua* ham from Zhejiang Province and *Rugao* ham from Jiangsu Province.) Xuanwei ham is named after its producing region, Xuanwei City of Yunnan Province, in southwestern China. Xuanwei ham is also recognized as Yunnan ham. Xuanwei ham is characterized by its rose-red color, delightful flavor, and pleasant taste. The ham is typically shaped like the *pipa*, a Chinese musical instrument. The clear red color of the meat is due to high levels of oxymyoglobin, nitrosylmyoglobin, muscle hemoglobin, and met-myoglobin. The fat is bright white. The pleasant flavor is a consequence of the partial decomposition of meat and fat to aldehydes, alcohols, esters, amino acids, and other organic compounds. The meat is tender and has a fragrant aroma.

H. Aixiang (✉)
Faculty of Food Science and Technology, Yunnan Agricultural University,
Kunming 650201, P.R. China
e-mail: aixianghuang@126.com

S. Sirisansaneeyakul
Department of Biotechnology, Faculty of Agro-Industry, Kasetsart University,
Bangkok 10900, Thailand
e-mail: sarote.s@ku.ac.th

Y. Chisti
School of Engineering, Massey University, Private Bag 11 222,
Palmerston North, New Zealand
e-mail: Y.Chisti@massey.ac.nz

© Springer Science+Business Media New York 2016
K. Kristbergsson, J. Oliveira (eds.), *Traditional Foods*, Integrating Food
Science and Engineering Knowledge Into the Food Chain 10,
DOI 10.1007/978-1-4899-7648-2_16

Traditionally, Xuanwei ham was produced using rear legs of *Wujin* breed of pigs that are bred in Xuanwei district. Wujin pig is a typical local breed that is usually reared outdoors. The pigs are fed mainly on corn flour, soybean, horse bean, potato, carrot, and buckwheat. Wujin breed grows slowly, but its meat is of a high quality. The quality of the meat is one reason for the fame of Xuanwei ham. Xuanwei ham processing begins in winter. The best pork legs are selected, and blood is completely pressed out. Then, the legs are rubbed repeatedly with salt and dried. The legs are allowed to cure for more than 6 months.

Xuanwei ham has a long storage life. Under ambient conditions it can be stored for 2 or 3 years. The ham is highly nutritious and easily digested. According to historical records, Xuanwei ham has been associated with many health benefits. Newly produced Xuanwei ham has a good flavor and is conveniently packaged, ready to eat, and a choice gift.

Xuanwei ham dates back to the fifth year of Yongzheng times in the Qing dynasty (1727 AD). The ham has been popular for hundreds of years (Yu et al. 2005). In 1909, Pu Zaiting, a folk entrepreneur, established "Xuanhe Ham Industry Company Limited." The company sent technicians to Shanghai, Guangzhou, and Japan to study advanced food processing technology. Xuanwei ham industry had become quite developed by 1915. Xuanwei ham won a gold medal at the Panama World Fair in 1915. It was one of the first famous local products of Yunnan to enter the world market. In 1918, canning equipment brought from the United States facilitated the establishment of "Xuanhe Canned Ham Industry Company Limited" to produce canned ham. The product was mainly exported overseas. In 1923, the famed states-man Sun Yat-sen tasted Xuanwei ham at the National Food Exhibition held in Guangzhou. Mr. Sun wrote of Xuanwei ham "yin he shi de," or "eat well for a sound mind." For many years, Xuanwei ham has enjoyed great popularity around the world. It has been exported to Hong Kong, Macao, Japan, Southeast Asian countries, and some European nations. In 1939, canned Xuanwei ham was being produced by four companies. The total production was 600,000 cans per year.

Xuanwei ham industry developed significantly after the Peoples Republic of China was established in 1949. The Municipal Authority of Kunming City built a factory for producing cans for packaging Xuanwei ham. Since the 1980s, fast growing foreign swine breeds (e.g., Duroc, USA; Landrace, Denmark; York, UK) have been bred with the Wujin breed to obtain pigs with improved growth rates and higher lean meat content than the Wujin breed.

Ensuring a high quality of Xuanwei ham and meeting consumer expectations while using modern industrial methods of production remains challenging. Research continues on the effects of processing methods on the quality of ham. Much of this work is funded by Yunnan Scientific Department, Yunnan Educational Department, and Xuanwei City Local Government, to promote continued development of Xuanwei ham production in China.

In 1999, the total production of Xuanwei ham was 13,000 tons, and the ham was being produced by 38 large producers. In March 2001, Xuanwei ham gained the status of a regional brand that has been protected by the Peoples Republic of China. As a consequence of this, the Chinese standard GB 18357-2003 for Xuanwei ham was officially issued. By 2004, the production of primary Xuanwei ham had reached 20,750 tons, and the manufacturing and packaging technologies for this product had greatly improved. Both canned and bag packed ham are available. The packaging is well designed and of a high quality. Improved packaging preserves flavor and has enhanced the popularity of this product with the consumers.

16.1.2 Present Status

Increased production of Xuanwei ham has been made possible by the introduction of crossbred pigs that grow more rapidly than the traditional Wujin breed. Crossbreeds such as Wujin×Duroc, York×(Wujin×Duroc), and DLY (Duroc×(Landrace×York) are being increasingly used. Studies suggest that the breed of pig does not significantly influence the final product so long as the breed has at least 25 % of Wujin blood (Yang and Lu 1987).

Pig production is big business in Xuanwei City, and the city government takes it seriously. For example, in 2004 the city government loaned 120 million yuan to pig breeders. The city has 31 pig breeding facilities that each produces more than 3000 pigs annually. In addition, there are around 9600 smaller breeding facilities. Animal production in the city is backed by 356 animal hospitals. In 2004, around 1.2 million pigs were sold in Xuanwei City.

Nine new ham processing facilities are planned in Xuanwei City (Li 2005). Seven of these will have a capacity of 500 tons each per year. Three processing plants will each produce 1000 tons of Xuanwei ham per year. Some of the newer facilities are attempting to move away from the traditional practice of extended curing and seasonal processing of ham. Industrial production methods are being implemented, and the quality and consistency of the product has improved greatly compared to the past. Many different forms of convenience ham products are now on the market.

An increasing number of large producers has meant an increasing competition, and this has enhanced the quality of the product. Increasingly, new research on Xuanwei ham is being published, and this knowledge is helping to improve the quality of the product. The ham market has seen organized development with direct sales shops established in Beijing, Shanghai, Guangzhou, and other cities. The product is being exported to many more countries than in the past.

16.2 Production of Xuanwei Ham

Typical climatic characteristics of Xuanwei City are said to explain its production of the famed ham. Xuanwei City is located in a low-latitude plateau monsoon climatic area where the north sub-torrid zone, the southern temperate zone, and the mid-temperate zone coexist. Winter lasts from November to January, and spring

Fig. 16.1 Flow sheet for
production of Xuanwei
dry-cured ham

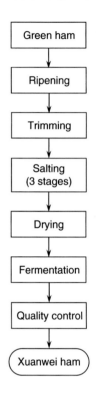

occurs from February to April. February, March, and April are sunny and clear, and
this leads to a low relative humidity in these months. From May to September, it is
overcast and rainy, and the relative humidity is comparatively high. Winter is the
best season for salting the ham, and early spring is suited to drying it. The rainy
season is good for ham fermentation.

A simplified flowchart for production of Xuanwei dry-cured ham is shown in
Fig. 16.1. Xuanwei ham is produced using the hind legs of pigs. The pigs are slaughtered at around 90–130 kg live weight.

The hind legs are cut off along the last lumbar vertebra and trimmed in an oval
shape that is similar to that of the Chinese musical instrument pipa. Blood is pressed
out of the meat by hand. The legs are held in a cold room at 4–8 °C and 75 % relative
humidity for 24 h for meat ripening (Huang et al. 2009). Ripened legs are known as
green ham (Fig. 16.2).

Green ham is then salted (Fig. 16.2). The salt used contains table salt (25 g/kg
leg) and sodium nitrite (0.1 g/kg leg). The salt is rubbed into green ham by massag-
ing with hands for 5 min. The salted ham is stacked in pallets and held in a cold
room at 4–8 °C, 75–85 % relative humidity, for 2 days. Salting procedure is then
repeated. This time the salt contains table salt (30 g/kg leg) and sodium nitrite
(0.1 g/kg leg). After a further 3 days of curing in the cold room, a third salting
occurs. In this stage of salting, the salt mixture contains 15 g table salt per kg leg and

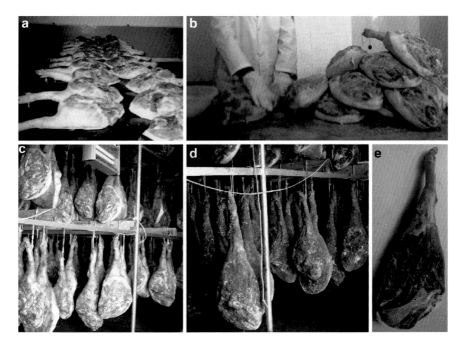

Fig. 16.2 Production stages of Xuanwei dry-cured ham: (**a**) green ham; (**b**) salting; (**c**) drying; (**d**) fermentation, or curing; and (**e**) the final product

sodium nitrite (0.1 g/kg leg). The salted green ham is stacked in pallets and held in a cold room under the above specified conditions for 3–4 weeks. During this period, ham piles are turned over four times.

A drying stage follows (Fig. 16.2). The salted legs are hung in a drying room held at 10–15 °C and 50–60 % relative humidity. Excess salt is brushed away and the legs are held for 40 days. The drying room has screened windows to permit ingress of fresh air, but pests such as flies are kept out. Maintaining the optimal drying temperature and relative humidity is important. For example, a high relative humidity can lead to spoilage, whereas too low a relative humidity leads to development of crust on the ham.

After drying, temperature is increased gradually to 25 °C, and the relative humidity is also increased progressively to 70 % to begin the fermentation process that lasts for at least 180 days (Fig. 16.2). Increased temperature and relative humidity benefit microbial and enzyme action that is responsible for producing the flavor of the ham. Flavor develops as a consequence of partial decomposition of lipids and proteins.

Producing Xuanwei ham requires care to prevent damage and loss associated with pests. The main pests include flies during ham curing and drying stages. Ham moth and mites (e.g., *Tyrophagus putrescentiae*) can cause damage during ham fermentation and storage stages. Flies such as *Piophila casei*, *Dermestes carnivorus* beetle, and mites occur if relative humidity exceeds 80 %. Controlling pests can be a challenge. Pests cause substantial loss each year. Extensive research has been carried out in preventing and treating mite infestations (Ma et al. 1997; Huang et al. 2002). Microorganisms such as *Streptomyces* strain s-368 help prevent and treat mite infestations.

The Chinese standard GB 18357-2003 (Chinese Bureau of Quality and Technology Control) has been developed for quality control of Xuanwei ham. The sensory requirements and physicochemical indices for the ham are summarized in Tables 16.1 and 16.2, respectively.

Table 16.1 Physicochemical quality indices

Physicochemical index	Required value
Lean meat content in ham (%)	≥ 60
Moisture in meat (%)	≤ 48
Sodium chloride in meat (%)	≤ 12.5
Sodium nitrite in meat (mg/kg)	≤ 4
Nitrogen trimethylamine (g/100 g)	≤ 1.3
Peroxide value in fat (meq/kg)	≤ 32

Table 16.2 Volatile compounds detected in Xuanwei ham[a,b]

Compound group	Compound
Aldehyde	2-Tridecenal
	13-Tetradecenal
	17-Octadecenal
	2-Pentenal
	2-Heptenal
	2,4-Heptadienal
	16-Octadecenal
	1,1-Diethoxyethane
	Hexadecyl oxirane
	1-Octene
	1-Nonene
	5-Decene
	1-Tetradecene
	3-Hexadecene
	8-Heptadecene
	5-Octadecene
	3-Ethyl-2-methyl-1,3-hexadiene
Alcohol	1-Nonanol
	6-Pentadecen-1-ol
	3-Furanmethanol
Ketone	6-Tridecanone
Acid	Hexanoic acid
Ester	Ethyloleate
Others	Anthracene

[a]All analyses were by GC/MS
[b]Found only in Xuanwei ham
Source: Modified from Huang (2006)

16.3   Physicochemical and Microbial Changes Accompanying Ham Processing

16.3.1   Color

The heme pigment that is chiefly responsible for the color of meat is myoglobin. Myoglobin can be converted to oxymyoglobin and metmyoglobin to produce different colors in flesh (Fig. 16.3) (Hemmings and Clermont-Ferrand 1992). To understand how this occurs, it is necessary to discuss the way oxygen reacts with myoglobin and hemoglobin. Myoglobin, oxymyoglobin, metmyoglobin, and their three hemoglobin equivalents are the common forms of the pigments that occur in meat. Oxymyoglobin is produced from myoglobin through binding of the latter with an oxygen molecule (Fig. 16.3).

In addition to oxygen, various other compounds can act as ligands. Ligation produces compounds that are associated with the characteristic pink color of cured ham. The color of the ham is an important attribute. Most consumers prefer a bright red color. Therefore, the oxidation of myoglobin to metmyoglobin should be avoided, as it imparts a brown color to red meat, a principal cause of consumer dissatisfaction. Significant research has focused on color development in ham. Huang et al. (2005) suggest that the addition of 100 mg/kg $NaNO_2$ to ham could improve the color of cured ham without exceeding the residual nitrite level specified by the Chinese dry-cured ham standard.

Addition of nitrate and/or nitrite to meat products converts the myoglobin into nitrosylmyoglobin, a stable red compound. During curing or cooking, nitrosylmyoglobin changes into pink-reddish nitrosohemochromogen. The reaction between nitrite and myoglobin has been studied extensively (Hemmings and Clermont-Ferrand 1992).

Because of food safety concerns that are sometimes associated with nitrite, materials that can substitute for nitrite are of potential interest. Wakamatsu et al. (2004) studied the Zn-porphyrin complex that contributes to bright red color in Parma ham

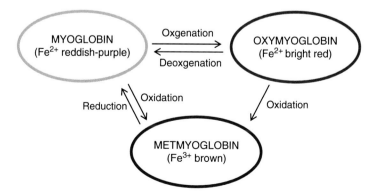

Fig. 16.3 The relationship among myoglobin, oxymyoglobin, and metmyoglobin. Based on Zheng and Yang (1989)

and suggested that the bright red color in Parma ham is caused by Zn-protoporphyrin IX in which the iron in the heme of myoglobin has been replaced by Zn. Lipophilic stable red pigment in Parma ham has been found to increase with aging (Parolari et al. 2003). Some bacteria isolated from meat products apparently convert myoglobin to the desirable red myoglobin derivatives (Arihara et al. 1993; Morita et al. 1996). Nitrite replacement in Xuanwei ham requires further study.

16.3.2   Chemical Constituents

16.3.2.1   Lipids

Lipid breakdown components are chiefly responsible for the flavor of ham. In addition to being precursors of flavors, lipids act as carriers for flavor compounds. Triglycerides and phospholipids are the main lipids that contribute to ham flavors. During curing, lipids undergo mainly hydrolysis and oxidation. Lipolysis is an enzymatic reaction that hydrolyses the lipids to release free fatty acids. Endogenous lipases present in meat and microbial lipases contribute to lipolysis. The compounds produced by lipids oxidation may react with other compounds via Maillard reaction to generate further flavors.

Oxidation of free fatty acids is an autocatalytic process. The first products of oxidation are free radicals that lead to formation of hydroperoxides and peroxides that undergo other reactions. Among the numerous products of fat degradation, low molecular weight compounds such as aldehydes and ketones are particularly associated with flavors. Both oxidation and microbial metabolic action contribute to flavor development. Unstable aldehydes seldom accumulate and are often oxidized to carboxylic acids (Rosset et al. 1977).

Yang and coworkers (2005) studied lipolysis of intramuscular lipids during processing of traditional Xuanwei ham. The lipid content of the *biceps femoris* muscle of Wujin crossbreed pigs was found to be 45 g kg^{-1} fresh meat (Yang et al. 2005). The glycerides and free fatty acids represented 25.3 % and 2.3 % of the total lipid content, respectively. These results are consistent with values reported for French white pigs (Buscailhon et al. 1994).

Among glycerides, the monounsaturated fatty acids (MUFA) were by far the most abundant (54.7 % of the total fatty acids), followed by saturated fatty acids (SFA, 34.2 %) and a small proportion of polyunsaturated fatty acids (PUFA, 10.7 %). Oleic, palmitic, stearic, and linoleic acids were the main fatty acids in glycerides (49.4 %, 22.2 %, 10.4 %, and 9.6 %, respectively). A high proportion of these fatty acids has also been reported in intramuscular glycerides in fresh meat of Iberian pigs (Tejeda et al. 2002; Petrón et al. 2004).

Among phospholipids, the PUFA are the most abundant (42.1 %), followed by SFA (36.8 %) and MUFA (21.1 %). These results are consistent with reports for French white pigs (Buscailhon et al. 1994; Coutron-Gambotti et al. 1999), but differ from data that has been reported for Iberian pigs (Cava et al. 1997; Martín et al. 1999). For the latter, SFA were the most abundant. This difference is probably related with differences in animal breeds and feeding regimes.

Amounts of phospholipids and free fatty acids change significantly during Xuanwei ham processing, but glyceride levels do not change much (Yang et al. 2005). There is a close relationship between the decrease in phospholipid content and the increase in free fatty acid content observed during processing, indicating that free fatty acids come mainly from phospholipids. This is consistent with other studies of changes in fat content and composition during processing of dry-cured hams (Buscailhon et al. 1994; Martín et al. 1999). After 12 months of ripening, a 77 % reduction in the total quantity of phospholipids has been observed compared with raw ham. During the first 4 months of processing, the percentage of PUFA decreased and that of MUFA increased, but there was no subsequent change. The amount of SFA did not change during the entire period of processing. The ratio of $n-3$ (C18:3 and C22:5) to $n-6$ (C18:2 and C20:4) fatty acids increased significantly during the first 4 months of processing. This was mainly associated with a significant decline in the content of linoleic acid. The trend of change in phospholipids was similar to that of the other fatty acids. Overall, the PUFA were degraded by 87.7 %. Lesser levels of SFA (76.9 %) and monounsaturated (56.3 %) fatty acids were degraded (Yang et al. 2005). These results were consistent with similar observations for Iberian dry-cured ham (Martín et al. 1999).

16.3.2.2 Proteins

Degradation of proteins also contributes to the flavor and taste of ham. Both endogenous enzymes and microbial enzymes contribute to proteolysis (Martín et al. 2004; Zhao et al. 2005a). Proteolysis produces small peptides and free amino acids. Amino acids may be further converted to pyruvate. Ammonia is one of the degradation products of proteins. During curing, the concentration of free amino acids increases. These amino acids contribute to flavor both directly and indirectly via conversion to volatile compounds. Several months of ripening is required to enable flavor development. The fungal population that is allowed to grow on the surface of some types of dry-cured hams may play a role in proteolysis (Martín et al. 2004).

Some studies attribute proteolysis mostly to endogenous enzymes in view of the low microbial counts found inside the ham (Toldra et al. 1988) and the low detected microbial enzyme activity levels (Molina and Toldra 1992). Zhao and coworkers (2005b) studied the changes in dipeptidyl peptidase (DPP) activity during processing of Jinhua ham. The activity of muscle DPP I decreased from an initial value of around 8608 to about 1842 U g^{-1} before aging. During aging, the activity gradually increased from around 1842 to 12,196 U g^{-1} at the end. Unlike the activity of DPP I, the activity of muscle DPP IV decreased continuously, and about 11 % of the initial potential activity was left at the end of processing. The results showed that endogenous enzymes were active during the entire period of ham processing.

The proteolytic microorganisms present in ham include lactobacilli (Reuter and Langner 1968), micrococci, streptococci, and yeasts (Wang et al. 2006). Molina and Toldra (1992) reported on the proteolytic activity of *Pediococcus pentosaceus* and *Staphylococcus xylosus*, as two major microorganisms that were isolated from Spanish Serrano dry-cured ham. Both microorganisms lacked endopeptidase activity.

However, *P. pentosaceus* showed strong leucine and valine arylamidase activities, while *S. xylosus* showed a very weak leucine arylamidase activity.

Xu et al. (2003) reported on the proteolysis activities of whole cells (cell), cell extracts (CE), and a combination of these (cell+CE) for *L. plantarum* 6003, on muscle sarcoplasmic protein extracts. Both the cell fraction and the mixture of cells and extract caused proteolysis, but the extract alone was inactive. The mixture of cells and extract was more active than the cells alone.

The contribution of some fungi to proteolysis in dry-cured ham has been studied (Martín et al. 2004). The ham was inoculated with nontoxigenic strains of *Penicillium chrysogenum* (Pg222) and *Debaryomyces hansenii* (Dh345), selected for their proteolytic activity on myofibrillar proteins. By 6 months of inoculation, the fungi were shown to cause higher proteolysis when compared with non-inoculated control ham. After 6 months, the presence of proteolytic strains among the wild fungal population on non-inoculated control hams led to similar levels of proteolysis as observed with inoculated ham. However, surface inoculation of fully dry-cured ham with stains Pg222 and Dh345 led to higher levels of most free amino acids in the external muscle. Furthermore, the concentration of some of the more polar free amino acids (i.e., Asp, Glu, Ser, and Gln) in inoculated ham was higher in the external muscles compared with the internal meat. Further work is required on the potential of selected fungi for modifying the organoleptic properties of dry-cured ham.

16.3.2.3 Microflora

Microorganisms are found in all stages of processing of dry-cured ham (Rodríguez et al. 1994; Martín et al. 2004). Microbiology of Western hams has been investigated extensively. Molds are usually thought of as dominant microorganisms on the surface of ham and affect ham quality. Molds are involved in the ripening of dry-cured meat products where they have a positive effect on flavor and appearance (Lücke 1986). Bacteria such as *Staphylococcus* and *Micrococcus* may not be recovered at the end of maturation of Iberian dry-cured ham, whereas yeasts are predominant during this stage (Rodríguez et al. 1994). The surface yeast population of Iberian ham has been shown to be useful for estimating the progress of maturation. *Candida zeylanoides* is the dominant yeast in the early stages, while *Debaryomyces hansenii* is dominant in mature ham (Núñez et al. 1996). Yeasts have been suggested to contribute to curing possibly through their proteolytic or lipolytic activity (Saldanha-da-Gama et al. 1997).

In studies of natural microflora of Xuanwei ham, Wang et al. (2006) showed that the *Streptomyces* bacteria were the dominant strains that accounted for almost half of the isolated *Actinomycetes*. Although few studies exist on the effects of *Actinomycetes* on ham quality, bacteria, except for some micrococci and lactobacteria, are usually thought to be harmful to human health and cause ham spoilage (Rodríguez et al. 1994; Losantos et al. 2000; Portocarrero et al. 2002; De Martinis and Freitas 2003). *Aspergilli* and *Penicillia* appear to flourish on the surface of Xuanwei ham from June to August. Eight species of *Aspergillus* were found. *A. fumigatus* was the dominant species and accounted for one third of all *Aspergilli*. *A. flavus* and *A. versicolor* were present. Other species occurred occasionally. Four species of *Penicillium* were found. Roughly equal levels of *P. lanate*, *P. divaricate*, and *P. velutin* were isolated, while *P. monoverticillate* appeared at a lesser frequency. Generally, high values of relative humidity benefit mold growth on the surface of ham (Arnau et al. 2003). Some molds may improve the quality of ham, but other molds may produce unwanted mycotoxins (Creppy 2002; Martín et al. 2003).

Yeasts appear to be the dominant fungi that are associated with Xuanwei ham (Wang et al. 2006). Yeasts may account for more than 50 % of the total microorganisms on mature ham at any given time. This predominance of yeasts is quite common on different types of hams and other pork-based products such as several types of bacon where the concentration of yeasts ranges from 10^3 to 10^9 cfu/g of fat. Proteolytic and lipolytic activities of yeasts appear to be desirable (Saldanha-da-Gama et al. 1997). Yeasts are predominant microorganisms near the end of maturation period also in Iberian dry-cured ham (Rodríguez et al. 1994; Núñez et al. 1996).

Humidity and temperature largely determine the microbial species that occur at different times during processing (Arnau et al. 2003). During the rainy season in Xuanwei district, the temperature and humidity are higher than in other seasons. Molds appear almost exclusively on the surface of the ham. They rarely occur in the interior. The quantity of *Penicillia* and *Aspergilli* is high in May when the temperature and humidity are high. These fungi peak in July or August. The peak concentrations of spores of *Aspergilli* and *Penicillia* have been recorded at 7×10^5 cfu/g and 9×10^4 cfu/g, respectively. Usually, molds begin to grow in May and became well established by June. Spores are formed in August and September. The quantity of spores falls off gradually after September.

Growth of bacteria and *Actinomycetes* does not seem to depend on humidity in the curing room. Levels of bacteria are generally lower than levels of yeasts. According to Wang et al. (2006), yeasts on ham multiply exponentially from the beginning of the salting stage to reach a peak in April, and then the numbers drop and stabilize to around 2×10^7 cfu/g. Yeast levels within the ham showed similar variation as the surface yeasts. According to Wang et al. (2006), yeasts account for 60–70 % of the total microbial population on the surface of ham. The concentration of yeasts in ham muscle can peak at anywhere from 1×10^6 to 3×10^6 cfu/g. In some cases, no molds have been found growing on the surface of good-quality ham; therefore, some researchers believe that molds do not play a direct role in determining the quality of the dry-cured ham (Jiang et al. 1990; Wang et al. 2006), but an opposing view also prevails (Martín et al. 2003).

According to the traditional view, high-quality Xuanwei ham must have "green growth" (i.e., molds) on it. However, fungi such as *Penicillia*, *Fusarium*, and *Aspergilli* are known to produce mycotoxin in foods such as dry-cured Iberian ham (Núñez et al. 1996; Cvetnić and Pepeljnjak 1997; Brera et al. 1998; Erdogan et al. 2003). More than 15 % of the mold strains examined were found to produce mycotoxins in Xuanwei ham (Wang et al. 2006). The toxins penetrated to a depth of 0.6 cm in the ham muscle. Because most of the fungi that occur on ham have not been examined for producing mycotoxins, contamination with toxins might be more prevalent than is realized.

16.4 Future Developments

16.4.1 Processing Techniques

In view of its significance, the local government of Yunnan Province has invested in research and further development of Xuanwei ham industry. Beginning in the 1980s, imported swine breeds were crossbred with local Wujin pigs to improve growth rates and increase lean content of the meat. Yunnan Agricultural University and Yunnan Animal Husbandry Institute have cooperated in studying the relationship between the Xuanwei ham quality and the use of crossbreeds. This work established that the quality was not significantly influenced if the animals contained 1/4 local Wujin blood.

Based on market research, the consumers mostly desire less salty meat products because of health concerns. A low salt intake is recommended for reducing hypertension (Morgan et al. 2001). Indeed, the producers commonly use less salt than in the past in processing of Xuanwei dry-cured ham (Lu and Yang 1988).

Some of the microorganisms isolated from the traditional Xuanwei dry-cured ham have been studied for their influence on quality of the final product (Huang et al. 2003). Traditional ham features are retained in hygienically processed ham that has the following attributes: 47.4 % (w/w) moisture, 6.2 % (w/w) NaCl, 28.8 % (w/w) protein, 13.3 % (w/w) fat, and 66.4 mg/100 g TVB-N (i.e., total volatile basic nitrogen, an index of freshness) value. In comparison with this, traditional Xuanwei ham (lean tissue) contains 8.8 % (w/w) NaCl, 30.4 % (w/w) protein, and 10.9 % (w/w) fat (Jiang et al. 1990).

The traditional technology of making Xuanwei ham relies on natural conditions. Ham is usually salted in winter when the temperature varies between 1.5 and 13 °C and the relative humidity ranges from 58 to 72 %. From February to April, it is dry in Yunnan Province, and most of dry curing occurs during this period. During May to September, in the rainy season, the temperature gradually increases from 10 to 25 °C, and the relative humidity increases progressively from 40 to 70 %. These conditions benefit microbial and enzyme action and contribute to ham fermentation.

Traditional processing methods are often incompatible with industrial processing and consumer demands. Therefore, studies have been carried out on using modern processes to produce Xuanwei ham (Huang et al. 2005; Qiao et al. 2006). For example, new processing methods and salting ingredients can reduce the time required for salting to 21 days compared with 28 days that are needed in traditional processing (Huang et al. 2005). Use of starter cultures for ham inoculation offers other possibilities for process improvements. Some companies have begun using industrial processing machinery in making Xuanwei ham. The product is now available as canned ham and ready-to-eat vacuum-packed ham.

16.4.2 Microflora

Microorganisms associated with Xuanwei ham have attracted substantial research interest (Jiang et al. 1990). For example, yeasts have been found to be important to the development of the cured ham quality traits (Jiang et al. 1990). Li et al. (2003)

suggest that the presence of salt-tolerant *Micrococci*, *Staphylococci*, and molds are highly correlated with flavor development in Xuanwei ham. Similar claims have been made for the Italian-type dry-cured ham (Hinrichsen and Pedersen 1995). In contrast with Li et al. (2003), Wang and coworkers (2006) showed that the quality of ham correlated with the dominance of yeasts. Furthermore, according to Wang et al. (2006) mycelial fungi such as *Penicillium* and *Aspergillus* that flourish on the surface of ham may not actually contribute much to its quality characteristics.

The quality of ham inoculated with yeasts that occur on traditionally produced ham has been found to be superior to traditional ham in terms of color, fragrance, flavor, and the amino acid profiles (Wang et al. 2006). According to Wang et al. (2006), with the use of yeast inocula and controlled fermentation conditions of humidity and temperature, a molds-free ham can be produced that is comparable to the traditional product. Unpublished data of Huang et al. suggest that whether molds contribute to Xuanwei ham fermentation remains to be resolved. Potential health hazards from uncontrolled fungal populations can be minimized by using nontoxigenic strains as starters in dry-cured hams.

16.4.3 Physicochemical Properties

The physical and chemical properties of dry-cured ham are important determinants of its quality (Jiang et al. 1990; Careri et al. 1993). The lean portion of Xuanwei ham contains 30.4 % protein, 10.9 % fat, 10.3 % amino acids, 42.2 % moisture, and 8.8 % salt (Jiang et al. 1990). The whole ham contains 17.6 % protein, 29.1 % fat, 5.6 % amino acids, 24.8 % moisture, and 3.3 % salt (Jiang et al. 1990). Many essential elements are present in the ham as are some vitamins. The ham is particularly rich in vitamin E (45 mg/100 g). The characteristic bright red color of Xuanwei ham is mainly attributed to oxymyoglobin and myoglobin. The flavor and taste are associated with the presence of various amino acids and volatile organic compounds. Xuanwei ham quality standard GB 18357-2003 requires the ham to meet the specifications shown in Table 16.2.

The volatile substances present in Xuanwei ham have been extensively studied (Qiao and Ma 2004; Yao et al. 2004). Seventy-five compounds were tentatively identified in the volatile fraction. The compounds identified included hydrocarbons, alcohols, aldehydes, ketones, organic acids, esters, and other unspecified compounds (Table 16.2).

16.5 Concluding Remarks

Traditionally produced Xuanwei ham is highly popular. Advances are being made in transforming the traditional production process to a more modern basis. Among important aims are achieving a hygienic product of consistently high quality, reducing the length of the production period, and achieving production from pig breeds that grow more rapidly than the breed that has been used traditionally.

Microorganisms and endogenous enzymes are believed to be the key factors involved in the production of dry-cured hams (Rodríguez et al. 1998; Martín et al. 2004; Zhao et al. 2005a). Substantial but sometimes contradictory knowledge has emerged on microbiology of processing of Xuanwei ham. Work has been done also on the contribution of endogenous enzymes to the development of various quality attributes of dry-cured ham (Toldra and Etherington 1993; Zhao et al. 2005a); nevertheless, a full transformation of Xuanwei ham production to a modern well-defined and controlled process requires continuing effort.

References

Arihara K, Kushida H, Kondo Y, Itoh M, Luchansky JB, Cassens RG (1993) Conversion of met-myoglobin to bright red myoglobin derivatives by *Chromobacterium violaceum*, *Kurthia* sp., and *Lactobacillus fermentum* JCM1173. J Food Sci 58:38–42

Arnau J, Gou P, Comaposada J (2003) Effect of the relative humidity of drying air during the resting period on the composition and appearance of dry-cured ham surface. Meat Sci 65:1275–1280

Brera C, Miraglia M, Colatosti M (1998) Evaluation of the impact of mycotoxins on human health: sources of errors. Microchem J 59:45–49

Buscailhon S, Gandemer G, Monin G (1994) Time-related changes in intramuscular lipids of French dry-cured ham. Meat Sci 37:245–255

Careri M, Mangla A, Barbieri G, Bolzoni L, Virgigili R, Parolari G (1993) Sensory property relationships to chemical data of Italian-type dry-cured ham. J Food Sci 58:968–972

Cava R, Ruiz J, López-Bote C, Martín L, García C, Ventanas J, Antequera T (1997) Influence of finishing diet on fatty acid profiles of intramuscular lipids, triglycerides and phospholipids in muscles of the Iberian pig. Meat Sci 45:263–270

Chinese Bureau of Quality and Technology Control (2003) Chinese Standard GB18357-2003 Xuanwei ham

Coutron-Gambotti C, Gandemer G, Rousset S, Maestrini O, Casabianca F (1999) Reducing salt content of dry-cured ham: effect on lipid composition and sensory attributes. Food Chem 64:13–19

Creppy EE (2002) Update of survey, regulation and toxic effects of mycotoxins in Europe. Toxicol Lett 127:19–28

Cvetnić Z, Pepeljnjak S (1997) Distribution and mycotoxin-producing ability of some fungal isolates from the air. Atmos Environ 31:491–495

De Martinis ECP, Freitas FZ (2003) Screening of lactic acid bacteria from Brazilian meats for bacteriocin formation. Food Control 14:197–200

Erdogan A, Gurses M, Sert S (2003) Isolation of moulds capable of producing mycotoxins from blue mouldy Tulum cheeses produced in Turkey. Int J Food Microbiol 85:83–85

Hemmings B, Clermont-Ferrand ATT (1992) In: Girard JP (ed) Technology of meat and meat products. Ellis Horwood, New York

Hinrichsen L, Pedersen SB (1995) Relationship among flavor, volatile compounds, chemical changes, and microflora in Italian-type dry-cured ham during processing. J Agric Food Chem 43:2932–2940

Huang AX (2006) Study on the formation of quality traits and microbial roles of Yunnan dry-cured ham. Dissertation, Southwestern University

Huang AX, Lu ZF, Ge CR, Chen T, Wang Y (2002) The main pest in Yunnan ham processing and storage. Storage Process 6:21–22, In Chinese

Huang AX, Hu YJ, Ge CR, Lu ZF (2003) The study of dry-cured meat with the characteristics of traditional dry-cured ham. J Food Sci 4:82–84, In Chinese

Huang AX, Ge CR, Chen ZD, Li HJ (2005) The influences of different processing techniques on Yunnan ham. J Food Ferment Ind 4:138–140, In Chinese

Huang AX, Sirisansaneeyakul S, Chen Z, Liu S, Chisti Y (2009) Microbiology of Chinese Xuanwei ham production. Fleischwirtschaft Int 24(3):64–67

Jiang DF, Duan RL, Ma P (1990) The study of chemical compounds and microflora in Xuanwei style ham. J Yunnan Univ 3:71–75, In Chinese

Li ZY (2005) The industrial development of Xuanwei ham. In: The proceedings of the 5th Chinese Congress of Meat Science, pp 1–6

Losantos A, Sanabria C, Cornejo I, Carrascosa AV (2000) Characterization of *Enterobacteriaceae* strains isolated from spoiled dry-cured hams. Food Microbiol 17:505–512

Lu ZF, Yang HS (1988) The study on reduction of salt in Yunnan ham. J Meat Poult Egg 6:24–25, In Chinese

Lücke FK (1986) Microbiological processes in the manufacture of dry sausage and raw ham. Fleischwirtschaft 66:1505–1509

Ma P, Jiang DF, Wei RC, Wang ZX, Yang ZS, Zhang LQ, Wang LZ (1997) Studies on the methods of the prevention and cure mite calamity of ham. J Yunnan Univ 19:414–416

Martín L, Córdoba JJ, Ventanas J, Antequera T (1999) Changes in intramuscular lipids during ripening of Iberian dry-cured ham. Meat Sci 51:129–134

Martín A, Córdoba JJ, Benito MJ, Aranda E, Asensio MA (2003) Effect of *Penicillium chrysogenum* and *Debaryomyces hansenii* on the volatile compounds during controlled ripening of pork loins. Int J Food Microbiol 84:327–338

Martín A, Córdoba JJ, Nunez F, Benito MJ, Asensio MA (2004) Contribution of selected fungal population to proteolysis on dry-cured ham. Int J Food Microbiol 94:55–66

Molina I, Toldra F (1992) Detection of proteolytic activities. J Food Sci 57:1308–1310

Morgan T, Aubert J, Brunner H (2001) Interaction between sodium intake, angiotensin II, and blood pressure as a cause of cardiac hypertrophy. Am J Hypertens 14:914–920

Morita H, Niu J, Sakata R, Nagata Y (1996) Red pigment of Parma ham and bacterial influence on its formation. J Food Sci 61:1021–1023

Núñez F, Rodríguez MM, Bermúdez ME, Córdoba JJ, Asensio MA (1996) Composition and toxigenic potential of the mould population on dry-cured Iberian ham. Int J Food Microbiol 32:185–197

Parolari G, Gabba L, Saccani G (2003) Extraction properties and absorption spectra of dry cured hams made with and without nitrate. Meat Sci 64:483–490

Petrón MJ, Muriel E, Timón ML, Martín L, Antequera T (2004) Fatty acids and triacylglycerols profiles from different types of Iberian dry-cured hams. Meat Sci 68:71–77

Portocarrero SM, Newman M, Mikel B (2002) *Staphylococcus aureus* survival, staphylococcal enterotoxin production and shelf stability of country-cured hams manufactured under different processing procedures. Meat Sci 62:267–273

Qiao FD, Ma CW (2004) Characteristics analysis of specific quality and forming reasons of traditional Xuanwei ham. J Food Sci 8:55–61, In Chinese

Qiao FD, Ma CW, Yang HJ, Song Y, Li MT (2006) Physico-chemical changes and standardized technology in Xuanwei ham processing. J Food Sci 2:136–140, In Chinese

Reuter G, Langner HJ (1968) Entwicklung der Mikroflora in schnell reifender deuter Rohwurse and analoge quantitative Aminosäure Analyse ber einer Salami. Fleischwirtschaft 2:170–176

Rodríguez M, Núñez F, Córdoba JJ, Sanabria C, Bermúdez E, Asensio MA (1994) Characterization of *Staphylococcus* spp. and *Micrococcus* spp. isolated from Iberian ham throughout the ripening process. Int J Food Microbiol 24:329–335

Rodríguez M, Núñez F, Córdoba JJ, Bermúdez E, Asensio MA (1998) Evaluation of proteolytic activity of microorganisms isolated from dry cured ham. J Appl Microbiol 85:905–912

Rosset MR, Liger P, Roussel-Ciquard N (1977) La flavour de la viande. Serie Synthese Bibliographique 14, CDIUPA

Saldanha-da-Gama A, Malfeito-Ferreira M, Loureiro V (1997) Characterization of yeasts associated with Portuguese pork-based products. Int J Food Microbiol 37:201–207

Tejeda JF, Gandemer G, Antequera T, Viau M, García C (2002) Lipid traits of muscles as related to genotype and fattening diet in Iberian pigs: total intramuscular lipids and triacylglycerols. Meat Sci 60:357–363

Toldra F, Etherington DJ (1993) Cathepsins B, D, H and L activities in the processing of dry-cured ham. J Sci Food Agric 62:157–161

Toldra F, Richo E, Flores J (1988) Examination of cathepsins B, D, H and L activities in dry-cured ham. Meat Sci 59:531–538

Wakamatsu J, Nishimura T, Hattori A (2004) A Zn-porphyrin complex contributes to bright red color in Parma ham. J Food Sci 67:95–100

Wang XH, Ma P, Jiang DF, Peng Q, Ya H (2006) The natural microflora of Xuanwei ham and the no-mouldy ham production. J Food Eng 77:103–111

Xu XL, Xu WM, Zhao GH (2003) Study on proteolysis activities of L. plantarum on muscle protein extracts (I). Food Sci 24:56–61, In Chinese

Yang HS, Lu ZF (1987) The influence of different genotype of pig on Yunnan ham quality. Sci Technol Yunnan Agric Univ 1:30–32, In Chinese

Yang HJ, Ma CW, Qiao FD, Song Y, Du M (2005) Lipolysis in intramuscular lipids during processing of traditional Xuanwei ham. Meat Sci 71:670–675

Yao P, Qiao FD, Yan H, Ma CW (2004) Isolation and identification of volatile compounds of Xuanwei ham. J Food Sci 2:146–150, In Chinese

Yu ZS, Jiang WX, Qiu CY, Fan MY (2005) The history development of Xuanwei ham. In: The proceedings of the 5th Chinese Congress of Meat Science, pp 264–265

Zhao GM, Zhao GH, Wang YL, Xu XL, Huan YJ, Wu JQ (2005a) Time-related changes in cathepsin B and L activities during processing of Jinhua ham as a function of pH, salt and temperature. Meat Sci 70:381–388

Zhao GM, Zhao GH, Xu XL, Peng ZQ, Huan YJ, Jiang ZM, Chen MW (2005b) Study on time-related changes of dipeptidyl peptidase during processing of Jinhua ham using response surface methodology. Meat Sci 69:165–174

Zheng JS, Yang CJ (1989) Principles and application of food storage. China Financial and Economic Press, Beijing, p 47, In Chinese

Chapter 17
Selected Viennese Meat Specialties

Elias Gmeiner, Helmut Glattes, and Gerhard Schleining

17.1 Introduction

Meat is one of the most naturally occurring, well balanced, and easily obtained and digested packages of proteins, essential fatty acids, and source of iron and vitamins. Depending on the age, species, and living condition of the slaughtered animal, the meat has a different composition. But roughly it can be told that meat consists of approx. 15–22 % of proteins, 3–15 % of fat, 1–5 % of minerals and carbohydrates, and 65–75 % of water.

Red meat has a higher concentration of iron than white meat, because myoglobin, an oxygen-binding protein, has an iron(II) ion bound on the active site of the molecule.

Proteins are broke down in the stomach during digestion into essential and nonessential amino acids. These amino acids are used as precursors to vitamins and also for the anabolism and modification of proteins produced naturally in the body.

Fat is an important quality factor. The juiciness and the aroma of meat and its products are largely determined by the fat content. Fat is for the intake of indispensable fat-soluble vitamins. Most essential polyunsaturated fatty acids have to be taken through diet. Single and especially polyunsaturated fatty acids can lower the blood cholesterol level, which is increased by saturated fatty acids.

Vitamins belong to the regulator substances in the human body. Vitamins cannot be synthesized in sufficient quantities by an organism; they must be obtained from the diet in the form of vitamins or their precursors, which are modified and transformed into the particular chemical compound. Meat is a very good provider of vitamins from the B group and vitamin A (liver).

E. Gmeiner • H. Glattes • G. Schleining (✉)
Department of Food Sciences and Technology, BOKU, Muthgasse 18, Wien 1190, Austria
e-mail: gerhard.schleining@boku.ac.at

© Springer Science+Business Media New York 2016
K. Kristbergsson, J. Oliveira (eds.), *Traditional Foods*, Integrating Food
Science and Engineering Knowledge Into the Food Chain 10,
DOI 10.1007/978-1-4899-7648-2_17

Through the metabolism of carbohydrates, the human body gains ATP as an energy carrier. An overrun of carbohydrates can be stored momentary in the liver and muscle tissue as glycogen. If the supply of carbohydrates is bigger than the usage, the surplus is converted into fat and stored.

Minerals are only available in small amounts; the functions are mainly the upkeep of body structures and body functions such as nervous and muscular conduction (Hoffmann 2000).

17.2 Wiener Schnitzel–Viennese Schnitzel

17.2.1 History

The origins and history of the Viennese Schnitzel is not easy to clarify; therefore, legends were developed which are nowadays in common parlance seen as the true origins of the Viennese Schnitzel.

The term schnitzel is related to the south German word for slice, which stands for a little piece of cut meat. Crumbed veal escalope first appeared in the eighteenth century; the first known recipes are from the year 1768.

A frequently quoted legend has it that the Viennese of the Habsburg Monarchy got to know the Viennese Schnitzel through Field Marshal Earl Joseph Wenzel von Radetzky. He was sent to upper Italy in 1848 to quench revolts against the Habsburg Monarchy. There he came across the "Cotoletta alla Milanese," a cutlet dipped into beaten eggs and braised in clarified butter. This is how it was mentioned at least in a document from the Viennese National Archives, where he was cited.

So the "Cotoletta alla Milanese" had been seen as the forerunner of the Viennese Schnitzel. The legend of Field Marshal Radetzky as the bringer of the Viennese Schnitzel had been disproven by the historian Richard Zahnhausen in 2001 and also years before by folklorist Günter Wiegelmann in 1967. One cannot rule out the possibility that there are earlier influences from Italy.

It is said that gold is the beginning of all crumbed things. So it was customary in the Lombardy of the fifteenth and sixteenth centuries to cover meat with gold platelets in order to demonstrate one's prosperity during a meal and also because gold had the attribute of being healthy. As this culinary decadence took overhand in the Late Medieval, it was prohibited in 1514. As a consequence, an optical alternative was needed; the solution to this was coating the meat with a golden crust of bread crumbs. Historians suggest that it was already in common to gild meat in the Byzantium. From there on, the Jews delivered this custom of gilding meat to the Moors in Spain. Further it went onward to Italy.

As you can see the Viennese Schnitzel's origins lie elsewhere. Fact is that since the end of the nineteenth century, you can find recipes for crumbed escalope in every cookery book, and it was already in the Baroque era (1575–1770) tradition in Vienna to crumb vegetables, meat, fish, or poultry.

Traditionally Viennese Schnitzel was served with parsley, potatoes, lettuce, potato salad, or cucumber. A slice of lemon was always a fix part of the dish. The reason for this is that back in the nineteenth and twentieth centuries, sometimes the hygienic conditions were not the best and the lemon juice was used to cover the taste of slightly rotten meat, burned bread crumbs, and rancid fat. Nowadays it is also served with rice, French fries, or fried potatoes.

Currently the Austrians are serving two types of Viennese Schnitzel, either made of veal or pork, but the Codex Alimentarius Austriacus defines a Viennese Schnitzel such as only made of veal.

17.2.2 Ingredients and Preparation

In the following the preparation is described based on *Dippelreither* (2004):
For four servings, you need:

500 g of veal
2 eggs
1 tbsp. milk (1 tablespoon ~ 15 mL)
Salt
Breadcrumbs
Wheat flour
Lard, butter, or oil
Lemon wedges or halves

17.2.2.1 The Meat

- Flank of veal, Fricandeau (lean topside), or veal shoulder (thicker, less tender, more fat and strings).
- Veal cut into thin (5–8 mm) and even pieces at right angle to the fibers; the smallest enlargement of the Schnitzel should not be less than 10 cm; incisions (1–2 cm if it is a big piece) should be around the edges; otherwise the meat will curve (Mayer 2003).
- Fat and strings removed as good as possible.

17.2.2.2 Plating

- Plate the meat gently with the smooth side of the hammer (cover the meat with a plastic film, so your clothes and kitchen will stay clean and some of the outgoing meat juice will be resorbed; the meat will stay juicy).
- Make sure that the veal is not thinner than 4 mm and avoid making holes into the meat.

17.2.2.3 Coating

- Wheat flour
- Beat eggs with a fork, do not use a blender (eggs would lose adhesion power), and add some milk or optionally oil (coat of breadcrumbs becomes juicier)
- Breadcrumbs

Dip both sides of the escalope into flour (gently shake off surplus), then into the beaten eggs (let it drain a bit), and finally with slight pressure into the breadcrumbs (shake off surplus).

Mostly the escalope is salted before it gets covered and baked. The reason not to do so is that sodium chloride abstracts water, an effect also known as osmosis. As a matter of fact, the meat dries out. So it is highly recommended to salt the Schnitzel just before serving.

17.2.2.4 Baking

Heat plenty of fat (traditionally lard; the escalope needs to swim) in a suitable pan. Put the escalope into the hot fat 170–180 °C—otherwise the coating would stick to the ground—and brown. Repeatedly move the pan; the coating on the upper side of the escalope will get in contact with the hot fat and so the coating elevates a bit. This is called *fliegende Panier* (*flying breading*); it is a trademark of the Viennese Schnitzel.

Then carefully turn it over with a meat fork after 1–2 min (it depends on the size), fry until done, and lift it out of the pan with a shovel. Leave the Schnitzels to drain and dab with kitchen paper to remove surplus fat (Fig. 17.1).

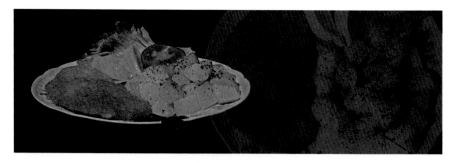

Fig. 17.1 Viennese Schnitzel with parsley potatoes (photo by Pfanner M. ©2011)

17.3   Tafelspitz: Boiled Beef

17.3.1   History

Boiled beef is not only used as a Viennese household specialty but also for Austrian restaurants. "Tafelspitz" derived its name from the piece of meat being used, the pointy and tender silverside of the beef. In the United States, the meat cut is also known as the tri-tip. The meat is usually from a young ox; best if it is a well-aged and well-hung piece of beef.

Austrian butchers gave almost every muscle of beef a separate name. Joseph Wechsberg, a Czech writer, journalist, musician, and gourmet, wrote: "the one who couldn't speak competently at least about a dozen different pieces of boiled beef, couldn't call his self a part of the Viennese community, no matter how much money he earned or whether he had the title of a court counselor" (Höbaus 2011).

While a lot of dishes of the classical Viennese cuisine come from the crown lands like Hungary, boiled beef is an original Viennese specialty. Already in the first half of the nineteenth century, boiled beef was a fix part of the imperial cuisine, as it was apparent from the menu of the imperial officers from the year 1836. Back then boiled beef was served daily with varying side dishes. The final prominence attained the dish during the regency of Emperor Franz Joseph I. He was known for his thrift and modesty. For the imperial household, simple diet was sufficient, like boiled beef plus side dishes. The side dishes were usually stew, a plate full of fresh shredded horseradish, young onions, old bread to soak up the dip, and three deciliter of wine. "The serving lesson text books of the vocational schools" from the year 1912 state: "A good piece of boiled beef, which belongs to his favourite dishes, is never missing on the private dining table of his majesty" (*Hohenlohe*2005).

At that time there was plenty of beef available because of cattle and fattened ox from Hungary, which were bought and imported to Austria. The meat of the cattle had the tenderness, juiciness, and condiment which made the boiled beef in those days to a real specialty.

Starting from the imperial court, this rather simple dish found its way through the servants into the middle-class households where Viennese citizens tried to imitate the eating habit of the emperor.

But who invented this dish we all know as "Tafelspitz?" The legend says that the Tafelspitz was invented in the famous Hotel Sacher in Vienna.

In the Austro-Hungarian Monarchy, during the occasion of a meal with the serving military, the emperor was always served first and also determined the end of the meal. Emperor Franz Joseph was known as someone who ate pretty fast and only a mouthful. Therefore, the soldiers had to eat somewhere else after "finishing" their meal at the imperial court. The famous Anna Sacher really had sympathy for them and created a dish that can simmer for hours and even improve its taste with time— the Tafelspitz.

Between Austria and Hungary, there was an agreement regarding the import of pork from Serbia. After cancellation of this contract, the result was a higher import of beef from Hungary and a higher consumption of beef products in the Austrian-Hungarian empire.

17.3.2 Ingredients and Preparation

In the following the preparation is described based on *Plachutta* and Plachuta (*2008*):

For 6–8 servings at a cooking time of 3–3½ h, you need:

1 onion in skin, cut into half
Approx. 2000 g Tafelspitz
Approx. 3.5 L water
10–15 peppercorns, black
250 g carrots, yellow carrots, celery, and parsley root in same amounts and peeled
½ stick of leek, peeled and washed
Stock
Salt
Chives cut into fine rolls

Halve onions and fry the cut areas without fat in a pan covered with aluminum foil until they become dark brown. Wash meat in lukewarm water and let it drain. Bring the water to boil in a large pot, put the meat in, and let it simmer. Skim occasionally and add peppercorns and the onion halves. Add carrots, yellow carrots, celery, and parsley root approx. 25 min before finishing and if desired also stock. Let it simmer until it is finished. Remove cooked, tender meat from the pot, season the soup with salt, and strain it through a sieve. Cut the meet into finger-thick slices across the grain, pour soup over it, and sprinkle with salt and chives.

Traditionally it is served with fried potatoes, apple–horseradish, and chive sauce, but dill string beans and cream spinach also fit pretty good.

To check whether the meat is already done, take a long needle and sting carefully. As soon as the needle enters easily, the meat is ready to serve (Fig. 17.2).

Fig. 17.2 Tafelspitz with fried potatoes, spinach, chive sauce, and apple–horseradish (photo by Pfanner M. ©2011)

17.4   Backhendl: Fried Chicken

17.4.1   History

In former times poultry was symbol for luxury and a good life. For the well-heeled poultry was a real culinary delight at auspicious occasions. For example, as Emperor Friedrich I. Barbarossa held court council in Mainz in the year 1184, even two houses were needed to store the poultry.

As already mentioned below, crumbing things does not have its origin in the Viennese Schnitzel but in the past in the fifteenth and sixteenth centuries or even at an earlier time in the Byzantium. Fact is that crumbing things were already pretty popular in Vienna even before the Viennese Schnitzel became famous. Fried chicken belonged to these specialties in the era of Biedermeier. Chicken or general poultry were the embodiment of prosperity. An old wisdom of ancient times says that "Chicken was served in peasant households, only if the chicken or the peasant was ill" (Agrarmarkt Austria 2011). Initially, with the emancipation of the serfs in the nineteenth century, the fried chicken became a part of the Viennese kitchen for all classes.

> Historical records numeralise the devotion of the Viennese with 22.000 fried chickens per year, and that in only two inns named "Wilder Mann" and "Paperl" in the Prater. 12.000 chicken and 4.000 geese were eaten in the Prater on Whitsunday of 1852 the traditional confirmation day. (Hohenlohe 2005)

Thus, fried chicken can be postulated as one of the most famous and eaten dishes in the nineteenth century in Austria. The oldest known recipe for fried chicken is from 1719. It was mentioned in the Salzburg cookbook written by Conrad Hagens.

17.4.2   Ingredients and Preparation

In the following, the preparation is described based on *Plachutta* (*2008*):
For 6–8 servings, you need:

2 chickens (approx. 1400 g), emboweled
Salt
Approx. 100 g wheat flour
2 eggs
Approx. 150 g breadcrumbs
Plenty of oil or butter
Parsley
2 lemons

Wash the chicken, and dry and chop them into four pieces. The spine and neck must be removed. The skin can be removed, but it is not compulsory. Cut the inside of the leg through to the bone. Salt strongly and dip into the flour, then into the

Fig. 17.3 Backhendl with potato salad (photo by Pfanner M. ©2011)

whisked eggs, and finally into the breadcrumbs. Pour plenty of oil into a deep pan and heat. Fry the chicken pieces approx. 6 min—do not forget to shake the pan slightly—turn the pieces around and fry until finish (golden coat). Take the fried chicken carefully out of the pan and drip off on a paper towel.

Traditionally it is served with potato salad, iceberg lettuce, or butterhead lettuce and garnished with parsley leaves and lemon slices.

Chickens from organic and adequate animal housing are better in meat structure and taste than conventional poultry. If the skin and the bones are removed, it will shorten the baking time and ease the consumption (Fig. 17.3).

References

Agrarmarkt Austria (2011) Geschichte. Augenweide und Gaumenfreude. http://www.rund-ums-gefluegel.at/index.php?id=geschichte. Accessed 25 Mar 2011

Dippelreither R (2004) Das Schnitzelkochbuch, Alte und neue Rezepte. Leopold Stocker Verlag, Graz/Stuttgart, pp 8–14

Höbaus E (2011) Traditionelle Lebensmittel in Österreich. http://www.traditionelle-lebensmittel.at/article/articleview/71563/1/26099/. Accessed 11 Mar 2011

Hoffmann P (2000) Lexikon der Lebensmittel, Wegweiser zur gesunden Ernährung, pp 261, 263, 264, 270, 283

Hohenlohe M (2005) Viennese cuisine, cook and enjoy. Pichler Verlag, Wien, pp 30, 36, 55

Mayer V (2003) …dass es über den Teller ragt. In: Kulinarischer Almachnach, Klett-Cotta, Stuttgart

Plachutta E, Plachutta M (2008) Plachutta, Meine Wiener Küche. Christian Brandstätter Verlag, Wien, pp 91, 94

Chapter 18
Dried Norse Fish

Trude Wicklund and Odd Ivar Lekang

18.1 History

In Norway, fish has been dried and used for local market or export for centuries. The good supply of fresh raw material and suitable climatic conditions has made it possible to make this into a reliable and economical success. Fishery and handling of fish have had a great influence on the settlements along the Norwegian coastline. Dried fish was one of the first Norwegian export products, and the Vikings were known to trade dried fish with other West European countries. The trade of dried fish is still very important for the Norwegian export market (Renaa and Notaker 2008). The stockfish has been exported around the whole world. It is especially popular in many Mediterranean countries, where it is included in dishes served during the fasting period. The Spanish and Portuguese seafarers brought this raw material with them on their journeys and introduced it to other cultures in other parts of the world. Dried fish has long traditions in the Caribbean cuisine as well as in some West African and South American countries.

T. Wicklund (✉)
Department of Chemistry, Biotechnology and Food Sciences, Norwegian University of Life Sciences, Postbox 5003, Aas 1432, Norway
e-mail: trude.wicklund@nmbu.no

O.I. Lekang
Department of Mathematical Sciences and Technology, Norwegian University of Life Sciences, Postbox 5003, Aas 1432, Norway

© Springer Science+Business Media New York 2016
K. Kristbergsson, J. Oliveira (eds.), *Traditional Foods*, Integrating Food Science and Engineering Knowledge Into the Food Chain 10,
DOI 10.1007/978-1-4899-7648-2_18

18.1.1 Dried Fish: Raw Material and Traditional and Modern Processing

Drying is the world's oldest preservation method. The drying procedure is included in many processes for preservation of fish. Fish preserved this way can be stored for years. Drying can take place by hanging the fish on wooden racks (drying flake), placing the fish on the rocks or, today, a more modernized process using mechanized equipment. It is common to divide in stockfish that is dried and cliff fish that is salted prior to drying (Riddervold 1993).

18.1.2 Stockfish

Stockfish is unsalted fish, dried under the sun and wind in special wooden racks (sticks) (Fig. 18.1). The English name, stockfish, possibly comes from the wooden racks it is dried on or from the fact that the fish looks like a stick when dry. The drying method is cheap and has little demands for investments, and large quantities of fish can be hanged during a limited time. The fish is usually dried in the winter when the temperature is low, which gives a product less likely to be destroyed by bacteria,

Fig. 18.1 Drying of stock fish in Iceland. Photo by Kristberg Kristbergsson

Fig. 18.2 Today both whole fish, heads and backbones (after filleting) are dried. Photo by Kristberg Kristbergsson

moulds or insects. This is also when the major catches of white fish take place in Norway (Renaa and Notaker 2008). Today most of the stock fish production has moved to drying heads and backbones after filleting (Fig. 18.2).

The totally dominating raw material used is cod, but other white fish like pollock, haddock, ling and tusk can also be used.

The fish is prepared immediately after capturing. It is gutted, the head removed, and then the fish is either split along the spine or dried whole. By splitting the fish, the drying time will be shorter. The drying process lasts from February to May. This time of the year and the locations for drying give good drying conditions, preventing damage from bacteria, moulds and insects. The conditions for drying are very good in Lofoten (latitude 67/68°N), where high-quality dried fish is produced. The largest seasonal cod fishery in Norway is also taking place in this period, Lofotfishery, and a method that can conserve large quantities of fish in a short time is required. In the Lofoten area, there will always be a breeze and very little rainfall, and the temperature is usually just above 0 °C during the drying period. The stockfish will have a dry matter content of 10–15 % and a_w 0.7. The lower the moisture content, the better to prevent mould growth during storage.

If temperature goes too low during the drying period, the fish will freeze. This will have a negative effect on the product quality, as freezing will destroy the fibres in the fish. The drying conditions further south (Western Norway, latitude 60/62°N) are warmer and more humid, so in these areas salting of the fish before drying is more common (klippfisk/cliff fish).

The long and slow drying process allows a fermentation to take place during the drying process, slightly more for the fish that is not split before hanging. The drying period on the flakes is from 2 to 3 months. After this drying period, the fish needs to be matured for another 2–3 months, usually in airy conditions, but under roof (Lynum 2007).

Depending on raw material quality before drying, whether the fish has been exposed to frost or pests will be important for classification of the final product. Up to 30 different parameters are evaluated. The classification is subjective and specially trained people do this classification (Vraker). The best quality was usually exported at highest prices. Lower qualities can be used for animal feed.

18.1.3 Klippfisk/Bacalao (Cliff Fish)

The cliff fish (*klippfisk* in Norwegian, also popularly known as *bacalao* in Spanish, *bacalhau* in Portuguese and *baccalà* in Italian) differs from dried fish by the amount of salt used in addition to the drying process. Cod and saithe are common raw materials. The fish is split along the spine and the coarsest part of the spine is removed. A lot of salt is added to the fish that is then left on the rocks for the initial drying, until it becomes firm. It is then washed and salted again and then alternately pressed and further dried.

The traditional way of production was to dry the fish on the cliffs near the sea. In the morning it was distributed on the cliffs for sun drying, while at the end of the day, it was picked up again and put into stack for the night, to avoid taking up humidity. Many women would be doing this work along the Norwegian west coast. This is also how this fish was named. Today the drying is done indoors in special driers (Vollan 1956).

This way of processing the fish is a relatively new method in Norwegian history. The method became more common as salt became available and cheaper throughout all of Europe in 1500–1700.

Besides Norway, also Iceland and the Faroe Islands are big producers of cliff fish.

18.2 The Use of Dried Fish

18.2.1 Lutfisk

Lutfisk is a dish that has long traditions in Norway, but similar recipes are also found in other Scandinavian countries. The fish is first soaked in cold water for 5–6 days, changing water twice a day. The soaked stockfish is then soaked for another period in a lye solution. The quality of the lye is important for the product quality. The old tradition was to prepare lye from ash from certain hardwood (mainly birch). This contains more potassium (K_2CO_3) and gives the fish a more mellow treatment than using NaOH.

The normal quantity of NaOH used is 4g per litre of boiling water. For preparation of lye, 4 L of hardwood ash and 10 L boiling water are used. Impurities are removed, and the lye solution is cooled before use.

18.2.1.1 Lye Treatment

The ash has to be clean and the trays used for the treatment should be from either wood enamel or plastic. Metal trays can give unwanted colour and taste. The fish is soaked in the lye solution for 2–3 days or until it gets the right consistency—when the fish flesh can easily be penetrated by a finger. During this treatment, the fish flesh becomes jelly-like, which is the desired consistency. Too long incubation time can cause saponification. The pH value at the end of the period is 11–12, so the fish needs soaking in cold water (daily changing of water) for nearly 1 week before further preparation.

The fish can then be cooked and prepared for a delicious meal.

18.2.1.2   Cooking

The fish can be simply steamed and placed on a grate, adding salt and enough water for steaming.

The "lutfisk" can also be baked in the oven, just adding salt and placing in an oven for 40–50 min at circa 225 °C.

18.2.1.3   Serving

The way of serving varies between regions. In some places, the fish is served in "lefse" (soft flat bread), while it is more common on a plate with boiled potatoes. Other possible supplements are bacon, green pea stew, white sauce, mashed ruta-baga, brown goat cheese, syrup, mustard, butter or cured cucumber.

Aquavit and beer are common drinks with this meal. Sterling silver should never be used when serving lutfisk.

The season for lutfisk usually starts in November and continues through Christmas.

18.2.2   Stockfish (Bacalao)

The origin of the word *bacalao* is Spanish and means *cod*, while in Norway *bacalao* is the name of a special dish prepared mostly from clip fish, sometimes also from stock fish.

There is a large variation in local recipes. The basic ingredients for this recipe in Norway are clipfish, potatoes, onions, tomatoes, pimiento, olive oil and Spanish pepper.

In Portugal *bacalhau* is the national dish, cooked in a wide variety of ways (according to tradition, there are 1000 different ways of cooking *bacalhau*). Boiling with potatoes and cabbages gives the most common meal for Christmas Eve, a tradition as strong as having turkey in the USA for Thanksgiving. In 2013 it was estimated that 80 % of Portuguese households had *bacalhau* for Christmas dinner. Outside this season, there are many other more popular forms—baked, fried, boiled in milk and made into mash; the variations are almost endless. Portugal imports about 30–40 thousand tons of salted cod fish (*bacalhau*) per year, which adds to the local production, giving the Portuguese a per capita consumption of 5–7 kg/year of dried salted cod fish (in return, only around 5 % of the Portuguese have ever tried fresh, unsalted cod). About 3/4 of the *bacalhau* consumed in Portugal in 2014 was imported from Norway.

In Italy, both the undried fish and the salty clip fish are common raw materials in their dish *baccalà*.

Stockfish is also very popular in West Africa as a good supplement to their grain staple food.

In Mexico and Brazil, stockfish is included in their traditional Christmas dishes. *Bacalhau* represents 25 % of the total exports of goods from Norway to Brazil.

18.2.3 *Boknafisk (No English Translation Found)*

A special dish is made from a partly dried fish: If the fish is dried for a short time, only 1–2 weeks, it becomes dry on the outside and soft on the inside. This allows a more marked fermentation process to take place, and the fish will have a stronger smell and taste, slightly acidic. The fish flesh, though, should remain white. This is regarded as a delicacy in the western and northern part of Norway.

There are different ways of preparing this fish: It can be filleted and soaked in sea water for some hours before drying for 10–12 days, usually hanging on the wall; the whole fish can be salted for 2–3 days before drying (hanging on the wall) for 1–2 weeks, or it can then be soaked and prepared like any other boiled fish.

18.2.4 *Other Uses of Dried Fish*

As a **snack**, stock fish is only knocked to soften or split the fish fibres, and it is consumed dry.

References

Lynum L (2007) Fisk som råstoff. Fagbokforlaget, Bergen. ISBN 978-82-519-1254-9
Renaa SE, Notaker H (2008) Tørrfisk til begjær. Messel forlag, Oslo. ISBN 9788276310986
Riddervold A (1993) Konservering av mat, levende norske tradisjoner. Teknologisk forlag, Oslo. ISBN 82-512-0421-6
Vollan O (1956) Den norske klippfiskens historie. Øens forlag, Førde

Suggested Additional Reading (English)

http://www.visitnorway.com/en/what-to-do/food-and-drink/stockfish/
Kurlansky M (1997) Cod: a biography of the fish that changed the world. Walker, New York. ISBN 0-8027-1326-2

Chapter 19
Utilization of Different Raw Materials from Sheep and Lamb in Norway

Trude Wicklund

19.1 History and Tradition

Sheep were brought to Scandinavia during the Stone Age. Some Scandinavian breeds are still regarded as little developed since that time, especially the *Old Norwegian Sheep*. This breed represents one of the most primitive kinds of domestic sheep still present in Europe. The sheep are frugal and can graze on areas that are not easy to exploit for more intensive farming. Sheep were also highly valued by the Vikings for their wool, meat, and milk. This sheep was the common type in Norway until the eighteenth century, when new breeds were brought to the country and mixed with the old ones. The new breeds were better adapted for inland farming and are now the dominating varieties.

The *Old Norwegian Sheep* is well adapted to the climate along the coastline, and with some supplement, they can stay outdoors throughout the year. Similar breeds are also common in Iceland and the Faroe Islands where the climate is very similar to the Norwegian coastline. The climate here is mild during wintertime and cool during summertime, differently from the inland climate. Sheep as a domestic animal has been important for development of the cultivated landscape along the Norwegian coastline and on the islands, throughout the centuries. Today, the *Old Norwegian Sheep* constitutes only 1 % of the total number of sheep in Norway. There is an increased interest for this type of sheep and farming nowadays. Sheep grazing on outlying fields and in the mountain areas develops a typical aroma/ flavor of the meat. The feed available in these areas, such as special herbs and heather, contribute to this flavor.

T. Wicklund (✉)
Department of Chemistry, Biotechnology and Food Sciences, Norwegian University of Life Sciences, Postbox 5003, Aas 1432, Norway
e-mail: trude.wicklund@nmbu.no

© Springer Science+Business Media New York 2016
K. Kristbergsson, J. Oliveira (eds.), *Traditional Foods*, Integrating Food
Science and Engineering Knowledge Into the Food Chain 10,
DOI 10.1007/978-1-4899-7648-2_19

Meat from sheep has long traditions in the Norwegian cuisine. Many people were poor and all parts of the sheep were used as food. Typical dishes have been developed, but similar dishes can also be found in other cultures where sheep has been an important livestock.

To prolong shelf life after slaughtering, which is normally done in the autumn, drying and smoking have been very important (Fellows 2009; Hemmer et al. 2006). It is common to combine this with additional salting and smoking. Drying of meat that has not been salted is very rare and is forbidden in Norway today. Earlier, this has been a tradition in some cultures, but special precautions have to be taken both in the slaughtering procedure and in the pretreatment before drying. Good knowledge of physiological and microbiological conditions in meat and special drying conditions are crucial (Riddervold 1993). Smoking of meat is more common in the coastal region, where the climate is more humid and drying conditions have been traditionally poorer. Smoke will give a better protection against mold growth on the surface.

As the normal slaughtering season for sheep is in the autumn, many of the traditional dishes are consumed before and during Christmastime, in addition to other special occasions.

Examples of special produce from sheep are:

Smalahove—sheep head
Fenalår—salted and dried lamb thighs
Værballer—testicles

There might be variation between preparations from district to district depending on local traditions.

19.2   *Smalahove*: Sheep Head

Sheep has been a highly valuated raw material in the Norwegian cuisine. Every part of the animal should be used (Riddervold 1993). Earlier, the sheep head was regarded as an "everyday food." The slaughtering season still has strong traditions in some regions, and feasts are normally arranged in this season. In modern times, many of the dishes have become a delicacy, and today it is often served for special occasions.

New EU directives have changed the traditional use of sheep head in the production of *smalahove*. Due to concern about proliferation of BSE (e.g., Creutzfeldt–Jakob disease (CJD) and other related diseases (transmissible spongiform encephalopathies/prion diseases - TSE)), there are now restrictions in the use of sheep older than 12 months for this purpose (Greger 1997).

19.2.1   Preparation of the Head

- The head can be prepared skinned or not skinned.
- The skin and fleece of the head are torched. The torching can be done on an open fire or in a special oven using firewood from birch or alder. A stick is put through the nose. In earlier traditions, the ears were not removed before torching. In this way, the owner of the sheep could still be identified. Today, the ears are removed before torching.
- After cleaving the head and removing the brain, the head should be kept in clean and cold water until the next day and then salted for 2–3 days.
- After salting, the head can be further processed by smoking. Another tradition is to prepare without smoking.

The salting and smoking process will give a shelf life of some months if it is stored in a dry and cool place.

Farmers very often prepare their own *smalahove* in their local traditional way, although today, a more industrialized process is more common, and such producers are responsible for the majority of the market today.

Another part of the lamb that is usually produced in the same way is lamb rib, in Norwegian called *pinnekjøtt*. The name actually means stick meat and originates from the way of preparation of the dish—the meat is kept on sticks, usually from birch, when steamed. It is served similarly to *smalahove*. The raw material is the ribs of lamb, salted, and died the same way as *smalahove*. Smoking is optional.

19.2.2   Serving

The head is boiled/steamed for ca 3 h or until the meat loosens from the bone. Watering before boiling/steaming will extract more salt from the head. Boiling, instead of steaming, will also affect the product in the same way. Tradition or desire will decide which way to prepare it.

The one half of the head is usually served on a plate together with boiled root vegetables and potatoes (Fig. 19.1). As an everyday dish, fermented milk was served as a drink, while at feasts, beer and aquavit are common.

The eye (the muscle behind the eye) and the ear are normally eaten first. These parts contain most fat and need to be warm when eaten. A special curiosity is to marinate the eye in the aquavit for some seconds and then drink the whole in one sip. The cheek is also a delicious part of the head. Another delicacy is the tongue.

Fig. 19.1 Smalahove
served with carrot,
rutabaga, and potato
(photo: Trude Wicklund)

19.3   Fenalår: Cured Lamb Thigh

The cured lamb thigh has long traditions in the coastal regions of Norway (Christie 2007). The sheep or lamb should be of some size, as a small animal will give a dry and skinny product.

For curing, the meat can be put in brine, dry salt, or a mixture with salt and sugar. The curing time will depend on the temperature in the surroundings—cool but not below 0 °C—3 days per kg is common. By using dry salt, the water activity on the surface will decrease more rapidly. This will prevent microorganisms from growing on the surface. The best salt quality will be sea salt (Riddervold 1993).

Some people like the meat smoked, others prefer the non-smoked variety. If the meat shall be smoked, the thigh should be dried for a couple of days after salting. Cold smoking for 3–4 h is common, and the smoke from juniper gives a good taste to the product. Alder wood is also used for smoking (Riddervold 1993).

After the salting and possible smoking, the meat shall be dried. It is usually hanged in an airy and cool place (storehouses) for 3–4 months. The thigh can be covered by a cloth bag to avoid insects that ruin the product (Håseth et al. 2007; Pettersen 2007).

During the salting and drying process, the meat undergoes a fermentation process and becomes both tender and firm. In earlier times, the meat was dried more than it is today. The meat should be sliced very thinly for serving (Fig. 19.2). Today, sliced meat can be bought in the shops as well as the whole thigh. Different types of cured and dried meat (swine, deer, elk, and reindeer) can be served together with the cured lamb. Flat bread—made basically from barley—scrambled eggs, sour cream, butter, potato salad, and some vegetable salad are usually served with the meat. A homemade beer will complete the meal, especially in summertime.

Fig. 19.2 Fenalår (photo:
Trude Wicklund)

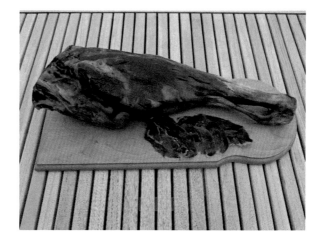

19.4   Værballer: Ram Balls

In a community with limited resources, all parts of the lamb or the sheep should be exploited. The use of testicles from sheep (in Icelandic language—*hrutspungar*) is an example of this. In other countries, testicles from ox or gems can also be used. This produce is not well known today and is regarded as a curiosity, but different variations have been reported from some countries in the Middle East, China, the USA, and Russia (Country Roads Catering 2015).

Raw sheep testicles feel like little balloons filled with jelly. When cooked, they are tender and mild and do not have any sheep meat taste or aroma. They have been compared to bratwurst (Grygus 2015).

For preparation, the testicles should be skinned and soaked in cold water before use. They are often sliced and fried either coated in a batter or not coated. The balls are cooked very quickly if sliced or fried only for a couple of minutes in hot oil. Other ways of preparing can be sautéed, sauced, fricasseed, poached, or roasted.

The testicles can also be fermented in whey (Icelandic style) then pressed into cakes or set in gelatin and served as a pâté. The consistency and taste are described as sour and spongy (Imsomboon 2014).

References

Christie H (2007) Fenalår—Historien om en norsk delikatesse og spesialitet. Kolofon, Oslo
Country Roads Catering (2015) http://www.bridgeandtunnelclub.com/bigmap/queens/menus/che-burechnaya.htm. Accessed 28 June 2015
Fellows PJ (2009) Food processing technology. CRC, Boca Raton. ISBN 978-1-84569-634-4
Greger M (1997) http://mad-cow.org, http://www.mad-cow.org/~tom/raymond.html#smalahove. Accessed 28 June 2015

Grygus A (2015) www.clovegarden.com, http://www.clovegarden.com/ingred/as_nutz.html. Accessed 28 June 2015

Håseth T, Thorkelsson G, Sidhu M (2007) Nordic products. In: Toldrá F (ed) Handbook of fermented meat and poultry. Blackwell, Oxford, pp 371–377

Hemmer E, Askim M, Karlsen H, Lynum L, Nordeng A, Nybraaten G (2006) Næringsmiddellære. Yrkeslitteratur as, Oslo

Imsomboon P (2014) http://modernfarmer.com, http://modernfarmer.com/2014/11/bizarre-fermented-foods/. Accessed 29 June 2015

Pettersen K (2007) www.matoppskrift.no, http://www.matoppskrift.no/artikkel/Fenalaaret-ekte-norsk#axzz3eMK7UeeG. Accessed 28 June 2015

Riddervold A (1993) Konservering av mat, levende norske tradisjoner. Teknologisk forlag, Oslo. ISBN 82-512-0421-6

Chapter 20
Muxama and *Estupeta*: Traditional Food Products Obtained from Tuna Loins in South Portugal and Spain

Jaime Aníbal and Eduardo Esteves

20.1 Importance of Tuna as Fishing and Food Resource

Tuna is a generic name for fish from the *Scombridae* family, which includes about 50 species, mostly from the *Thunnus* genus. Tuna is a very important commercial resource widely but sparsely distributed throughout the oceans of the world that generally occurs in tropical and temperate waters between about 45° north and south of the equator (FAO 2010). Because of high muscular activity, some tuna species display warm-blooded adaptations and can raise their body temperatures above surrounding water temperatures (Randall et al. 2002). This enables them to survive in cooler ocean environments and to inhabit a wider geographic range of latitudes than other kinds of fish.

The most important tuna species for commercial and recreational fisheries are yellowfin (*Thunnus albacares*), bigeye (*T. obesus*), bluefin (*T. thynnus*, *T. orientalis*, and *T. maccoyii*), albacore (*T. alalunga*), and skipjack (*Katsuwonus pelamis*) (FAO 2010). Between 1940 and the mid-1960s, the annual world catches of the five main marketed species of tunas rose from about 300,000 tons to about 1 million tons, most taken by hook and line. With the development of purse-seine nets, now the predominant gear, catches have risen to more than four million tons annually during the last few years (6.3 million tons in 2008). Of these catches, about 68 % are from the Pacific Ocean, 22 % from the Indian Ocean, and the remaining 10 % from the

J. Aníbal
CIMA—Centro de Investigação Marinha e Ambiental and Departamento de Engenharia Alimentar, Instituto Superior de Engenharia, Universidade do Algarve, Faro, Portugal

E. Esteves (✉)
Departamento de Engenharia Alimentar, Instituto Superior de Engenharia, Universidade do Algarve and Centro de Ciências do Mar CCMar—CIMAR Laboratório Associado,
Faro, Portugal
e-mail: eesteves@ualg.pt

© Springer Science+Business Media New York 2016
K. Kristbergsson, J. Oliveira (eds.), *Traditional Foods*, Integrating Food Science and Engineering Knowledge Into the Food Chain 10,
DOI 10.1007/978-1-4899-7648-2_20

Atlantic Ocean and the Mediterranean Sea. Skipjack makes up about 57 % of the catch, followed by yellowfin (27 %), bigeye (10 %), albacore (5 %), and bluefin (1 %). Purse seiners catch about 62 % of the world production, longline about 14 %, pole and line about 11 %, and a variety of other gears the remainder (FAO 2010).

In Portugal, landings of tuna were around 10,000 tons in 2009, which represented ca. 5 % of total national catches (189 thousand tons) (INE 2010). Apart from tunas caught in the Portuguese EEZ, landed fish are from fishing grounds in southwest Atlantic (242 tons of bigeye, 217 tons of yellowfin) and southeast Atlantic (87 tons of albacore). A total of 22 tons of yellowfin tuna are reported by INE (2010) as being produced via intensive aquaculture—an *Almadrava* (described below) still operating in waters off the Algarve was noted, where undersized tunas are fed (fattened) before being shipped to Japan. Most of the 12,000 tons of processed tuna are consumed canned. In fact, no explicit statistical records exist for its utilization or trade in the dry-salted (seafood) product categories of Portuguese statistics (INE 2010). In neighboring Spain, fresh seafood consumption varied annually around 550–620 thousand tons between 1987 and 2000; tuna's contribution to this consumption was 20–24 thousand tons per year (Navarro 2001).

In the south of Iberian Peninsula (Algarve and Andalucía), tuna was traditionally caught using a fishing apparatus named *Almadrava*, i.e., an offshore "maze" of bottom-fixed nets to imprison, capture, and hold the fish. Changes in the migratory patterns of tuna schools, arguably due to climate change (Dufour et al. 2010), were probably the main reason leading to declining catches and the disappearance of this kind of fishing method during the 1960s (Santos 1989). While operating and prospering, the numerous *Almadravas* established in Portuguese and Spanish waters supplied raw material for the canned tuna plants that became an important regional economic asset, creating significant employment (Santos 1989; Lã and Vicente 1993).

20.2 Traditional Processing of Tuna Loins

In the south of Portugal and Spain, tuna has been traditionally consumed in three different ways: fresh, as tuna steaks; canned, e.g., in olive oil; and cured, salted, and/ or salted-dried. Much like other seafood, fresh tuna spoils very fast due to its protein-rich composition and high summer temperatures (Sikorski 1990; Huss et al. 2004), especially without a proper refrigeration system. This system only became ubiquitous in the food industry in the last half century. Before, alternative methods were used to preserve (sea)food. Thence curing was an important process for preserving tuna for longer periods of time outside the fishing season (April to September) (Santos 1989).

At the end of the (traditional) quartering of tuna a.k.a. *ronqueamento* (Portugal) or *ronqueo* (Spain), about 50–60 % of a tuna can be utilized for the manufacturing of food products or preparation of meals, viz., *Muxama, Estupeta, Mormos,*

Rabinhos, Faceiras, Orelhas, Ventresca, Tarantela and *Sangacho, Espinheta, Tripa, Bucho*, and *Ovas* (Lã and Vicente 1993).

The word *Muxama* (Portugal) or *Mojama* (in common use in Spain) comes from the Arab *musama* and means "dry." The production of *Muxama* was developed by the Arabs more than a thousand years ago, based on a combination of salt preservation followed by drying.

The preparation of *Muxama* involves a series of stages/steps. After beheading, eviscerating, and fining, tunas are quartered in four big loins, two in the tuna's dorsal (upper) side and two other in the ventral (lower) side. The central parts in each of these loins are composed of red muscle with low lipid content that after being properly cut into rectangle parallelograms are used to produce *Muxama* (Fig. 20.1). The trimmed loins are generally 50 cm in length, 8 cm in width, and 3 cm in height (Lã and Vicente 1993).

Then, the tuna loins are piled/stacked in salt for 24 h and afterwards washed several times. After all residues of crystallized salts are eliminated from the loin's surface, they are dried at 14 °C and 60 % relative humidity for 10–12 days, depending on their size (Lã and Vicente 1993).

Muxama production has not changed much during the last centuries. Back in the Middle Ages (around the tenth century), the tuna loins were air-dried during summer, when the weather in the south Iberian Peninsula is hot and dry. Nowadays, legislation imposes the use of a "temperature- and humidity-controlled room/chamber." Given that tuna used in the production of *Muxama* is fished elsewhere and arrives frozen to the plants (Barat and Grau 2009), the use of "temperature- and humidity-controlled room/chamber" allows year-round production of *Muxama*.

On the other hand, *Estupeta*—a narrow piece of white, lipid- and fiber-rich muscle closely located to the dorsal loins described above—is cut from the loins and brined for at least 30 days (Lã and Vicente 1993) in a 10–25 % NaCl solution. Its commercialization is done in light-brine solution jars.

Fig. 20.1 Sagittal section of the body of a tuna. *Black rounded rectangles* represent (sections of) the muscle (loins) used in the production of *Muxama*

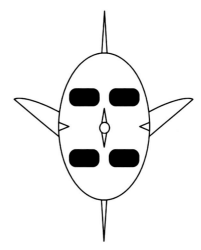

20.3   Methods of Consumption of *Muxama* and *Estupeta*

Although the productions of *Muxama* and *Estupeta* have been adapted to use new technologies, their consumption is still made as it was centuries ago. *Muxama* is cut into thin slices pretty much like ham and served as an appetizer. In contrast, because *Estupeta* is commercialized in brine solution, it has to be thoroughly washed under tap water until it becomes softer and free of salt residues. After "unweaving" the long and thin pieces of *Estupeta*, these can be added to tomatoes, onions, and sweet peppers to prepare a cold salad, seasoned with olive oil and vinegar.

As important as understanding how *Muxama* and *Estupeta* are processed and consumed is to ensure that they will be produced in the future. Traditional products tend to disappear when producers retire or pass away or when the production and trade is no longer profitable. Social and economic stakeholder and universities should look at regional traditional products as a priority issue, because they are part of our heritage and should be preserved as part of our culture. Moreover, they can play an important part in the sustainable development of populations and regions.

References

Barat JM, Grau R (2009) Thawing and salting studies of dry-cured tuna loins. J Food Eng 91:455–459

Dufour F, Arrizabalaga H, Irigoien X, Santiago J (2010) Climate impacts on albacore and bluefin tunas migrations phenology and spatial distribution. Prog Oceanogr 86(1–2):283–290

FAO (2010) The state of world fisheries and aquaculture. Food and Agriculture Organization of the United Nations, Rome, p 197

Huss HH, Ababouch L, Gram L (2004) Assessment and management of seafood safety and quality. Food and Agriculture Organization of the United Nations, Rome, p 266

INE (2010) Estatísticas da Pesca 2009. Instituto Nacional de Estatística, I.P., Lisboa, p 101

Lã A, Vicente L (1993) O Atum Esquecido. Escola Superior de Tecnologia, Universidade do Algarve, Faro, p 147

Navarro AL (2001) Nuevas tendencías en el consumo y la comercialización de los productos de la pesaca. Distribuición y Consumo, pp 33–53

Randall D, Burggren W, French K (2002) Eckert animal physiology: mechanisms and adaptations, 5th edn. W. H. Freeman, New York, p 736

Santos LFR (1989) A pesca do atum no Algarve, Loulé, p 61

Sikorski ZE (1990) Seafood: resources, nutritional composition and preservation. CRC, Boca Raton, p 256

Chapter 21
Salting and Drying of Cod

Helena Oliveira, Maria Leonor Nunes, Paulo Vaz-Pires, and Rui Costa

21.1 Introduction

Salting and drying of cod has been known since 1453 (Moutinho 1985). There are records of Portuguese salted fish already in the fourteenth century, but it was during the fifteenth century that salting of cod in the North Atlantic, Greenland, and Newfoundland began.

Dried and/or salted product has a long shelf life (months or even years) without any special packaging needs but it requires storage with controlled relative humidity and temperature. However, the objective of salting and drying fresh cod is no longer aimed at producing a shelf-stable product (with low moisture and high salt content) but to promote essential sensory properties (odor, color, texture, and a characteristic taste) that remain throughout desalting and cooking and make this product highly appreciated by consumers (Andrés et al. 2005a). Cod has also long been valued as

H. Oliveira
Division of Aquaculture and Upgrading, Portuguese Institute for the Sea and Atmosphere,
I.P. (IPMA, I.P.), Avenida de Brasília, Lisbon 1449-006, Portugal

ICBAS-UP, Abel Salazar Institute for the Biomedical Sciences, University of Porto,
Rua Jorge Viterbo Ferreira 228, Porto 4050-313, Portugal

CIIMAR/CIMAR, Interdisciplinary Center of Marine and Environmental Research,
University of Porto, Rua dos Bragas 289, Porto 4050-123, Portugal

CERNAS, College of Agriculture of the Polytechnic Institute of Coimbra,
Bencanta, Coimbra 3045-601, Portugal

M.L. Nunes
Division of Aquaculture and Upgrading, Portuguese Institute for the Sea and Atmosphere,
I.P. (IPMA, I.P.), Avenida de Brasília, Lisbon 1449-006, Portugal

CIIMAR/CIMAR, Interdisciplinary Center of Marine and Environmental Research,
University of Porto, Rua dos Bragas 289, Porto 4050-123, Portugal

© Springer Science+Business Media New York 2016
K. Kristbergsson, J. Oliveira (eds.), *Traditional Foods*, Integrating Food
Science and Engineering Knowledge Into the Food Chain 10,
DOI 10.1007/978-1-4899-7648-2_21

dried and/or salted product due to its nutritional value (high protein and low fat content) (Martínez-Alvarez et al. 2005a; Heredia et al. 2007). Before consumption, the product is subject to desalting, which involves soaking the fish in water, resulting in water uptake and salt leaching out of the muscle tissue. After desalting, the fish may be either consumed immediately, stored chilled for some days, or stored frozen and then used for the preparation of various cod dishes (Lauritzsen et al. 2004).

Dried and salted cod is highly appreciated in Latin countries (Martínez-Alvarez et al. 2005a; Muñoz-Guerrero et al. 2010) under the name bacalhau (in Portugal) or bacalao (in Spain) (Bjørkevoll et al. 2003). The particular texture and flavor make it very appreciated and, certainly, if a food could represent a country, Portugal would be historically and culturally typified by its hundreds of salted cod recipes from all over the country (Rodrigues et al. 2003). Even with the development of other preservation techniques (like freezing) and correspondent facilities, the gastronomic culture of Portugal maintains the need for a supply of dried salted cod (Brás and Costa 2010).

Portugal ranks as number one in the world for seafood consumption, together with Greenland, with values above 60 kg lwe *per capita*/year (lwe = live weight equivalent) (FAO 2012). About 40 % of this consumption refers to cod (*Gadus morhua*) in dried salted form. Portugal is the main consumer country of dried salted fish and has an important salting and drying industry of cod (Dias et al. 2003). Norway is the largest worldwide cod producer and Portugal by far the largest market for Norwegian dried salted cod (60 % of total export) (Dias et al. 2003; Nystrand 2012). The annual consumption of salt-cured products based on cod from the North Atlantic fisheries can be estimated to be more than 150,000 metric tons (Norwegian Export Council 2009), and another approximately 130,000 metric tons of salted cod are exported from Iceland (Statistics Iceland 2014).

Many factors, including the quality and condition of the raw material and the concentration, type, and quality of salt as well as the salting method used, are believed to affect the characteristics and quality of the final product (Thorarinsdóttir et al. 2004). Fresh raw materials and good manufacturing practices during salting, drying, and desalting are crucial to improve cod quality and shelf life (Pedro et al. 2002).

P. Vaz-Pires
ICBAS-UP, Abel Salazar Institute for the Biomedical Sciences, University of Porto, Rua Jorge Viterbo Ferreira 228, Porto 4050-313, Portugal

CIIMAR/CIMAR, Interdisciplinary Center of Marine and Environmental Research, University of Porto, Rua dos Bragas 289, Porto 4050-123, Portugal

R. Costa (✉)
CERNAS, College of Agriculture of the Polytechnic Institute of Coimbra, Bencanta, Coimbra 3045-601, Portugal
e-mail: ruicosta@esac.pt

21.2 Raw Materials

21.2.1 Fish

21.2.1.1 Species

In the production of dried salted cod, different species are used: *Gadus morhua*, *Gadus macrocephalus*, and *Gadus ogac*, being the most used in commercial terms and therefore the most studied. These species belong to the class of Actinopterygii (ray-finned fishes), to the order of Gadiformes, and to the Gadidae family. Gadids are characterized by having three dorsal fins and two anal fins, with the first dorsal behind the head; pelvic fins before pectorals; no spines; teeth present on vomer, usually with barbell; eggs without oil globules; no otophysic connection between swim bladder and auditory capsules; and up to about 2 m maximum length in *Gadus morhua* (Fig. 21.1). Members of the family are found in temperate and circumpolar waters, principally of the Northern Hemisphere and thus in the Arctic, Atlantic, and Pacific Oceans (Cohen et al. 1990).

Atlantic cod (*Gadus morhua*) are protected by strict management systems designed to limit overexploitation (Warm et al. 2000). The capture of cod has decreased gradually between 1970 and 2000, from approximately 3 to 1 million tons per year. In 2011, the global capture production of *Gadus morhua* was 1,049,666 tons. To compensate this decrease, the production of farmed cod was stimulated. A sharp increase has occurred in the production of farmed cod between 2000 and 2003, from 169 to 2565 tons. Such increase continued, and in 2011, the global aquaculture production of *Gadus morhua* was 16,126 tons (FAO 2011). However, the salting and drying industry did not adopt farmed cod due to the considerable sensory differences of the processed product when compared to the traditional product obtained from wild cod.

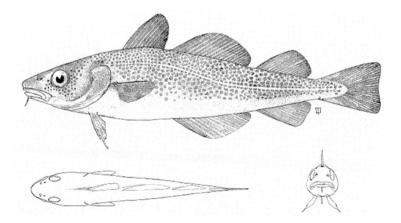

Fig. 21.1 Graphical representation of the species *Gadus morhua* (FishBase 2012)

21.2.1.2 Chemical Composition

Atlantic cod is classified as a lean species, with a lipid content of the body muscu-
lature representing less than 1 % of the muscle wet weight (Table 21.1) (Lambert
and Dutil 1997), and therefore considered negligible, as reported by Barat et al.
(2004c). The muscle is the main protein depot, while lipid reserves are primarily
stored in the liver ranging between 2 and 75 % of its weight (Lambert and Dutil
1997).

Cod composition is affected by gender, age, season, water temperature, type, and
abundance of available food (Love 1958; Ingolfsdóttir et al. 1998; Waterman 2001).

To process cod, the minimum acceptable quality parameters generally followed
by Portuguese processors are presented in Table 21.2.

Table 21.1 Nutritional data of cod (*Gadus morhua*)[a]

Nutritional data (/100 g)	
Energy value (kcal/kJ)	84.7/354.5; 82/343
Edible part (%)	79.2
Total fat (g)	0.4; 0.7
Saturated fat (g)	0.1
ω3 (g)	0.20; 0.21
ω6 (g)	0.005; 0.02
Cholesterol (mg)	43; 52
Protein (g)	17.8; 19
Vitamin A (µg)	3.8; 40 IU
Vitamin E (mg)	0.3; 0.6 (alpha-tocopherol)
Potassium (mg)	362; 413
Phosphorus (mg)	116; 203

[a]Reconstructed from Bandarra et al. (2004) and Nutrition data (2012)

Table 21.2 Quality parameters for raw cod processing

Chemical parameter	
Protein (g/100 g)	16–18
Total volatile basic nitrogen (TVB-N) (mg/100 g)	<25
Free amino acid nitrogen (FAA-N) (mg/100 g)	<85
Trimethylamine nitrogen (TMA-N) (mg/100 g)	<2.5
Moisture (g/100 g)	80–85

21.2.1.3 Freshness

Freshness is a controversial concept (Bremner and Sakaguchi 2000) but is one of the most important parameters of fish quality in most markets of fresh and lightly preserved fish. Therefore, the role of this attribute in cod processing has been widely studied. The aquaculture industry of farmed cod makes pre-rigor fish (which means highly fresh) more available for processing.

With respect to transport kinetics and weight yield, the results observed by Barat et al. (2006) revealed that the main influence of cod freshness occurs throughout the salting process, as the freshest raw material gives the lowest overall yield and the lowest salt uptake. Taking into account both the sensory quality and microbial status evaluations after desalting, the freshest raw material tended to give a firmer fish with less flakiness, while the oldest raw material gave higher flakiness (Barat et al. 2006).

Frozen fish may also be used for salting but only after being thoroughly thawed and inspected for suitability. Products exposed to repeated freezing and thawing cycles show protein denaturation and, consequently, evidence of considerable quality loss (Mackie 1993; Hurling and McArthur 1996).

21.2.1.4 Parasites

Some authors stated that the parasites only threaten human health if they are alive in the product (Sastre et al. 2000; Rodrigues 2006). Nevertheless, other studies have reported that even inactivated parasites can cause allergic reactions in some consumers (Fernández de Corres et al. 1996).

Parasite larvae in fish are located mainly in the intestines. Rodrigues (2006) mentioned that if fish is kept chilled with viscera after catching, rupture of viscera can happen and then parasites can migrate to the muscle and infect it. Therefore, the immediate removal of the viscera (gutting) after catching is essential to reduce the parasites in cod muscle (Rodrigues 2006). After that, a brief visual inspection of parasites on a light table (candling procedures) has been used by the fish processing industry (Levsen et al. 2005). However, this method is not totally effective (FDA 1999) because larvae are sometimes whitish making it difficult to differentiate from the cod muscle or they are located so deep in the tissue that it is not possible to detect (Rodrigues 2006).

Although salting may reduce the parasite hazard in fish, this process does not eliminate it nor minimize it to an acceptable level (FDA 2011). Rodrigues (2006) showed that the total number of parasites in salted cod was generally lower than in fresh cod due to manipulation and removal by drag with salt water of the parasites that are located superficially.

Freezing of fish at −20 °C or lower for 7 days or at −35 °C or lower for 15 h destroys 99 % of the larvae (Huss 1994; FDA 1999, 2011), and it has been suggested that surviving larvae may be so damaged that they do not represent a danger (Rodrigues 2006). Wootten and Cann (2001) reported that freezing fish at −20 °C for 60 h kills all worms. Nevertheless, the processing of salted cod very often does not imply a prior freezing and these products are considered unsafe (Rodrigues 2006). For this reason, a short period of freezing either of the raw material or the final product must be included in the processing as a means of parasite control (Huss 1994).

No matter how carefully fish is inspected by processors, caterers, and retailers, some worms will occasionally be observed in fish by the consumer. Furthermore, it should be emphasized that the presence of parasites in fish offered for sale does not imply carelessness or bad practice by the processor or retailer (Wootten and Cann 2001).

21.2.2   Salt

Generally, marine salt is used in salted fish production with sodium (Na) as the predominant component. NaCl is essential to produce the desired texture and flavor and guarantee the safety of the product (Ismail and Wooton 1992). Nevertheless, salt produced from marine sources can contain halophilic bacteria and mold, which survive in the salt and in the dry-salted fish and can contribute to spoilage (Codex Alimentarius 2003).

Other salt components affect the quality of salted cod. For example, salted cod produced by dry salting with salt containing magnesium chloride ($MgCl_2$) and/or calcium chloride ($CaCl_2$) showed a whiter appearance and was more opaque and firm than fish just salted with pure NaCl (Penso 1953; Rodrigues et al. 2005). Depending on the content of such salts, the drying process can take much more time. Calcium salts promote firmness of salted fish up to a concentration of 0.3 %, while higher concentrations make the fish excessively compact and firm (Beatty and Fougère 1957; van Klaveren and Legendre 1965). Additionally, the effect of salt concentration is dependent, in turn, on the pH of the medium (Martínez-Alvarez et al. 2005b).

Salt granulometry is also an important issue. The use of very fine salt granules could result in cluster formation, which is not favorable for ensuring the uniform distribution of salt on the fish. On the other hand, the use of very coarse salt granules can cause damage on the fish flesh throughout the salting process and may reduce the maturation rate (Codex Alimentarius 2003). To obtain a more efficient salting, a mixture of salts with different granulometry is often used. The smaller crystals dissolve more easily and the coarse crystals allow a better flow of liquids. Thus, besides dissolving and penetrating inside the fish tissue, salt will absorb water from the same tissue, allowing an effective conservation. Hence, the salt will induce both salting and drying of the fish simultaneously. A mixture of 1/3 of fine salt granules and 2/3 of coarse salt granules has been considered adequate (Rodrigues 2006).

The salt used for salting cod should possess a minimum sodium chloride content of 95 %, white color, no odor, and no impurities, and the particle size of 90 % of the total grains should pass through a sieve with a mesh of 8 mm (Codex Alimentarius 2009).

21.3 Processing

Traditional Portuguese cod processing comprises usually two main operations: salting and drying (Martínez-Alvarez et al. 2005b; Martínez-Alvarez and Gómez-Guillén 2005). Salting is a preservation method based on lowering the water activity (a_w) that inhibits bacterial growth and enzymatic spoilage. The a_w value of heavily salted cod generally ranges from 0.7 to 0.75 (Lupín et al. 1981; Gomez and Fernandez-Salguero 1993). Drying reduces the water content without dramatic effect on water activity. Between salting and drying, there is a curing stage.

Traditional curing methods and curing time for cod vary with regions and producer and/or market requirements. However, cod acquires its characteristic appearance, stockfish-like flavor, and firm consistency during this processing phase (Vilhelmsson et al. 1997; Jónsdóttir et al. 2011). In some regions, the process of heavy salting still relies on the main traditional stages reported some decades ago (Beatty and Fougère 1957; van Klaveren and Legendre 1965).

Before the main processing operations, cod is subjected to bleeding and gutting after catching and then is chilled or frozen onboard and stored. If frozen, the fish is defrosted before salting with water at a temperature lower than 18 °C (AIB 2010). At the processing unit, the fish is beheaded, trimmed, filleted, or split. This later step consists of cutting cod along the ventral median line and opening to remove the vertebral column for about 2/3 to 3/4 of its length (Di Luccia et al. 2005; AIB 2010). Immediately after splitting, fish is washed in running clean water or clean seawater to remove blood, and, according to the intended presentation, the "black membrane" (peritoneum) might be removed from the belly walls before salting (Codex Alimentarius 2003).

21.3.1 Salting

Salting is a process which basically involves the transfer of salt and water, between cod and its surroundings. The fish muscle takes up salt and loses water (Andrés et al. 2002). The water content of cod muscle is generally reduced from approximately 82 % to about 54 % during the traditional salt-curing process. Therefore, heavy salt-matured products have a water content of approximately 55 % (w/w) and a salt content of about 20 % (w/w) (Bjørkevoll et al. 2003).

Cod salting may be carried out by dry salting, wet salting, brine salting, brine injection, or a combination of these techniques (Codex Alimentarius 2003).

Traditional Portuguese salting of cod is done by dry salting (or kenching). In this process, cod is placed with the skin side down in stacks with dry salt crystals interspersed between the layers until fully cured (pile salting) (Barat et al. 2003) with 0.33 kg of salt per kg of fish (AIB 2010). The resultant brine, formed by salt in the water leached out from the fish tissue, is allowed to drain away continuously (Rodrigues et al. 2005). This stage takes usually 2–8 weeks, depending on the fish thickness and the degree of desired curing in the resulting products (Burgess et al. 1987; Rodrigues et al. 2005). Additionally, the use of overpressure on the fillets during dry salting can improve salting kinetics, and this is what occurs in stack salting due to the pressure exerted by the weight of the piled fillets and the salt crystals (Andrés et al. 2005a).

Wet salting (or pickling) is the process whereby cod is mixed with salt, layered alternately, and stored in watertight containers under the resulting brine, designated as "pickle" (Codex Alimentarius 2003), which is formed within 1 day (Lauritzsen et al. 2004). Pickle salting takes a week (5–7 days), and then cod is generally salt matured by kenching in stacks (dry salting) for at least 10 days, producing salt-cured cod (Bjørkevoll et al. 2003; Lauritzsen et al. 2004). The fish can be removed from the container and stacked so that pickle drains away (Codex Alimentarius 2003).

Brine salting (or brining) is carried out by immersing fish into a brine solution prepared with coarse salt and water (Codex Alimentarius 2003). The diffusion of salt into the muscle depends on several factors, such as the concentration and composition of the brine, ratio of brine to product, shape and thickness of the product, and duration of brining (Oliveira et al. 2012).

The salting method has an influence on the structural and mechanical properties of the fish muscle (Andrés et al. 2002; Barat et al. 2003; Thorarinsdóttir et al. 2004). The presence of high salt concentrations in muscle gradually increases the water-holding capacity (WHC), obtaining a maximum at an ionic strength of 1 M (about 5.8 % salt) (Offer and Knight 1988), but at higher ionic strengths, WHC decreases, apparently by a salting-out effect due to water binding by the salt and concurrent dehydration of the protein (Martínez-Alvarez et al. 2005b). Therefore, salting promoted changes in the muscle proteins (including some precipitation) resulting in changes in weight and WHC (Thorarinsdóttir et al. 2002, 2004; Martínez-Alvarez and Gómez-Guillén 2006) and in texture, such as increase in hardness (Barat et al. 2003; Gallart-Jornet et al. 2007). It is well known that high salt concentration, close to saturation (25 % w/w) in brining and dry salting, denatures the proteins and reduces their WHC (Gallart-Jornet et al. 2007), while for more diluted brines, protein denaturation is lower and an increase in the WHC due to salt uptake is observed (Thorarinsdóttir et al. 2002). During heavy salting of cod, the activity of cathepsin B/L remains unchanged, whereas the activity of acidic proteases declines as the salt concentration increases in the muscle (Stoknes et al. 2005).

In spite of the large influx of salt into the cod muscle during the curing process, weight is lost mainly due to extensive dehydration and protein leaching from the muscle (Barat et al. 2003; Lauritzsen et al. 2004). Beyond soluble muscle proteins, vitamins and free amino acids can also be lost throughout the process (Larsen and Elvevoll 2008). These losses should be reduced as much as possible (FAO 1981). Moreover, Lauritzsen et al. (2004) reported that most of the weight loss occurs in the kench-curing steps and the fillets placed in the lower parts of the stacks will be exposed to a heavier weight from the upper layers. As a result, the protein and liquid losses from the fillets may be increased even further. The same authors also mentioned that the lost protein ends up in the brine or in the dry salt used and represents both a waste of valuable fish protein and a possible environmental problem (Lauritzsen et al. 2004).

After dry salting, the salted fish is piled again but with inverse order and stored between 8 and 12 °C and with 80–85 % of relative humidity for at least 1 month where it loses salted solution (AIB 2010). After this step, the salted cod is called green salted cod.

Before drying, there is another step of dry salting but with less salt added and another inversion of order in the piles. This step must take at least another month at a temperature equal or lower than 4 °C and a relative humidity between 80 and 85 % (AIB 2010). In this period, curing continues with development of flavor and color changes. After this time, the cod is washed again to remove salt grains and then is dried.

21.3.2 Drying

Drying intends to reduce the moisture to values lower than 47 %, and, when it happens, dried salt-cured cod (klipfish) is obtained (Lauritzsen et al. 2004; AIB 2010). Recommended conditions for drying are drying discontinuously in forced air at temperatures between 18 and 21 °C and relative humidity of 45–80 % for 2–4 days, depending on the fish thickness (AIB 2010).

Some studies indicate other drying conditions. Linton and Wood (1945) proposed optimal conditions of salted cod drying: an air velocity of 1–1.25 m/s, a relative humidity of 45–55 %, and temperature at 26 °C (Jason 1965). Del Valle and Nickerson (1968) cite other authors with similar values. For lightly salted cod, the following conditions were proposed: an air velocity of 1.5–2.0 m/s, a relative humidity of 50–55 %, and a temperature of 30 °C (Linton and Wood 1945). Modern industrial practices vary; in Portugal, temperatures are kept below 25 °C, with air velocity and humidity held constant during the drying process. However, studies performed in Norway demonstrated varying conditions throughout the drying process, with higher temperature (20–26 °C) and air velocity (1.9–2.4 m/s) and lower relative humidity (50–35 %) (Barat et al. 2006).

The end point of drying in the Portuguese industry is usually determined by the moment at which the dried salted cod stays in a horizontal position when held with one hand on the loin (Brás and Costa 2010).

The changes occurring in drying are all related to water transfer. Water present on the salted cod surface evaporates and is transported by convection into the dried air. The water content decrease at the surface establishes a concentration difference with the inner zone, which constitutes the driving force for water migration from the inner zones to the outer surface (Oliveira et al. 2012). There is also a formation of an impermeable layer of salt and protein at the surface of the fish (Barat et al. 2004b). Del Valle and Nickerson (1968) suggested that the formation of such layer is the result of salt solution syneresis in the salted fish. They suggested that the muscle shrinkage was the cause of syneresis and the shrinkage is due to water loss and protein denaturation because of high salt content. On the other hand, Barat et al. (2004b) attributed the phenomenon to capillary action.

As a result of salted cod drying, there is also a slight salt loss. Barat et al. (2006) observed losses of 2–4 % salt from the surface resulting from the separation of sodium chloride crystals due to handling of cod.

In addition to the chemical changes previously discussed, other quality changes (color, texture, and microbiological changes) have been identified during drying. This process causes a yellowing of the salted cod that correlates well with the water content (Brás and Costa 2010). Other causes have been suggested for color change in salted fish and can be the reasons of this change during drying: protein denaturation due to low pH (Lauritzsen et al. 2004; Stien et al. 2005), oxidation of phospholipids (Anon 1967; Stien et al. 2005), and reactions resulting from the presence of iron or copper ions (Anon 1967).

Cod hardens with drying due to protein denaturation and reduction of protein hydration (Brás and Costa 2010). Nevertheless, the degree of hardness after drying depends on the salting method used. Brás and Costa (2010) obtained a lower firmness when cod was salted in brines with lower salt content, obtaining an inversely proportional relationship between firmness and water content beneath the surface of the dried fish. The dried fish firmness can also be attributed to a lower pH of the muscle or the presence of calcium or magnesium ions in the salt (Lauritzsen et al. 2004; Brás and Costa 2010).

The high salt content and low water content of dried salted cod result in water activities lower than 0.75, a value not favorable for the growth of halophilic bacteria. Beyond the development of these bacteria, other bacteria are halotolerant, particularly Gram-positive cocci which can survive in dried and salted products, and when facing optimum conditions, like in desalting, may grow and spoil the product (Rodrigues et al. 2003).

The a_w value decreases slightly during drying from 0.73 to 0.75 (unpublished values from our laboratory) to 0.70 (Rodrigues et al. 2003) when compared to the moisture value which decreases from close to 60 % (m/m; total basis) to less than 50 % (m/m; total basis). The relationship of water activity to water content of salted cod throughout drying and storage is given by isohalic isotherms (Doe et al. 1982).

The split fish thickness is a limiting parameter in the mass transfer rate when the transfer depends on the internal mechanisms like during drying of salted cod. It has been reported that a greater mass of fish takes longer to dry because it has a greater thickness (Del Valle and Nickerson 1968). Additionally, Brás and Costa (2010) measured a linear increase in drying time with the fish weight, up to 20 h per extra kg of fish above fishes of 0.5 kg.

The dried salted cod products are light and do not take too much space and are therefore suitable for transport and storage (Di Luccia et al. 2005). These products are sold mainly at retail level as unpacked split fish, without storage and/or desalting recommendations, and are generally consumed cooked after desalting, and in some cases raw, after being desalted or not (Rodrigues et al. 2003; Pedro et al. 2004).

AIB (2010) defined the quality parameters of dried and salted cod as presented in the Table 21.3, published in the Portuguese law DRE (2011).

Table 21.3 Quality parameters for dried and salted cod defined by Portuguese Association of Cod Processors (DRE 2011)

Parameter	
Protein (g/100 g)	≥26
Total volatile basic nitrogen (TVB-N) (mg/100 g)	≤35
Free amino acid nitrogen (FAA-N) (mg/100 g)	95–120
Trimethylamine nitrogen (TMA-N) (mg/100 g)	≤3.0
Moisture (%)	≤47 %
Sodium chloride (% g/100 g)	≥20.0
Mesophilic aerobic counts (CFU/g)	$<10^3$
Total coliform counts (CFU/g)	<10
Sulfite-reducing *Clostridia* counts (CFU/g)	<10
Staphylococcus aureus counts (CFU/g)	$<10^2$
Listeria monocytogenes	Negative

21.3.3 Storage

Storage conditions for dried and salted cod should prevent the uptake or loss of water from/to the surrounding environment and the growth of halophiles. To accomplish this and since packaging is limited to card boxes, the relative humidity (RH) in the storage space should be close to 70 % (Costa 2010) and the temperature would be preferably lower than 7 °C (Portuguese legislation 2005) but it can stand months at higher temperatures without spoilage. Therefore, the accurate control of humidity and temperature is crucial with regard to the growth of halophilic bacteria and the stability of quality parameters, since some microorganisms and enzymes are still active despite the high salt content and chilled storage conditions (Rodrigues et al. 2003; Pedro et al. 2004). When the environmental RH is higher than muscle a_w in fully salted muscle (0.75), the salted products absorb water and consequently gain weight, whereas the opposite has been observed when environmental RH is lower than the muscle a_w of the salted cod (Doe et al. 1982). Other methods using modified atmosphere (Aas et al. 2010) and super-chilling or freezing temperatures (Nguyen et al. 2011) have been studied during storage but generally without success.

Extremely halophilic archaea that caused reddening ("pink") at high storage temperatures and extremely xerophilic molds that caused brown spots ("dun") have been considered the only microbes growing in salt-cured and dried salt-cured cod (Bjørkevoll et al. 2003). However, Vilhelmsson et al. (1997) showed that the halotolerant bacteria found in high numbers in dried salted cod belong to *Staphylococcus arlettae/xylosus*. Doe and Heruwati (1988) also detected *Staphylococcus xylosus* in spoiled dried salted cod. Moreover, some of the industrial abuses may lead to a condition named sliming, characterized by the appearance of a semi-greasy, sticky, glistening layer of yellow-gray or beige color and a sour pungent odor not caused by extreme halophiles (van Klaveren and Legendre 1965).

Dried salted cod is considered a quite stable product. However, several problems (e.g., less interesting sensory characteristics, etc.) can arise due to deficient or improper preparation, packing, salting, or drying. The main preparation defects observed in dried salted cod are slits, clots, patches of liver and blood, and treacle. The presence of parasites or parasitic infections detectable by the naked eye and foreign bodies are considered preparation and presentation defects (Portuguese legislation 2005). The conservation defects reported in dried salted cod are the presence of reddening (pink), dun or brown spots (Huss and Valdimarsson 1990), unpleasant odor, patches of abnormal color, and sour cod (cod looks cooked on the ventral surface) (Portuguese legislation 2005).

21.3.4 Desalting

Desalting is a process usually done at home. The fish is cut into serving-size pieces, and then the pieces are desalted in tap water for at least 24 h at room temperature or under refrigeration with several changes of water. The process generally takes about 2 days although it depends on the thickness of the fish pieces.

Today the consumer trend is moving toward ready-to-cook products which have increased the demand of producing desalted frozen cod. The desalting procedure in industrial operations is similar to that performed by consumers at home (Barat et al. 2004c). However, commercial rehydration, freezing, and distribution often lead to a longer storage time between completed rehydration and consumption than traditional rehydration at home (Lorentzen et al. 2010). Before commercial desalting, salted cod is generally cut into pieces of different size and shape that should be commercially acceptable. It is common to find cod pieces in the market in Portugal corresponding to loin, fins, tail, and muscle with important mass differences depending on the fish part. This is a crucial practical aspect that has to be taken into consideration by the cod industry because the time needed for cod desalting will depend largely on the muscle zone (Andrés et al. 2005b).

The cod changes found during the desalting operation are well known; the protein/cod matrix is rehydrated resulting in a firmness decrease (Martínez-Alvarez et al. 2005a) of the muscle obtained by salting and, consequently, in an improvement of cod texture (Barat et al. 2004a). The cod matrix rehydration strongly affects the yield (defined as weight of final desalted cod divided by the weight of initial salted cod) and therefore increases the industrial economic benefits (Barat et al. 2004b, c). The NaCl content decreases to concentrations suitable for human consumption (Barat et al. 2004b, c), implying the proteins absorb water and there is an increase in WHC, contributing to the total cod weight increase (Barat et al. 2004a, c). Therefore, samples become more spongy and softer during desalting (Barat et al. 2004c). The cod desalting process is considered as simultaneous leaching and hydrating processes (Barat et al. 2004b). The analysis of the mass transfer phenomena in the desalting process indicates that the main components transferred were water and NaCl (Na$^+$ and Cl$^-$ ions) and, in smaller proportion, protein, resulting in protein-containing residual brine (Barat et al. 2004b; Muñoz-Guerrero et al. 2010).

Temperature control during desalting and storage is very important, among other reasons, due to the rapidity of microbial growth in cod once it is desalted (Pedro et al. 2002). Martínez-Alvarez et al. (2005a) reported that high temperatures increase the speed of desalting, but since the product is highly unstable, temperatures of 0–4 °C are preferable.

According to some experiments (Barat et al. 2004b, c, 2006; Andrés et al. 2005b), a cod-water mass ratio of 1:9 (w/w) is considered adequate to carry out the process but Muñoz-Guerrero et al. (2010) stated that a 1:7 cod-water ratio (w/w) is also suitable.

References

Aas GH, Skjerdal OT, Stoknes I, Bjørkevoll I (2010) Effects of packaging method on salt-cured cod yield and quality during storage. J Aquat Food Prod Technol 19:149–161

AIB (2010) Caderno de Especificações do Bacalhau de Cura Tradicional Portuguesa. Associação dos Industriais do Bacalhau, Gafanha da Nazaré, p 11

Andrés A, Rodríguez-Barona S, Barat JM, Fito P (2002) Note: mass transfer kinetics during cod salting operation. Food Sci Technol Int 8:309–314

Andrés A, Rodríguez-Barona S, Barat JM, Fito P (2005a) Salted cod manufacturing: influence of salting procedure on process yield and product characteristics. J Food Eng 69:467–471

Andrés A, Rodríguez-Barona S, Barat JM (2005b) Analysis of some cod-desalting process variables. J Food Eng 70:67–72

Anon (1967) Causas do amarelecimento do bacalhau salgado. Conservas de Peixe 257:17–20

Bandarra NM, Calhau MA, Oliveira L, Ramos M, Dias MG, Bártolo H, Faria MR, Fonseca MC, Gonçalves A, Batista I, Nunes ML (2004) Composição e valor nutricional dos produtos da pesca mais consumidos em Portugal, vol 11. Publicações Avulsas do IPIMAR/INIAP (Instituto Nacional de Investigação Agrária e das Pescas), Lisboa, p 103

Barat JM, Rodríguez-Barona S, Andrés A, Fito P (2003) Cod salting manufacturing analysis. Food Res Int 36:447–453

Barat JM, Rodríguez-Barona S, Andrés A, Ibáñez JB (2004a) Modeling of the cod desalting operation. J Food Sci 69:183–189

Barat JM, Rodríguez-Barona S, Andrés A, Visquert M (2004b) Mass transfer analysis during the cod desalting process. Food Res Int 37:203–208

Barat JM, Rodríguez-Barona S, Castelló M, Andrés A, Fito P (2004c) Cod desalting process as affected by water management. J Food Eng 61:353–357

Barat JM, Gallart-Jornet L, Andrés A, Akse L, Carlehög M, Skjerdal OT (2006) Influence of cod freshness on the salting, drying and desalting stages. J Food Eng 73:9–19

Beatty SA, Fougère H (1957) The processing of dried salted fish. Bulletin No. 112. Fisheries Research Board of Canada, Ottawa

Bjørkevoll I, Olsen RL, Skjerdal OT (2003) Origin and spoilage potential of the microbiota dominating genus *Psychrobacter* in sterile rehydrated salt-cured and dried salt-cured cod (*Gadus morhua*). Int J Food Microbiol 84:175–187

Brás A, Costa R (2010) Influence of brine-salting prior to pickle salting in the manufacturing of various salted-dried fish species. J Food Eng 100:490–495

Bremner HA, Sakaguchi M (2000) A critical look at whether 'freshness' can be determined. J Aquat Food Prod Technol 9(3):5–25

Burgess GHD, Cutting CL, Lovern JA, Waterman JJ (1987) Curado con Sal. In: Burgess CL, Cutting JA, Waterman JJ (eds) El pescado y las industrias derivadas de la pesca. Acribia SA, Zaragoza, pp 105–122

Codex Alimentarius (2003) Code of practice for fish and fishery products. CAC/RCP 52. FAO/WHO, Rome, p 134

Codex Alimentarius (2009) Code of practice for fish and fishery products, 1st edn. World Health Organization/Food and Agriculture Organization, Rome

Cohen DM, Inada T, Iwamoto T, Scialabba N (1990) FAO species catalogue. Gadiform fishes of the world (Order Gadiformes). An annotated and illustrated catalogue of cods, hakes, grenadiers and other gadiform fishes known to date. FAO Fisheries Synopsis No. 125, vol 10. FAO, Rome, p 442

Costa R (2010) Secagem do bacalhau. Atas do 1° Encontro Português de Secagem de Alimentos, Instituto Politécnico de Viseu 22 de Outubro, Viseu

Del Valle FR, Nickerson JTR (1968) Studies on salting and drying fish. 3. Diffusion of water. J Food Sci 33:499–503

Dias JF, Menezes R, Filipe JC, Guia F, Guerreiro V, Asche F (2003) Short and long run dynamic behaviour of the cod value for Norway-Portugal. In: Kalaydjian K (ed) Proceedings of the fifteenth annual conference of the European Association of Fisheries Economists (EAFE), Ifremer, Brest, France, 15–16 May

Di Luccia A, Alviti G, Lamacchia C, Faccia M, Gambacorta G, Liuzzi V, Spagna Musso S (2005) Effect of the hydration process on water-soluble proteins of preserved cod products. Food Chem 93:385–393

Doe PE, Heruwati ES (1988) A model for the prediction of the microbial spoilage of sun-dried tropical fish. J Food Eng 8:47–72

Doe PE, Hashmi R, Poulter RG, Olley J (1982) Isohalic sorption isotherms. I. Determination for dried salted cod (Gadus morhua). J Food Technol 17:125–134

DRE (2011) DIARIO DA REPUBLICA—2.ª SERIE, N° 68, de 06.04.2011, Pág. 15895. Despacho 6006/2011 de 6 de Abril (Portuguese law)

FAO (1981) The prevention of losses in cured fish. FAO fisheries technical paper, 219. pp 70–74

FAO (2011) Species fact sheets. Gadus morhua (Linnaeus, 1758). FAO Fisheries and Aquaculture Department. http://www.fao.org/fishery/species/2218/en. Accessed 25 Mar 2013

FAO (2012) The state of world fisheries and aquaculture (SOFIA). FAO Fisheries and Aquaculture Department, Food and Agriculture Organization of the United Nations, Rome, p 209. http://www.fao.org/docrep/016/i2727e/i2727e.pdf. Accessed 26 Aug 2013

FDA (1999) Anisakis simplex and related worms. In: Bad Bug Book. Introduction to foodborne pathogenic microorganisms and natural toxins handbook. Food and Drug Administration, Silver Spring. http://www.fda.gov/Food/FoodSafety/FoodborneIllness/FoodborneIllnessFoodbornePathogensNaturalToxins/BadBugBook/ucm070768.htm. Accessed 25 Mar 2013

FDA (2011) Parasites (Chapter 5). In: FDA, US. Food and Drug Administration. Fish and fishery products hazards and controls guidance, 4th edn. Government Printing Office GPO Access, Washington, pp 91–98

Fernández de Corres L, Audicana M, Del Pozo MD, Muñoz D, Fernández E, Navarro JA, García M, Diez J (1996) Anisakis simplex induces not only anisakiasis: report on 28 cases of allergy caused by this nematode. J Investig Allergol Clin Immunol 6:315–319

FishBase (2012) A global information system on fishes. http://www.fishbase.org/photos/thumbnailssummary.php?ID=69. Accessed 30 Jan 2013

Gallart-Jornet L, Barat JM, Rustad T, Erikson U, Escriche I, Fito P (2007) A comparative study of brine-salting of Atlantic cod (Gadus morhua) and Atlantic salmon (Salmo salar). J Food Eng 79:261–270

Gomez R, Fernandez-Salguero J (1993) Note: water activity of Spanish intermediate-moisture fish products. Revista Espanola de Ciencia y Technologia de Allimentos 33:651–656

Heredia A, Andrés A, Betoret N, Fito P (2007) Application of the SAFES (systematic approach of food engineering systems) methodology to salting, drying and desalting of cod. J Food Eng 83:267–276

Hurling R, McArthur H (1996) Thawing, refreezing and frozen storage effects on muscle functionality and sensory attributes of frozen cod (*Gadus morhua*). J Food Sci 61:1289–1296

Huss HH (1994) Assurance of seafood quality. FAO fisheries technical paper no. 334. FAO, Rome, p 169

Huss HH, Valdimarsson G (1990) Microbiology of salted fish. FAO Fish Tech News 10:1–2

Ingolfsdóttir S, Stefänsson G, Kristbergsson K (1998) Seasonal variations in physicochemical and textural properties of north Atlantic cod (*Gadus morhua*) mince. J Aquat Food Prod Technol 7:39–61

Ismail N, Wooton M (1992) Fish salting and drying: a review. ASEAN Food J 7:175–183

Jason AC (1965) Drying and dehydration. In: Borgström G (ed) Fish as food, vol III. Academic, New York, pp 1–54

Jónsdóttir R, Sveinsdóttir K, Magnússon H, Arason S, Lauritzsen K, Thorarinsdóttir KA (2011) Flavor and quality characteristics of salted and desalted cod (*Gadus morhua*) produced by different salting methods. J Agric Food Chem 59:3893–3904

Lambert Y, Dutil J-D (1997) Can simple condition indices be used to monitor and quantify seasonal changes in the energy reserves of Atlantic cod (*Gadus morhua*)? Can J Fish Aquat Sci 54:104–112

Larsen R, Elvevoll EO (2008) Water uptake, drip losses and retention of free amino acids and minerals in cod (*Gadus morhua*) fillet immersed in NaCl or KCl. Food Chem 107:369–376

Lauritzsen K, Akse L, Gundersen B, Olsen RL (2004) Effects of calcium, magnesium and pH during salt-curing of cod (*Gadus morhua* L.). J Sci Food Agric 84:683–692

Levsen A, Lunestad BT, Berland B (2005) Low detection efficiency of candling as a commonly recommended inspection method for nematode larvae in the flesh of pelagic fish. J Food Prot 68:828–832

Linton EP, Wood AL (1945) Drying of heavily-salted fish. J Fish Res Board Can 6d:380–391

Lorentzen G, Olsen RL, Bjørkvoll I, Mikkelsen H, Skjerdal T (2010) Survival of *Listeria innocua* and *Listeria monocytogenes* in muscle of cod (*Gadus morhua* L.) during salt-curing and growth during chilled storage of rehydrated product. Food Control 21:292–297

Love RM (1958) Studies on the North Sea cod. III.—effects of starvation. J Sci Food Agric 9:617–620

Lupín HM, Boeri RL, Moschiar SM (1981) Water activity and salt content relationship in moist salted fish products. J Food Technol 16:31–38

Mackie IM (1993) The effects of freezing on flesh proteins. Food Rev Int 9:575–610

Martínez-Alvarez O, Gómez-Guillén MC (2005) The effect of brine composition and pH on the yield and nature of water-soluble proteins extractable from brined muscle of cod (*Gadus morhua*). Food Chem 92:71–77

Martínez-Alvarez O, Gómez-Guillén MC (2006) Effect of brine-salting at different pHs on the functional properties of cod muscle proteins after subsequent dry-salting. Food Chem 94:123–129

Martínez-Alvarez O, Borderías J, Gómez-Guillén MC (2005a) Use of hydrogen peroxide and carbonate/bicarbonate buffer for soaking of bacalao (salted cod). Eur Food Res Technol 221:226–231

Martínez-Alvarez O, Borderías AJ, Gómez-Guillén MC (2005b) Sodium replacement in the cod (*Gadus morhua*) muscle salting process. Food Chem 93:125–133

Moutinho M (1985) História da pesca do bacalhau. Editorial Estampa, Lisboa

Muñoz-Guerrero H, Gutiérrez MR, Vidal-Brotons D, Barat JM, Gras ML, Alcaina MI (2010) Environmental management of the residual brine of cod desalting. Quantification of mass transfer phenomena and determination of some parameters on the residual brine important for its treatment by membrane technology. J Food Eng 99:424–429

Nguyen MV, Jonsson A, Thorarinsdóttir KA, Arason S, Thorkelsson G (2011) Effects of different temperatures on storage quality of heavily salted cod (*Gadus morhua*). Int J Food Eng 7(1)

Norwegian Export Council (2009) Statistics Norway. www.ssb.no. Accessed 16 Jan 2013

Nutrition data (2012) Nutrition facts. Fish, cod, Atlantic, raw. Condé Nast websites. http://nutritiondata.self.com/facts/finfish-and-shellfish-products/4041/2. Accessed 30 June 2013

Nystrand BT (2012) Value chain analysis of dried and salted cod from Norway to Portugal. Report No. MA 12-03. Møreforsking Marin, p 24. http://www.moreforsk.no/default.aspx?menu=794&id=975. Accessed 30 Jan 2013

Offer G, Knight P (1988) The structural basis of water holding in meat. In: Lawrie R (ed) Developments in meat science, vol 4. Elsevier, London, pp 63–171

Oliveira H, Pedro S, Nunes ML, Costa R, Vaz-Pires P (2012) Processing of salted cod (*Gadus* spp.): a review. Compr Rev Food Sci Food Saf 11:546–564

Pedro S, Magalhães N, Albuquerque MM, Batista I, Nunes ML, Bernardo MF (2002) Preliminary observations on spoilage potential of flora from desalted cod (*Gadus morhua*). J Aquat Food Prod Technol 11:143–150

Pedro S, Albuquerque MM, Nunes ML, Bernardo MF (2004) Pathogenic bacteria and indicators in salted cod (*Gadus morhua*) and desalted products at low and high temperatures. J Aquat Food Prod Technol 13:39–48

Penso G (1953) Salage. Les produits de la pêche. Vigot Frères, Paris

Portuguese legislation (2005) Decreto-Lei n.° 25/2005 de 28 de Janeiro. Diário da República—I Série-A. N.° 20, p 696–703

Rodrigues MJ (2006) Estudos de qualidade em bacalhau. PhD Thesis, Instituto de Ciências Biomédicas Abel Salazar, Universidade do Porto, p 205

Rodrigues MJ, Ho P, López-Caballero ME, Vaz-Pires P, Nunes ML (2003) Characterization and identification of microflora from soaked cod and respective salted raw materials. Food Microbiol 20:471–481

Rodrigues MJ, Ho P, López-Caballero ME, Bandarra NM, Nunes ML (2005) Chemical, microbiological, and sensory quality of cod products salted in different brines. J Food Sci 70:M1–M6

Sastre J, Lluch-Bernal M, Quirce S, Arrieta I, Lahoz C, Del Amo A, Fernández-Caldas E, Marañón F (2000) A double-blind, placebo-controlled oral challenge study with lyophilized larvae and antigen of the fish parasite, *Anisakis simplex*. Allergy 55:560–564

Statistics Iceland (2014) Available from: http://www.statice.is/ Accessed on Oct 6 2015.

Stien LH, Hirmas E, Bjørnevik M, Karlsen Ø, Nortvedt R, Rørå AMB, Sunde J, Kiessling A (2005) The effects of stress and storage temperature on the color and texture of pre-rigor filleted farmed cod (*Gadus morhua* L.). Aquacult Res 36:1197–1206

Stoknes IS, Walde PM, Synnes M (2005) Proteolytic activity in cod (*Gadus morhua*) muscle during salt-curing. Food Res Int 38:693–699

Thorarinsdóttir KA, Arason S, Geirsdottir M, Bogason SG, Kristbergsson K (2002) Changes in myofibrillar proteins during processing of salted cod (*Gadus morhua*) as determined by electrophoresis and differential scanning calorimetry. Food Chem 77:377–385

Thorarinsdóttir KA, Arason S, Bogason SG, Kristbergsson K (2004) The effects of various salt concentrations during brine curing of cod (*Gadus morhua*). Int J Food Sci Technol 39:79–89

van Klaveren FW, Legendre R (1965) Salted cod. In: Borgström G (ed) Fish as food, vol III. Academic, London, pp 133–163

Vilhelmsson O, Hafsteinsson H, Kristjánsson JK (1997) Extremely halotolerant bacteria characteristic of fully cured and dried cod. Int J Food Microbiol 36:163–170

Warm K, Nelsen J, Hyldig G (2000) Sensory quality criteria for five fish species. J Food Qual 23:583–601

Waterman JJ (2001) The cod. Torry Advisory Note No. 33. Torry Research Station, Ministry of Technology. http://www.fao.org/wairdocs/tan/x5911e/x5911e00.htm. Accessed 30 Jan 2012

Wootten R, Cann DC (2001) Round worms in fish. Torry Advisory Note No. 80. Torry Research Station, Ministry of Agriculture, Fisheries and Food. http://www.fao.org/wairdocs/tan/x5951e/x5951e00.htm. Accessed 29 Nov 2013

Chapter 22
Brazilian Charqui Meats

Massami Shimokomaki, Carlos Eduardo Rocha Garcia, Mayka Reghiany Pedrão, and Fabio Augusto Garcia Coró

Salted meat products in Brazil can be broadly divided in two classes: first, the intermediate moisture meat products (IMMP), with shelf life of months at room temperature (charqui (*charque*) meat itself and its derivative jerked beef are the main representatives of this family), and, second, non-IMMP with shelf life of days presenting a relatively low content of salt and high moisture and high water activity (aw). *Carne-de-sol*, sun meat, is the representative of this kind of meat product and known under various names as *carne-de-sertão*, *carne-do-ceará*, *carne serenada*, *carne-de-viagem*, *carne-mole*, and *carne-do-vento*, all very popular in the northeastern region of Brazil.

Charqui meat (CH) also known as *carne seca*, dry meat, has its name etymologically derived from the Quechuan language of the Andes region from the word *ch'arki* and the salted meat manufactured from llama meat. CH consumption in Brazil has its roots closely related to the country's own history. During colonial times, it was consumed largely in the northeastern region, and it is nowadays consumed all over the country due in particular to migration movements. Because of an

M. Shimokomaki (✉)
Federal University of Technology—Paraná
Campus Londrina, Av. dos Pioneiros, 3131, Londrina, Paraná 86036-370, Brazil

Graduate Program in Animal Science, Department of Veterinary Medicine Preventive, Londrina State University, Rodovia Celso Garcia Cid—Pr 445 Km 380, s/n—Campus Universitário, Londrina, Paraná 86057-970, Brazil
e-mail: mshimo@uel.br

C.E.R. Garcia
Department of Pharmacy, Paraná Federal University,
Av. Pref. Lothário Meissner, 632, Curitiba, Paraná 80210-170, Brazil

M.R. Pedrão • F.A.G. Coró
Federal University of Technology—Paraná
Campus Londrina, Av. dos Pioneiros, 3131, Londrina, Paraná 86036-370, Brazil

© Springer Science+Business Media New York 2016   291
K. Kristbergsson, J. Oliveira (eds.), *Traditional Foods*, Integrating Food
Science and Engineering Knowledge Into the Food Chain 10,
DOI 10.1007/978-1-4899-7648-2_22

extreme dry season in 1780, animals were few in number locally; thus, the production moved down to Pelotas city in Rio Grande do Sul state, transforming economically this region, and CH was its main food product. There is a report stating its importance as exportable product for over 150 years in particular to Cuba, Uruguay, and even the USA (Marques 1992). Pelotas no longer produces so much CH, and the production is concentrated in the southwest regions, in particular in the São Paulo state. It is an ingredient of one of the most popular dishes known as "feijoada" cooked with black beans and other meat products.

The total charqui consumed is approximately 206 thousand ton/year, a consumption of 1.0–1.5 kg/capita equivalent to app. to US$1.000 mi annually (Brazilfocus 2007). According to the Brazilian legislation, CH should contain 40–50 % moisture and 10–20 % salt (Brasil 1962) and 0.75 the final value of its water activity ranking it as an IMMP (Torres et al. 1994). On the other hand, jerked beef (JB) is officially characterized by Brazilian legislation by having maximum moisture of 55 %, sodium nitrite of 50 ppm, salt concentration of 18 %, and final a_w value of 0.78 and should be vacuum packed and technologically is an improvement from CH (Brasil 2000; Shimokomaki et al. 2003).

Despite the high consumption, CH and JB have their production based on traditional technologies applying a heavy salting and drying under the sun associating to the wind therefore reaching the intermediate a_w values being microbiologically safe products. Recently, it was reported that some fermentative halophylic bacteria were present in the processing, in particular *Staphylococcus xylosus* and *carnosus*, and it has been suggested that enhancing those bacteria as starter culture would improve flavor and taste (Pinto et al. 2002).

The charqui meat preparation, essentially, has its processing starting from deboning the carcass, followed by wet salting when the ingredients are incorporated and evenly distributed into the meat accelerating the curing and stabilizing its color. The whole piece is immersed in concentrated brine approximately 20–25°B containing 1–2 % potassium or sodium nitrite for hours allowing the solution to diffuse into the meat. Nowadays, this brine solution is injected into the meat accelerating this first step of salting followed by tumbling the meat in order to guarantee the uniform distribution. Thereafter, meat samples are subsequently submitted to dry salting with rock salt overnight. The meat pieces are stacked into piles separated from each other by a layer of coarse marine salt, approximately 5 mm thick. Daily, throughout four subsequent days, the meat is restacked, and the uppermost meat pieces are repositioned on the bottom of the new piles activity known as *tombo*. After restack and rinsing excess salt from the meat surfaces, samples are hung in a stainless steel rail under the sun and wind for an approximately 8 h period when the temperature reaches 40–45 °C. At night and in rainy days, samples are collected and piled in a concrete floor and covered with tarpaulin, and finally the product is either vacuum packed for jerked beef or no vacuum for charqui meat although in markets today, vacuum-packed CH can also be found (Shimokomaki et al. 2003). The fermentation process may occur at two particular steps, while at the *tombo* phase when the piled meat is kept still, and during the drying period when the samples were collect at night and piled (Fig. **22.1**).

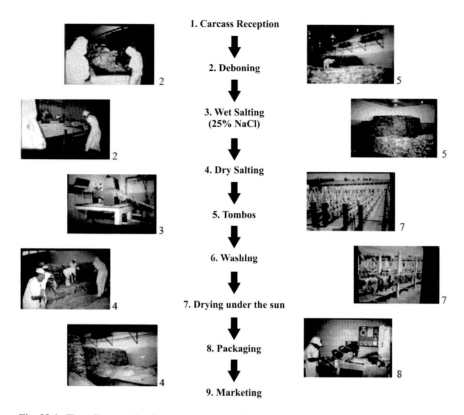

1. Carcass Reception
↓
2. Deboning
↓
3. Wet Salting
(25% NaCl)
↓
4. Dry Salting
↓
5. Tombos
↓
6. Washing
↓
7. Drying under the sun
↓
8. Packaging
↓
9. Marketing

Fig. 22.1 Flow diagram showing charqui or jerked beef processing consisting of *1*, raw material; *2*, deboning; *3*, brine injection; *4* and *5*, dry salting and *tombos*; *6*, washing the meat piece surface; *7*, sun dry in stainless steel rail; *8*, packaging; and *9*, marketing (Shimokomaki et al. 2003)

For centuries, production of either *carne-de-sol* or charqui meats was considered a prehistorical practice. In fact these products are similar to European meat products in particular those of the Iberian Peninsula brought into South America by settlers from the Colombian period onwards. It seems however that CH meats were originally processed in America because these kinds of products are not found in Europe. Although timid yet, there are attempts to modernize this manufacture processes in order to make them feasible to be standardized by applying quality management tools, implementing HACCP, ISOs, etc. This is necessary to make them exportable products. JB is an example of this trend and is an excellent food product derived from the hurdle technology theory (Leistner 2000) to obtain a safe product (Shimokomaki 2006). Salting, curing, drying, water activity, and vacuum packaging are hurdles sequentially applied in order to hinder microbial growth, and the chemical oxidation deterioration is inhibited by the presence of nitrite which acts both as botulism inhibitors and as lipid antioxidant (Shimokomaki et al. 1998). Despite the harsh conditions of processing, its biological value evaluated with experiments in rats resulted in high protein efficiency ratios, high net protein utilizations, and high nitrogen balances, thus showing a high biological value and also high true digestibility, with net protein utilization similar to casein (Garcia et al. 2001).

Lastly, charqui meats are at a turning point today for production from ancestral practices to modern technologies justified by the food safety demanded by the markets and also by financial incentives.

References

Brasil, Ministério da Agricultura e Abastecimento. Departamento Nacional de Inspeção de Produtos de Origem Animal (1962) Regulamento de inspeção industrial e sanitária dos produtos de origem animal. Rio de Janeiro

Brasil, Ministério da Agricultura, Pecuária e Abastecimento. Instrução Normativa n° 22 de 31 de julho de 2000. Ministério da Agricultura, Pecuária e Abastecimento (2000) Regulamento técnico de identidade e qualidade de carne bovina salgada curada dessecada ou jerked beef. Brasília

Brazilfocus (2007) Consumo de charque. http://www.datamark.com.br/. Accessed 15 Sept 2007

Garcia FA, Mizubuti IY, Kanashiro KY, Shimokomaki M (2001) Intermediate moisture meat product: biological evaluation of charqui meat protein quality. Food Chem 75(4):405–409

Leistner L (2000) Shelf stable product and intermediate moisture foods based on meat. In: Beuchat LB, Rockland L (eds) Water activity theory and application to food. Marcel Dekker, New York, pp 295–328

Marques AF (1992) Destino do charque produzido no Rio Grande do Sul. In: A economia do charque (Culinária do charque. O charque nas artes). Martins Livreiro Ed., Porto Alegre, p 188

Pinto MF, Ponsano EHG, Franco BDGM, Shimokomaki M (2002) Charqui meats as fermented meat products: role of bacteria for some sensorial properties development. Meat Sci 61(2):187–191

Shimokomaki M (2006). Charque, jerked beef e carne de sol. In: Shimokomaki M, et al (eds). Varela, São Paulo, pp 47–62

Shimokomaki M, Franco BDGM, Biscontini TM, Pinto MF, Terra NN, Zorn TMT (1998) Charqui meats are hurdle technology meat products. Food Rev Int 14(4):339–349

Shimokomaki M, Youssef E, Terra NN (2003) Curing. In: Caballero B et al (eds) Encyclopedia of food sciences and nutrition, 2nd edn. Elsevier, St. Louis, pp 1702–1707

Torres EAFS, Shimokomaki M, Franco BDGM, Landgraf M (1994) Parameters determining the quality of charqui, an intermediate moisture meat product. Meat Sci 38(2):229–234

Part V
Traditional Beverages

Chapter 23
German Beer

Martin Zarnkow, Christopher McGreger, and Nancy McGreger

23.1 Introduction

Beer has been man's constant companion from the dawn of agriculture and sedentism through the rise of civilization and into the modern age of industry and biotechnology. The word "civilization" refers to the progression over time that communities undergo toward increasing complexity, as small clan-based settlements slowly become cities, where social stratification, urban development, and eventually writing emerge. There is no doubt that beer played a central role in this evolution. The so-called Neolithic Revolution or demographic shift toward agriculture and sedentism was not limited to one area of the world but occurred independently in a number of regions. It did, however, happen first in the ancient Near East in what is known as the Fertile Crescent approximately 11,000 years ago. Though there is currently no direct evidence for brewing there at that time, it seems quite plausible that the hunter-gatherers of the Fertile Crescent began brewing beer before grain was domesticated, which would make beer a very ancient beverage indeed. Prior to the discovery of the brewing process, ancient humanlike creatures, species of Homininae from which humans and their ilk are descended, had already evolved the ability to digest alcohol due to their propensity for eating rotting fruit—and much later honey diluted with water (Hornsey 2012). Because malting and brewing are much more complicated processes, beer would have first appeared on the scene after wine and mead. In order to malt cereal grains, their moisture content has to be increased to the point that the plantlets inside of each kernel are induced to

M. Zarnkow (✉)
Research Center Weihenstephan for Beer and Food Quality, Technische Universität München, Munich, Germany
e-mail: martin.zarnkow@tum.de

C. McGreger • N. McGreger
McGreger Translations and Consulting, Munich, Germany

© Springer Science+Business Media New York 2016
K. Kristbergsson, J. Oliveira (eds.), *Traditional Foods*, Integrating Food
Science and Engineering Knowledge Into the Food Chain 10,
DOI 10.1007/978-1-4899-7648-2_23

commence growth, thereby releasing enzymes which degrade the large molecules found in the endosperm. These kernels are then dried, crushed, and subsequently added to water—known as the mash—to form a suspension, where further enzymatic processes occur to then create an aqueous solution free of the husks, known as wort, from the molecular remnants of the complex compounds, which were broken down during the processes of malting and mashing. Beer is defined as any fermented beverage created using cereal grains, but the majority of modern beer styles are produced using only water, barley malt, hops, and yeast (Zipfel and Rathke 2008). Although beer was produced for millennia in Mesopotamia and Egypt, countries in Western and Central Europe, such as Great Britain, Belgium, Germany, Austria, and the Czech Republic, are now best known as traditional beer-brewing nations.

Biotechnology is said to have originated with brewing. The first beer was most likely produced using einkorn wheat as an accompaniment to ancient hunter-gatherers' seasonal gatherings, of which feasting was an integral part. The beer would have been relatively low in alcohol, somewhat sour, and probably spiced with fruit or herbs. Microbes not only made water safe to drink due to the alcohol they produce, but the probiotic benefits imparted by a mixture of microbes, a topic which up until recently had not been given enough attention by scientists, would have been yet another advantage. These early grain-based beverages were perhaps brewed using bread dough, since the earliest beers for which there is evidence were brewed in such a manner. Therefore, early bread and beer may have simply been the liquid and dry forms of principally the same food. In ancient Sumer, in the land of Inanna (Ishtar), Gilgamesh and Enkidu, and the first cities, by approximately 3500 BC, the profession of brewer was born. Their expertise allowed them to gauge the quality of the raw materials and adjust the malting and brewing processes accordingly. Because beer was a staple in the ancient Near East, professional brewers were essential in Mesopotamia and Egypt. Figure 23.1 shows a photo of drinkers enjoying spontaneously fermented beer in sub-Saharan Africa, which bears a striking resemblance to beer drinking in ancient Mesopotamia (inset).

The fundamental difference between beer and other fermented foods is that it relies on converted starch from cereal grains as the source of sugar to feed the microbes, which ultimately transform it from wort to beer. Enzymes solubilize the starch before the microbes, usually yeast, can consume it during fermentation. Nothing has changed since man first began experimenting with fermenting sugars derived from cereals: as much fermentable extract is brought into solution as possible in the malting and brewing processes. The amount of extract dissolved in water is referred to as "original gravity." The methods developed by early Neolithic farmers allowed them to break the starch down to the point that it would become soluble, which involved grinding grain, making dough, and baking bread. Therefore, it is impossible to say which is older: baking bread or brewing beer. As mentioned above, they may ultimately be two forms of the same food. By heating a mixture of flour and water, the starch undergoes gelatinization. This unravels the tightly packed helical macromolecule, thus greatly facilitating enzymatic degradation of the starch into smaller molecules, one being glucose and molecules made of glucose, such as

Fig. 23.1 Communal drinking of locally brewed beer in Uganda (Photo: Michael Eberhard). *Inset*: Mesopotamian beer drinkers using straws (Berlin, Vorderasiatisches Museum, Inv.-Nr. VA 522; Drawing: D. Hinz)

maltose and maltotriose. Once these smaller molecules are available, the microbes are then able to convert them into carbon dioxide, alcohol, and heat—at least if they are brewers' yeast. Other microbes, such as bacteria, may produce other fermentation by-products, like lactic acid. The type of beer being brewed specifies what types of ingredients are selected and which processes are employed. In the ancient Near East, the lack of combustible materials (wood, animal dung) forced brewers to malt and mash at ambient temperatures; however, given the harsh midday summer sun, "ambient" was actually quite warm (50 °C in the shade). Even the desert heat is not sufficient to reach the optimal temperatures of the enzymes required to break down glucose polymers in the mash. For this reason, the duration of the mashing process was increased to several days (Zarnkow et al. 2006). The polymers of glucose, known as amylose and amylopectin, are degraded by naturally occurring enzymes known as amylases. These enzymes can be found in plants and animals and even in humans. Our saliva contains amylase, which is the reason that some carbohydrates, like bread, taste slightly sweet when we chew them. Some brewing methods actually make use of the fact that human saliva contains amylase. The brewers chew the grain, be it millet, corn, cassava, etc., and spit it back out into a container. The salivary amylase breaks down the starch, which then is fermented into beer by means of spontaneous fermentation. One such beverage is called *chicha* in Latin America.

The farming of domesticated cereal grains spread in all directions away from their points of origin in the Near East into climate zones that were conducive to their cultivation. Brewing did not take hold in Europe, at least not immediately, due to the challenges presented by the cooler and wetter climate. However, once it was discovered that fire could be utilized to produce malt, countries in the grain belt of Europe started brewing beer, and it quickly became a staple, providing much needed nutrition for a growing population. Southern Europeans tended to drink wine, but in a belt from the Black Sea to the British Isles, where grapes cannot be cultivated and cereal grains grow well, beer was the beverage of choice. As little as 150 years ago, untold numbers of small breweries were still in existence in these regions of Europe. The capacity of these breweries was quite small and their equipment, very primitive. The only opportunity brewers had to acquire knowledge was through experience handed down. Therefore, a master brewer learned to reliably brew good beer by gaining skills and knowledge as an apprentice and through many years of practicing his profession. This knowledge was steeped in tradition and was passed down in guilds or other vocational organizations to apprentices who, in turn, would become master brewers. Only after the onset of the Industrial Revolution, when equipment and technology enabled brewers to produce large volumes of beer, did these tiny breweries begin to disappear. England was the first country to industrialize, and pale ale was the first mass-produced beer; lager followed. In fact, most of the styles beer drinkers are accustomed to today arose during this period of industrialization. In a modern brewery, large volumes of beer are brewed on state-of-the-art equipment. Maltsters and brewers have optimized their raw materials and developed finely tuned production methods through the application of scientific principles. Production processes in most breweries, even in smaller ones, are completely automated.

In Bavaria in the year 1516, an edict was issued, declaring that only barley malt, hops, and water were allowed in beer. Though the Code of Hammurabi is older by thousands of years and also regulated the production of beer, the Bavarian law is the oldest consumer protection law still in effect. It is known as the *Reinheitsgebot*. At the time, yeast were unknown as such and were therefore not included in the *Reinheitsgebot* but were mentioned in a decree shortly afterward in 1551. The law, of course, only applied to the commoners of Bavaria. The aristocracy brewed any kind of beer they wished, and it came to light that they rather enjoyed wheat beer, known as Weißbier in German. The Dukes of Bavaria retained this singular privilege for centuries, until Georg Schneider I acquired the rights to brew wheat beer from King Ludwig II of Bavaria. Only since 1919 has the *Reinheitsgebot* been legally binding in all of what is now Germany.

Before the yeast is added and fermentation sets in, on average, the unfermented beer (wort) has an extract content (original gravity) of 11–12 %. Once the fermentable sugars have been consumed by the yeast, the extract content reaches 3.7–4.0 % by weight or approximately one-third of the original gravity. According to EU regulations, the alcohol content must be declared on beer labels in percent by volume. In order to calculate the volume of alcohol in beer, the volumetric percent is calculated using the percent by weight and the density of alcohol. Beer generally contains circa 4.7–5.0 % alcohol by volume. The extract which remains unfermented by the yeast amounts to approximately 4 % and consists of a number of substances, including carbohydrates, proteins, tannins, minerals, and vitamins. An average beer with an original gravity of 12 % contains around 450 kcal per liter.

By varying the original gravity of the wort, one can alter the strength of the finished beer. At one end of the spectrum, there are reduced calorie, low-alcohol beers and also Berliner Weisse with original gravities of 7–8 % and an alcohol content of approximately 3 % by volume, while at the other end, there are beer styles such as doppelbock, barley wine, and some Belgian Trappist ales with original gravities of 18–21 % and up to 10 % alcohol by volume. A few unconventionally produced beers have been known to reach 40 % alcohol by volume, but these are rare. Nonalcoholic beers have been brewed for some time now using a variety of methods, which include distillation, interrupted fermentation, and reverse osmosis (Meussdorffer and Zarnkow 2014).

Modern beer is generally golden to light amber in color, but this is a relatively recent phenomenon. Up until coke was invented, a fuel derived from coal, most malt was darker than it is now. Coke is produced from coal by driving off the impurities and thus its offensive odor, leaving almost pure carbon. Coke was employed for the first time to make malt in seventeenth-century England. This allowed for the malt to be kilned in such a way that it produced a more lightly roasted malt and ultimately led to the creation of what by the early eighteenth century became known as "pale ale." By the early nineteenth century, London breweries were already producing pale ale; however, shortly afterward when brewers in Burton upon Trent began doing so, pale ale greatly increased in popularity. The sulfate-rich water of Burton combined with the lightly kilned malt and a substantial dose of hops produced a beer with a clean, crisp, bitter flavor. As the British Empire expanded, so did the production of pale ale. Stronger, hoppier ale was shipped out to India to nourish and quench the thirst of the British troops stationed there, which resulted in a beer style now known as India pale ale or IPA. Of course, today malt is kilned using indirect heat so that no fuel emissions pass through the malt; however, before this was possible, most beer brewed using kilned malt would have exhibited at least a slightly smoky flavor. Prior to the invention of water treatment, early brewing centers were known for unique styles of beer, in part because the mineral content of the brewing water—or *liquor* as it is known among brewers—was a crucial factor in determining how darkly roasted the malt needed to be: the higher the alkalinity of the water, the darker the roast. Roasting malt was one way of increasing the acidity of the mash and the finished beer. This is why London and Munich are famous for porter and dunkles, respectively, and why Plzeň (Pilsen) is famous for pale lager. The "lager revolution" began in Bohemia in 1842, when a Bavarian brewmaster traveled to Plzeň to counsel the brewers there on methods for producing better tasting beer. Using local ingredients, including the region's extremely soft water, and existing equipment, pilsner was born. Since then, this beer style's popularity has yet to wane.

Pilsner is a "bottom-fermented" beer, while pale ale and most other beers that came before it are "top-fermented." This distinction refers to the type of yeast and the methods employed on the cold side of the brewing process, that is, during fermentation and maturation. Though this is not always the case, bottom-fermenting yeast perform best at much cooler temperatures, around 7–10 °C, while top-fermenting yeast generally prefer 18–25 °C. Bottom-fermenting yeast strains are so named, because once primary fermentation begins to trail off, they have a tendency to settle at the bottom of the tank. The opposite is true of top-fermenting yeast, which rises to the top after most of the extract has been consumed. These yeast strains are also known as top-cropping yeast, because they are harvested at the surface of the young, fermenting beer and repitched in a fresh batch of wort. Though many of the beers of Central Europe are bottom-fermented, such as pilsner, export, Märzen, helles, and dunkles, there are very old beer styles which have retained top-fermentation, like Berliner Weisse, Southern German Weißbier, Altbier in Düsseldorf, and Kölsch in Cologne. A number of Belgian beer styles and all native British beer styles, like mild ale, stout, porter, and bitter, are top-fermented as well. Cold, bottom-fermentation first appeared in medieval Europe and is generally associated with the monastery breweries of Central Europe. Prior to that time, all beer was fermented at warmer temperatures with a mixture of microbes, among them top-fermenting yeast.

In Germany, according to the *Vorläufiges Biergesetz*, the law derived from the *Reinheitsgebot*, malt may be produced from the following cereal grains: wheat, rye, oats, spelt, einkorn, triticale, and emmer. However, these types of malt are only allowed in top-fermented beer. Most of these types of malt play a negligible role in the brewing industry with the exception of wheat; however, craft brewers especially are becoming aware of their potential. "Organic" malt—or cereals grown without the use of pesticides and according to the laws governing the production of such foods—is gaining in popularity, due to increasing public awareness and the so-called Slow Food movement.

23.2   Malt

All beer is produced using some portion of malt, because it contains the enzymes necessary to degrade the larger molecules in the mash, a mixture of crushed malt and brewing water or liquor, so that the yeast can assimilate them. Malt contains sufficient quantities of these enzymes so brewers sometimes add a portion of other ingredients known as "adjuncts." These ingredients generally do not contain enzymes but serve as a source of carbohydrates to supplement what is in the malt. These adjuncts can be unmalted cereal grains, corn, or rice. If the proportion of adjuncts in the mash becomes too high, then enzymes derived from fungus or bacteria can be added to target the macromolecules present in the particular adjunct employed.

Malt is produced by allowing cereal grains to sprout. This means that the natural germination process, as it would normally occur when grain is sown, is induced and allowed to take its course under carefully controlled conditions (Fig. 23.2).

Brewers play a significant role in determining how barley varieties are selected and cultivated. The germinative energy is a key factor in this selection and must be close to 100 %. Brewers in Europe, that is, those who tend to avoid the use of adjuncts, prefer barley with a relatively low protein content, somewhere between 9 and 11 %. Once it has been harvested, the barley is thoroughly cleaned and then stored for a time. The malting process commences with the steeping of the grain, in which the moisture content of the barley rises to approximately 38 % within 24 h.

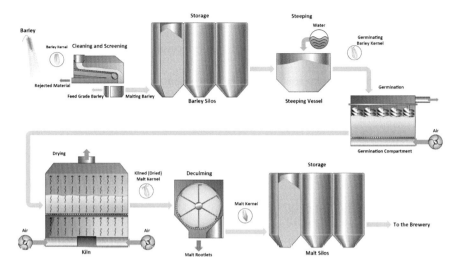

Fig. 23.2 A schematic representation of the malting process (Gesellschaft für Öffentlichkeitsarbeit der Deutschen Brauwirtschaft e. V.)

This is when germination begins. The rootlet develops first, followed by the acrospire. The tiny plantlet releases and activates enzymes, which begin degrading the endosperm, because until the first leaf can extend out and open up to absorb sunlight to initiate photosynthesis, the plantlet requires low molecular weight substances, mostly sugars and amino acids, for the development of new cells and tissues. The rigid structure of the endosperm that preserved and protected the kernel up to this point has to be broken down as well. To achieve this, another enzyme complex is required. These enzymatic processes cause the grain to soften. After only 1 day of steeping and 6 days of germination, the green, living cereal grains have been sufficiently "modified," meaning that the process must now be terminated. If it were to continue, the plantlet would consume the entire endosperm and then grow into a stalk of barley. However, maltsters and brewers have their own designs on the endosperm, and therefore, the plantlet is deactivated and the endosperm preserved in its present state by slowly drying and curing the kernels.

Large cylindroconical vessels are employed for steeping the grain. Germination has been traditionally carried out on the floor of malthouses, where the green malt had to be manually turned two or three times per day. In modern malthouses, germination vessels can reach capacities of 150–300 t.

Germination generally continues for 6 days at temperatures of 14–18 °C. The moisture content of the green malt rises gradually over this time to 44–48 %. Once the malt is sufficiently modified, it can be moved to the kiln. Over a period of approximately 20 h, air at temperatures between 50 and 65 °C is pushed through the malt. This process is known as "withering." Once a certain moisture content is reached, the temperature of the air is increased to approximately 80 °C for light malt and 95–105 °C for darker malt. This latter stage is known as "curing." After the kilning process is complete, the "culms" or rootlets are removed and the malted grain is cleaned. The malting process is very energy intensive; however, with the use of energy recovery and recycling systems, energy usage can be cut by approximately 50 % (Narziss 1999).

Malt is often called the soul of beer and for good reason—it is the foundation upon which a beer is built. The type of malt ultimately determines what style the beer is brewed.

23.3 Brewing

This stage in the process of creating beer involves bringing the compounds in the endosperm of the grain, produced during the precisely controlled process of malting, into an aqueous solution by mixing crushed grain with liquor or brewing water. As mentioned before, this liquid is called "wort" and forms the substrate for the downstream process of fermentation. Both the hot and cold sections of the brewing process are depicted in Fig. 23.3.

The enzymatic processes that occurred during germination are continued in the brewhouse where high molecular weight compounds are further degraded and solubilized. This is possible because during withering and curing in the kiln, maltsters are careful to prevent the denaturation of the enzymes that the plantlet has produced,

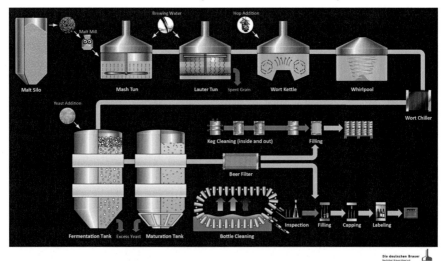

Fig. 23.3 A schematic representation of the brewing process (Gesellschaft für Öffentlichkeitsarbeit der Deutschen Brauwirtschaft e. V.)

though the plantlet itself does not survive. The malt is cleaned again at the brewery, before being milled.

The resulting grist is mixed with the liquor and is generally subjected to a series of temperature rests: approximately 50 °C for the so-called protein rest, around 62 °C for degradation of starch to maltose (malt sugar), and finally 70–72 °C for degradation of high molecular weight dextrins. In order to curtail the activity of the enzymes, a final rest is held at 75–78 °C. Some beers, particularly British ale, are brewed with a greatly simplified mashing process, which requires that the maltster modifies the malt even further prior to handing it over to the brewer. The mashing process can last from 120 to 180 min, depending upon the levels of enzymes in the malt and the degree of "modification" as well as the beer style being brewed.

After the mashing is completed, the grain must be separated from the liquid, that is, the wort must be separated from the malt solids. This can be achieved using a variety of methods and equipment. Most common is the lauter vessel, which is equipped with a screen upon which the solids can rest without touching the bottom of the vessel. The husks of the barley form a filter bed on top of this false bottom, and after recirculating for a short time, the wort runoff becomes clear. Brewers refer to the wort that comes off the bed once recirculating has finished as the "first runnings." It can possess an extract content of 18 %. Before the liquid level drops too far, liquor at approximately 75 °C is sprayed over the grain solids forming the bed, in order to rinse out the remaining extract. The collected wort should—once it is well mixed—reach a target concentration of 10–11 %, depending upon the desired original gravity and the duration of boiling. The spent grains are removed from the lauter tun once the remaining liquid has been drained off and are then generally fed to farm animals.

The collected wort is then moved to the wort kettle, where it is generally boiled for 60–90 min. An herb known as hops is added at this point. The flower is the portion of the hop plant that is used by brewers. The flowers resemble the cones of conifers but are much smaller. In the past, the whole cones were added directly into the wort kettle. Hops are now generally employed in a pelletized form or in the form of extract. Both have been very valuable to brewers, because they retain their freshness much longer and can be stored over extended periods without deteriorating as whole hop cones do. The percentage of bitter substances present in the hops and the quantity added to the wort kettle determine the bitterness of the finished beer. Many hop varieties are cultivated in different growing regions around the world. The bitterness and aroma of the beer are largely determined by both the variety and provenance of the hops. The bitter substances in the hops become soluble as the wort is boiled. Precipitation of the high molecular weight proteins and evaporation of undesirable aroma compounds are also desirable outcomes of boiling the wort. Any remaining enzymes derived from the malt are denatured during the boiling process, and the wort is effectively sterilized. Boiling the wort represents a very energy-intensive process step in the brewing process, even given the fact that modern boilers and wort kettles have been designed to be rather energy efficient. Wort kettles are manufactured with highly effective evaporative surfaces, and heat exchangers may be installed to recover energy at critical points in the brewery. Through implementation of such energy-saving measures and installation of the necessary equipment, only approximately 50 % of the energy expended in the wort boiling process using conventional methods is required for the same process in a modern brewhouse (Back 2005).

The boiled wort must undergo another separation step due to the addition of hops and due to solids formed during the boiling process. The wort is pumped tangentially into a vessel known as a whirlpool. The name of the vessel is descriptive of its function. Through the circular motion of the hot wort, the so-called hot break material (protein) and spent hops form a mound of sediment or "cone," as it is known, in the center of the vessel. The sediment remains on the floor of the vessel, while the wort is pumped out through a heat exchanger, which cools it to the temperature required for fermentation. Cold liquor or some type of food grade coolant runs countercurrent to the hot wort in the heat exchanger. The cold liquor is heated for the next brewing cycle to approximately 80 °C (Narziss 1992).

23.4 Fermentation–Maturation–Storage

The wort has now entered the cold section of the brewing process. Upon exiting the heat exchanger, the wort must be aerated. Though oxygen is often described as the number one enemy of the brewer, the yeast requires oxygen at this stage in order to reproduce. As brewers say, the yeast is "pitched" or introduced into the wort, where after a short interval of getting its bearings and cellular reproduction, it commences fermentation. The yeast obtains amino acids and other substances (e.g., higher fatty acids) it needs from the wort to create new cells. Fermentation primarily involves ingesting the sugars present in the wort and turning them into alcohol and carbon dioxide. The yeast also produces by-products during fermentation, thereby playing a considerable role in determining the flavor profile of the finished beer. As already discussed, yeast is commonly divided into two groups: top-fermenting and bottom-fermenting strains. However, within these broad categories, they are grouped into smaller categories, such as English ale, Belgian ale, German ale, etc. Even each individual strain of yeast within these smaller categories produces its own flavor and aroma profiles, which can vary according to substrate, temperature, pH, and other conditions. Therefore, yeast, as only one raw material used in the production of beer, imparts a dynamic and multifaceted dimension to the brewing process. Along with the intricacies of aroma and flavor derived from the many types of malt, the numerous varieties of hops, and the effects water has on the process, one can understand how brewers will never exhaust the great well of creativity provided by these four ingredients, especially since brewers are also known to experiment with other microbes and raw materials. Sometimes yeast can produce undesirable flavors in beer as well. These are known as off-flavors. Compounds known as vicinal diketones are known among these substances, and one of them, called diacetyl, evokes an unrefined, butter-like flavor and aroma in the beer. Diacetyl and its precursor are eliminated from the beer during the maturation period that follows primary fermentation. Acetaldehyde and volatile compounds containing sulfur also lend an unrefined, "green" flavor to beer. Thankfully, during maturation, they are scrubbed out by the carbon dioxide bubbles rising to the surface. The length of fermentation and maturation is dependent upon temperature. The process can last up to seven days at 9–11 °C for the more slowly bottom-fermented lagers. Top-fermenting yeast, on the other hand, tends to work more quickly at higher temperatures and can attenuate the wort in as little as 3 days. Bottom-fermented beer is referred to as "lager" due to the longer period of maturation and cold storage immediately following primary fermentation. Top-fermented beers are generally "conditioned" for a short time following primary fermentation. For bottom-fermented beers, a cooler lagering stage is required for achieving the requisite protein stability and a more rounded flavor. Toward the end of primary fermentation, the yeast usually flocculates out allowing it to be collected, and then it is rinsed, sieved, and stored until reused. Many bottom-fermented beers undergo an extended lagering period, up to 6–8 weeks. During this time, the yeast ferments the residual extract, the "green" beer flavors and aromas disappear, the beer clears, and carbon dioxide dissolves in the beer at a pressure selected by the brewmaster. Modern brewers often need to expedite the process and have optimized the cold side of brewing to curtail the time required to reach the end result through selection of a suitable yeast strain, by using modern propagation methods and by carefully maintaining the collected yeast.

Formerly, relatively small, open fermentation tanks constructed of wood were standard. These were later replaced by rectangular vats with capacities of up to 600 hL made of aluminum or stainless steel. A variety of fermentation tanks have been developed over the years, some horizontal, some vertical; however, many made collecting yeast difficult. Cylindroconical tanks have a capacity of up to 6500 hL as fermentation tanks, and up to 8000 hL as lager tanks have been highly successful since their introduction. The height of the beer in the tank can reach 12–18 m. The ratio of the diameter of the tank to the height of the beer in the tank should be 1:2 to 2.5 for optimal function. Cylindroconical tanks allow brewers to maintain the precise temperatures required for fermentation. Old lager cellars filled with row after row of wooden barrels would not have been as conducive to this type of fine temperature control. In this way, a precisely defined procedure can be developed according to temperature, pressure, and the timely removal of yeast sediment. This results in uniform beer quality and consistent flavor characteristics over time. At this stage of the process, a microbrewery or a brewpub, for instance, would generally serve their beer directly from the serving tank. It is fresh, often unfiltered, and how most of humanity has consumed its beer throughout the ages, in a naturally cloudy form. For modern brewers, however, additional steps are required in the process before customers can enjoy their beer. The mature beer enters the next step in the process: filtration.

23.4.1   Filtration

The objective of the filtration process is to eliminate any turbidity or haze that may be present in the beer. This originates from proteinaceous material, residual bitter substances, or yeast still in suspension. Filtration can also be employed as a means for the removal of beer-spoiling microbes. Thus, brewers refer to this stage as "stabilization." In the past, beer was filtered over pulp filter sheets made by pressing cotton or cellulose fibers together. After each filtration cycle, they were removed from the filter, washed, and pressed together again. Yeast cells still in the beer and any beer-spoiling microbes were able to pass through this type of filter sheets. In order to eliminate this from the beer, particularly the beer-spoiling microbes, sheet filters with very small pores were inserted downstream in the process. These filters are referred to as "sterile filters."

Today, filtration is performed using diatomaceous earth or kieselguhr. Diatomaceous earth is found in large deposits in the USA, France, and Germany. It is extracted, ground, sieved, and sometimes sintered. It is classified into various grades according to defined particle sizes. The diatomaceous earth is slurried with water or beer and then dosed into the beer upstream from the filter itself. This mixture coats either filter sheets or metal candle filters. The so-called "pre-coating" step creates a filter bed on the surfaces of the filter, which continues to be constantly dosed with diatomaceous earth over the course of the filtration. This enables the filter to operate from 7 to 14 h. However, if under-modified malt was used to brew the beer or if mistakes were made at some point during the production process, this could wreak havoc in the filtration process. In such cases, filter operations would be cut short, and the filter would be cleaned, sterilized, and pre-coated again. Used diatomaceous earth is difficult to dispose of, and regeneration, although possible, is very expensive. Due to these issues, a substitute for diatomaceous earth has been sought for many years. Crossflow filtration with membranes of a defined pore size represents a potential replacement for conventional filtration but as yet not yielded entirely satisfactory results. Membranes with defined pore sizes can also be employed for sterile filtration (Narziss 2005).

Tunnel or flash pasteurization can also reduce the microbial load in beer. Tunnel pasteurizers heat the canned or bottled beer slowly to the required temperature and then cool it slowly; however, this kind of pasteurization tends to age the beer prematurely and is therefore not attractive as a means for stabilizing beer. Thin layer or flash pasteurizers heat the beer at a set pressure to temperatures in the range of 68–74 °C, where it is held for 30–60 s. The pasteurized beer is then run countercurrent to the beer entering the pasteurizer, while the pressure is reduced, in order to cool it back down. Unlike tunnel pasteurization, flash pasteurization has proven to be an effective means for stabilizing foods and beverages without diminishing quality.

23.4.2 Filling and Packaging Beer

23.4.2.1 Barrels

At the beginning of the twentieth century, the vast majority of the beer produced in Europe was filled in wooden barrels. Unfortunately, barrels were too large and unwieldy for private consumers, who were often forced to take large flagons or pitchers to the nearest brewery in order to collect beer for home consumption.

In the past, wooden barrels were often lined with hot pitch and then allowed to cool. The layer of pitch on the barrels had to be renewed at regular intervals. However, barrels filled with beer destined for export to distant locations had to be pitched prior to each use, in order to hinder infection of the beer. Unpitched wooden barrels are now making a comeback among craft brewers, who find that aging beer in them imparts a unique flavor, particularly if they reuse barrels from wineries or distilleries. As time progressed, barrels were manufactured using other materials, such as aluminum alloy or stainless steel. These quickly replaced the old wooden barrels, but retained their form. For centuries, beer was served directly from a barrel through a tap connected to a valve under atmospheric pressure, which was rather unfavorable since the beer became flat within a few hours. Traditional British ale continues to be served unfiltered and unpasteurized, though it is treated with finings to bring about precipitation of any particles or yeast creating turbidity, from casks in which a secondary fermentation takes place in the pub. A hand pump, known as a beer engine, is employed to pull the still living beer up from the cellar and into the glass at the bar. Cask ales are delicate and require attentive maintenance and are therefore unappealing to some pub owners. Beer is now most often served under CO_2 pressure, which keeps the beer fresher for longer if maintained and operated properly. The introduction of cylindrical kegs represented an important innovation, because the fitting for tapping the beer is integrated into the container itself. Kegs were named after the older English wooden drums and helped to ensure higher quality of the beer. One potential drawback lies in the fact that the cleanliness of the entire beverage dispensing system lies with the gastronomic establishment where the beer is served. Sometimes this can be lacking. Kegs can be automatically cleaned, sterilized, pressure-tested, and refilled on a single filling line at a brewery.

23.4.2.2 Bottling

Today, 80 % of all the beer produced is packaged in bottles or cans. Modern bottling lines can fill 3000–60,000 bottles per hour. Although it is possible to operate the bottle washer as well as the depalletizing and packing equipment at higher speeds, the size of fillers is restricted to 60,000 bottles per hour for quality reasons.

23.5   Ingredients in Beer

23.5.1   *Brewing Liquor*

The importance of the brewing liquor, or the water used to brew beer, is evident when one considers that beer is 90 % water. Brewing liquor is subject to all of the regulations that apply to drinking water in Germany. However, breweries place much higher demands on their brewing liquor than the authorities do on public drinking water. The hardness of the water is extremely important for the brewing process and is determined by the quantity of calcium and magnesium salts, while the ratio of carbonate to non-carbonate hardness is also important. For example, Stuttgart's water originates from Lake Constance, where the water is very soft, whereas water from the Munich area possesses very high carbonate hardness. Hard water requires special treatment. This is generally carried out using membrane filter systems, which are easy to operate and "filter out" the salts responsible for hardness. Many breweries have their own natural springs or wells which supply brewing water.

23.5.2   *Malt*

High-quality malting barley is selected for producing malt. Plant breeders in Germany and other European countries submit 10–25 new varieties each year for registration at the Federal Plant Variety Office. New varieties are tested for 3 years at different locations. From an agricultural standpoint, essential traits include high yield and sufficient resistance to lodging as well as to disease. The most important malting and brewing qualities are high extract content coupled with low protein, a high limit of attenuation, and abundant enzyme capacity. Above all, the barley must possess adequate enzyme capacity for adequate modification of the cell walls during malting.

23.5.3   *Hops*

Brewers make a distinction between aroma and bittering hops according to the amount of substances present in the flowers or hop cones, which lend bitterness to the finished beer. The primary constituents of these substances are collectively known as α-acids. Aroma hops possess fewer α-acids and often more β-acids. The characteristic bitterness of these acids is not perceptible, until they are oxidized or polymerized. Hops exhibit very specific aroma profiles, which are linked to their genetic makeup. This not only affects the beer aroma but its bitterness as well and helps to round out the flavor. Bittering hops possess low amounts of β-acids and soft resins but ample amounts of α-acids. Furthermore, substances found in hop oils are capable of lending a somewhat harsher note to the beer. The polyphenol content of hops is important, because they contain substances with antioxidant effects. Antioxidants are important for health reasons (same as red wine). Research into hop cultivation is focused on developing varieties containing substances valued by brewers which can also fulfill agricultural requirements. Ideally, hop varieties must be bred which are not only characterized by a high yield but possess a tolerance to or even a resistance for disease in order to minimize pesticide application. This reduces the amount of pesticide residues in the environment, which are already quite low for cultivation of hops and barley in Germany. In addition to the individual varieties, the growing regions occupy an important role. In Bavaria, these regions consist of the Hallertau, Hersbruck, and Spalt; in Baden-Württemberg, the Tettnang; and, in central Germany, the Elbe-Saale region. Saaz hops from the Czech Republic are also of great significance due to their high quality. US hop growers are quite competitive on the international market, producing very interesting varieties. Hops are dioecious plants, indicating that the male and female flowers develop on separate plants. In Germany, only the hop cones of the unfertilized female plants are used in the brewing process. If female plants are fertilized and produce seed, yield is higher; however, even though there are more cones and they are larger and heavier, the quality of the hops suffers due in part to an increase in their lipid content. Furthermore, the lupulin content is also lower in hops with seeds. For this reason, male hop plants are not allowed in the hop-growing regions of Germany, but this is not the case in many places where hops are grown in the world.

23.5.4 Beer Yeast

Both bottom- and top-fermenting yeast exert a strong influence on the flavor profile of every beer style. Yeast had not yet been discovered in medieval times. The most plausible method for fermenting wort at the time was most likely to simply allow spontaneous fermentation to take its course. The ambient environment of the fermentation cellar and especially that of the wooden fermentation vats would have contributed significantly to the "microbial flora" which would eventually colonize the wort (as is still the case with spontaneously fermented "lambic" beers in Belgium). The beers made in this way would have fluctuated greatly, rendering the beer undrinkable at times, if spoilage microbes gained the upper hand during the course of fermentation. The brewers of this period learned to save the "sediment" at the bottom of each batch to inoculate in the next. In this manner, yeast strains became adapted to specific conditions present in each respective brewery. However, infections with "wild yeast" and acid-producing bacteria were a constant problem, often resulting in spoiled beer especially during warmer periods. Approximately 150 years ago, yeast was discovered as the fermenting organism in beer. Two scientists, Emil Christian Hansen from Denmark and Robert Koch from Germany, invented a method for propagating pure yeast strains by isolating single cells under a microscope and allowing them to reproduce under sterile conditions. This represents the advent of yeast propagation as we know it. In doing so, the two men performed a great service for the field of brewing, as brewers were now able to select the yeast best suited for their purposes, thereby achieving a more consistent product quality. Infected or weak yeast could be discarded and replaced with fresh yeast from continuous propagation.

Today, "yeast banks" in Germany and other countries have a wide range of strains with well-documented characteristics in their inventory, which can be retrieved at any time.

We have thus come full circle. Beer is a natural product, which, through scientific and technological advances made over the years, may be produced at a consistently high level of quality (Narziss 1992, 1999, 2005; Back 2005).

References

Back W (2005) Ausgewählte Kapitel der Brauereitechnologie. Fachverlag Hans Carl, Nürnberg

Hornsey I (2012) Alcohol and its role in the evolution of human society. Royal Society of Chemistry, London

Meussdorffer F, Zarnkow M (2014) Das Bier: Eine Geschichte von Hopfen und Malz. C.H. Beck, Munich

Narziss L (1992) In: Bierbrauerei D (ed) Die Technologie der Würzebereitung, vol 402, 7th edn. Ferdinand Enke Verlag, Stuttgart

Narziss L (1999) In: Bierbrauerei D (ed) Die Technologie der Malzbereitung, vol 466. 7th edn. Ferdinand Enke Verlag, Stuttgart

Narziss L (2005) Abriss der Bierbrauerei, vol 7. Wiley-VCH, Weinheim

Zarnkow M et al (2006) Interdisziplinäre Untersuchungen zum altorientalischen Bierbrauen in der Siedlung von Tall Bazi/Nordsyrien vor rund 3200 Jahren. Technikgeschichte 73(1):3–25

Zipfel W, Rathke K-D (2008) Lebensmittelrecht. C.H. Beck, Munich

Chapter 24
"Pálinka": Hungarian Distilled Fruit

Zsuzsanna László, Cecila Hodúr, and József Csanádi

24.1 Introduction

Pálinka is a traditional Hungarian spirit drink produced exclusively by the alcoholic fermentation and distillation of fleshy fruit or grape marc. The word "pálinka" originates from a Slovak term "pálit" which means to distil.

According to EU regulations, pálinka is a form of brandy distilled exclusively from Hungarian fruit with strictly no sugar or flavorings added. Pálinka is distilled at less than 86 vol%, and the distillate has aroma and taste that are derived from the distilled raw material (Anon 2010).

However, brandies are produced all over the world, but none are similar to pálinka's flavor and character. More than 100 pálinka varieties all have the characteristics of the Carpathian Basin's fruit varieties, climate, and soil.

Traditionally, the homemade pálinka was made from windfall, spoiled fruits, but the quality of this spirit was not high.

Historically, the pálinka distillers (tandem with breweries) have operated from the seventeenth to the eighteenth centuries. Home distillation was forbidden, but regardless has been popular from the eighteenth century.

Pálinka was produced in larger scale from the nineteenth century. From 1836, a tax on pálinka was introduced, and then from 1850, distillation of the spirit became a state monopoly. Recently, in 2010 home distillation of 50 L spirit/person/year has been permitted (Anon 2011).

Z. László • C. Hodúr (✉) • J. Csanádi
Faculty of Engineering, Institute of Process Engineering, University of Szeged,
Moszkvai str. 5-7, Szeged 6725, Hungary
e-mail: zsizsu@mk.u-szeged.hu; hodur@mk.u-szeged.hu

© Springer Science+Business Media New York 2016
K. Kristbergsson, J. Oliveira (eds.), *Traditional Foods*, Integrating Food
Science and Engineering Knowledge Into the Food Chain 10,
DOI 10.1007/978-1-4899-7648-2_24

24.2 Raw Materials for Pálinka

Pálinka can be produced from any wild or farmed fruit growing in Hungary. The fruit grown in Hungary's various regions gives every Hungarian pálinkas there distinct characteristics. There are as many types of pálinka as there are types of fruit grown in Hungary. The most common fruits for making pálinka are apricots, pears, plum, cherry, and apple.

Apricot is one of the most popular pálinka varieties, because it is not subtle and has a very characteristic and taste which are easy to recognize. The apricot pálinka has a distinctive apricot flavor, slightly yellow in color. It should be stoned (removal of pits) (similar to other stone fruits) before making a mash to avoid hydrogen cyanide production. The expected alcohol yield[1] is 4–5 L/100 kg.

Plum is a traditionally popular raw material of pálinka. It has a relatively high (4.5–5.5 L/100 kg) alcohol yield. Pits are sometimes left in the mash for extra flavor, although this is strictly controlled by the EU because of minute traces of cyanide. The Szatmár and Békés regions are best known for their plum pálinka.

Cherry has a high sugar content, and its alcohol yield is extremely high (5–7 L/100 kg). It should also be stoned to remove the pits. For cherries this can be done effectively by machines.

The best known **pear** pálinka varieties are the Williams pear and Kieffer pear (from Kiskunhalas). To produce good pear pálinka, it is essential that fruit should be gathered before it is completely matured, and it should be processed after ripening. Its alcohol yield is 3–4 L/100 kg.

Quince is a variety of the pear or apple but with much lower sugar content and a bitter taste but still considered a valuable fruit. It is not eaten as is but is processed into jam, jelly, or pálinka. However, its unique flavor and aroma come through strongly in pálinka. Its water content is low; thus, 30–40 % water addition is necessary during fermentation. The alcohol yield is 2–2.5 L/100 kg.

Berries are also used as raw materials for pálinka, but these fruits are subtle, so their processing is relatively expensive. The most popular berries in Hungary are raspberry, red currants, black currants, and strawberries.

Pálinkas made from **wild fruits** are very rare and considered a genuine delicacy. Wild apples are distilled in the Őrség and wild cherries in the Zemplén and Hegyköz hills, while wild pears can be found in larger quantities in the undulating forests and meadows of northern Hungary. The blackthorn is also suitable for making pálinka as it grows sweet after the winter frosts; excellent pálinka can also be made from elderflower, rose hips, and dogwood.

---

[1] Traditionally, in Hungary the alcohol yield is expressed as hectoliter degree (hlf) which means the volume of 100 % ethyl alcohol.

24.3 Processing of Fruits for Production of Pálinka

Careful selection of the right fruit is crucial to making high-quality pálinka. The fruits gathered in their original form are not fit to processing. The fruit must first be cleaned and then the pits removed from fruits containing pits to ensure the absence of traces of cyanide in the final product. The fruits are then mashed to create a jam-like base, known as "cefre," which is fermented at low temperature for 10–15 days. The mixture of fruit and alcohol produced from the mash is then distilled. Once distilled, the pálinka is aged, usually in casks, to allow the flavors to mature (Békési and Pándi 2005). Some pálinkas are stored in wooden casks to add flavor and soften the alcohol by allowing it to breathe. In some cases, fruit is added to the cask or bottle to increase the fruitiness and add sweetness, but in this case, the term "pálinka" cannot be used for the final product. After distillation and aging, the pálinka is diluted to between 37.5 and approximately 60 % with distilled water or springwater.

24.3.1 Mashing

The gathered fruits are **graded** and leafs, boughs, and spoiled fruits removed. In some cases, the fruits are contaminated by soil or soil-originated microorganisms which may cause acrolein contamination of the final product. The acrolein is a toxic substance produced as a fermentation by-product from glycerin by soil bacteria. The soil and pesticide residues must be removed by **washing**. This step is often missed from traditional homemade fruit processing. In modern technologies spray diffusers combined with conveyor or in larger factories fruit washing machines (tumbler machines, brush washer, or fluidization washers) are used.

The fruits containing pits, mainly peach or apricot, must be **stoned**. The stone inner contains amygdalin, which may cause bitter taste of pálinka. The amygdalin may decompose to the bitter benzyl aldehyde and the very toxic hydrogen cyanide. The hydrogen cyanide content of the pálinka must not exceed 40 mg/kg. About 20 % of the pits may be retained to promote the formation of the so-called stone flavor. After stoning or often in the same step, the fruit is crushed and **mashed** to ensure that yeasts reach the sugar content of the mash.

In the traditional way, the pit-containing fruits were stoned and crushed by hand; later stoners, choppers, and extractors were used. The fruit digested in this way was placed in ferroconcrete or metallic containers where it was fermented.

In modern closed processing systems, the stoning and crushing are done in stainless steel equipment. The fruit is moved from spout to stoner and crusher by a feeder screw. Because of the high temperature of the mash (which may be as high as 40 °C in the summer), the mash is cooled to 18 °C by tube heat exchanger to avoid unwanted fermentation. The chilled mash is pumped to a preparatory container where it is further cooled and mixed; the pH is adjusted to 2.8–3.2 which is the ideal pH value. Yeast cells are present in the air and on the skin of the fruit, but the wild yeasts may cause unwanted fermentation producing acetone, lactic acid, and butyric acid. Modern practices call for special yeasts and pectin-decomposing enzymes to prevent the unwanted components to form and to ensure the optimal formation of aromas and alcohol yield.

24.3.2 Fermentation

Fermentation of the sugary mash is initiated by yeast and enzymes. Sugar is transformed into alcohol and carbon dioxide. The fermentation is carried out in an anaerobic environment. The ideal temperature for the fermentation process is between 14 and 16 °C, and the process takes between 10 and 15 days. The gases (CO_2) leave through pipes. In modern processes, the whole process of fermentation is computer controlled (Biacs et al. 2010).

Traditionally, the fermentation tanks (casks, barrels) were clamped or sprayed with water to cool to the optimal fermentation temperature.

Modern processes use stainless steel tanks set in cool cellars or climatized buildings where the temperature is kept between 5 and 10 °C in the winter and at a maximum of 18–20 °C in the summer, or tanks have jackets for circulating cold water. The fermentation tanks are closed to ensure the absence of oxygen and to prevent the outflow of carbon dioxide. Tanks are usually equipped with high-pressure spray jets for cleaning. The high-efficiency cleaning process is also computer controlled.

Fermentation can be operated in batch or continuous mode, but the traditional batch fermentation is more frequent. Fermentation has three phases: pre-fermentation, main fermentation, and pro-fermentation. In the pre-fermentation phase, the inverting of sugar begins, and then after 2–3 days, the process passes into the main fermentation phase. During main fermentation of the mash bubbles, yeasts convert sugar to alcohol with CO_2 and heat production. After 5–6 days, the pro-fermentation phase begins and a CO_2 layer forms on the surface.

The fermentation is an anaerobic process, but the traditional fermentation was carried out in open tanks. In this case, the supernatant mash "hat" must have been removed from the tank, because the mash surface in contact with air is a good medium for (aerobic) microorganisms. In modern, closed processes, this is not a problem because of the carbon dioxide layer on the surface of the mash. The alcohol content of the mash is relatively low, 4–8 %.

24.3.3 Distillation

The mixture of fruit and alcohol produced from the mash is then distilled to concentrate the alcohol, aroma, and flavors. The mash is poured into a still and heated to 78 °C, the boiling point of ethanol, ensuring the absolute purity of the alcohol. As the mash heats up, the amount of water in the vapor grows.

The pure distillate with the characteristic flavor of the fruit must be separated from the "head" and "tail" fractions containing ethyl acetate, fossil oil, and other toxic forms of alcohol. The alcohol content of the first drops may exceed 90 % by volume (Nagygyörgy et al. 2010). The tail also contains alcohol, but it has unwanted aromas and flavors which make the pálinka unacceptable. Traditionally, the still was heated until the distillate was diluted to around 50 % alcohol by volume, which allowed bitter flavors of tail to get into the pálinka.

There are two types of still, the traditional "kisüsti" (a pot still usually made from copper) and the column still. In the case of pot still, distillation is a maximum permitted capacity of 1000 L. This type of still differs from a column still in that it requires the pálinka to be distilled twice. The first distillate has about 20–30 % alcohol content, but it can be concentrated in the second step to 55–60 %.

Modern column stills are able to effectively separate the fractions of the pálinka using a single distillation, with several levels of plates with a larger surface area. In this case, the presence of copper in the pot is important, because the copper has catalytic effect during distillation, mainly in preventing sulfur compounds to get into the pálinka. This method gives a little bit lighter, aromatic distillate with different character from pot still distillates.

24.3.4 Aging and Storage

After distillation, the pálinka must be aged, usually in metal casks, to allow the flavors to mature. Some pálinkas are stored in wooden (usually oaken or mulberry) casks for several months to add flavor and soften the alcohol by allowing it to breathe. In some cases, fruit is added to the cask or bottle; in this case, the pálinka is aged with a minimum of 10 % fruit for 3 months, increasing the fruitiness and adding sweetness to the pálinka. After distillation and aging, the pálinka is diluted to between 37.5 and 60 % with distilled water.

24.4 Hungarian Pálinka Specialities

In the Carpathian Basin and the Great Plains of eastern Hungary, the climate is suitable for a wide variety of fruit species, and each region has its own pálinka specialty. There are currently six regions producing fruit varieties that have been officially recognized for their unique qualities:

Plum pálinka from Szatmár
Apricot pálinka from Kecskemét
Plum pálinka from Békés
Apple pálinka from Szabolcs
Apricot pálinka from Gönc
Sour cherry pálinka from Újfehértó

References

Anon (2010) Raw materials. http://palinkaoldal.hu/index.php. Accessed 22 Feb 2010
Anon (2011) The history of the palinka. http://www.brillpalinkahaz.hu/apalinka. Accessed 25 Feb
    2011
Békési Z, Pándi F (2005) Pálinkafőzés. Mezőgazda Kiadó, Budapest
Biacs P, Szabó G, Szendrő P, Véha A (2010) Élelmiszer-technológia mérnököknek. SZTE Mérnöki
    Kar, Szeged
Nagygyörgy L, Popovics A, Martin A (2010) Pálinkakészítési alapismeretek, e-learning lectures,
    ver 1.0. http:// czidro.eu/wessling/palinka_keszites/. Accessed 25 Feb 2011

Chapter 25
Tokaji Aszú: "The Wine of Kings, the King of Wines"

Cecilia Hodúr, József Csanádi, and Zsuzsa László

25.1 Introduction

Tokaji (Hungarian of Tokaj) is the name of the wines from the region of Tokaj-Hegyalja in Hungary. The name Tokaji (which is a protected designation of origin) is used for labelling wines from this wine district. This region is noted for its sweet wines made from grapes affected by noble rot, a style of wine which has a long history in this region. Tokaj is mentioned in the official national anthem of Hungary. Tokaji is a "Hungaricum", a term used to refer to uniquely Hungarian products, especially cuisine.

Since 2007, only authorised wine producers from the Hungarian wine region of Tokaj-Hegyalja are permitted to use the Tokaj brand name.

25.2 History of Aszú Wine

According to Hungarian legend, the first aszú (wine using botrytised grapes) was made by Laczkó Máté Szepsi in 1630. However, mention of wine made from botrytised grapes had already appeared in the *Nomenklatura* of Fabricius Balázs Sziksai, which was completed in 1576. A recently discovered inventory of aszú predates this reference by 5 years. When vineyard classification began in 1730 in the Tokaj region, one of the gradings given to the various terroirs cantered around their potential to develop *Botrytis cinerea*.

In 1703, Francis Rákóczi II, Prince of Transylvania, gave King Louis XIV of France some Tokaji wine from his Tokaj estate as a gift. The Tokaji wine was served at the

---

C. Hodúr (✉) • J. Csanádi • Z. László
Faculty of Engineering, University of Szeged, Szeged, Hungary
e-mail: hodur@mk.u-szeged.hu

© Springer Science+Business Media New York 2016
K. Kristbergsson, J. Oliveira (eds.), *Traditional Foods*, Integrating Food
Science and Engineering Knowledge Into the Food Chain 10,
DOI 10.1007/978-1-4899-7648-2_25

French royal court at Versailles, where it became known as Tokay. Delighted with the precious beverage, Louis XV of France offered a glass of Tokaji to Madame de Pompadour, referring to it as "Vinum Regum, Rex Vinorum" ("Wine of Kings, King of Wines"). This famous line is used to this day in the marketing of Tokaji wines.

Tokaji wine became the subject of the world's first appellation control, established several decades before Port wine and over 120 years before the classification of Bordeaux. Vineyard classification began in 1730 with vineyards being classified into three categories depending on the soil, sun exposure and potential to develop noble rot, *Botrytis cinerea*, first class, second class and third class wines. A royal decree in 1757 established a closed production district in Tokaj. The classification system was completed by the national censuses of 1765 and 1772.

25.3 Technology

Aszú is the world-famous wine that is proudly cited in the Hungarian national anthem. It is the sweet, topaz-colour wine that was formerly known throughout the English-speaking world as Tokay.

The original meaning of the Hungarian word *aszú* was "dried", but the term aszú came to be associated with the type of wine made with botrytised (i.e. "nobly" rotten) grapes.

25.3.1 Raw Materials: Grapes

Only six grape varieties are officially approved for Tokaji wine production:

1. Furmint
2. Hárslevelű
3. Yellow Muscat
4. Zéta (previously called Oremus—a cross of Furmint and Bouvier grapes)
5. Kövérszőlő
6. Kabar (a cross of Hárslevelű and Bouvier grapes)

Furmint accounts for 60 % of the area and is by far the most important grape in the production of aszú wines. Hárslevelű stands for another 30 %. Nevertheless, an impressive range of different types and styles of wines are produced in the region, ranging from dry whites to the Eszencia, the world's sweetest wine.

The area where Tokaji wine grapes are traditionally grown is a small plateau, 457 m (1500 ft) above sea level, near the Carpathian Mountains. The soil is of volcanic origin, with high concentrations of iron and lime. The location of the region has a unique climate, beneficial to this particular viniculture, due to the protection of the nearby mountains. Winters are bitterly cold and windy; spring tends to be cool and dry, and summers are noticeably hot. Usually, autumn brings rain early on, followed by an extended Indian summer, allowing a very long ripening period. The Furmint grapes begin maturation with a thick skin, but as they ripen the skins become thinner and transparent. This allows the sun to penetrate the grape and evaporate much of the liquid inside, producing a higher concentration of sugar. Other types of grapes mature to the point of bursting; however, unlike most other grapes, Furmint will grow a second skin which seals it from rot. This also has the effect of concentrating the grape's natural sugars. The grapes are left on the vine long enough to develop the "noble rot" (*Botrytis cinerea*) mould.

Noble rot is the benevolent form of a grey fungus, *Botrytis cinerea*, affecting wine grapes. Infestation by *Botrytis* requires moist conditions, and if the weather stays wet, the malevolent form, "grey rot", can destroy crops of grapes. Grapes typically become infected with *Botrytis* when they are ripe, but when then exposed to drier conditions become partially rinsed, and the form of infection brought about by the partial drying process is known as noble rot. Grapes when picked at a certain point during infestation can produce particularly fine and concentrated sweet wine.

Grapes then are harvested, sometimes as late as December (and in the case of true Eszencia, occasionally into January).

Typical yearly production in the region runs to a relatively small 10,028,000 L (2,650,000 gal).

25.3.2   The Traditional Technology

The process of making aszú wine is as follows.

Aszú berries are individually picked; it is the first step of the grape harvest. These berries used to be collected into wooden butts in the vineyard then poured into huge vats in the cellar and used to trample the grapes into the consistency of paste— known as aszú dough. The berries are pressed and its liquid has extremely high sugar content (350–800 g/L); it has raw essences. Extra quality wine can be made from this by a very slow fermentation process. This "wine" is used as a quality enhancer for the other wines of the vintage, and in earlier times it was used as a medicine.

Today the berries are collected into boxes, and instead of trampling, the aszú dough is prepared by a masticator or screw press. The most important element of this operation is to make the dough a homogenous dough and not to disrupt the seeds of the grapes.

The next step is extraction. The dough is extracted or leached by must or wine from the same age group. The leaching time is 1–1.5 days during which sugar, acid and other components are liberated. This operation step is finalised by pressing.

The classical fermentation vessel for aszú wine is the "gönci" cask (Fig. 25.1). This is a big wooden barrel (136 L). The concentration of aszú was traditionally defined by the number of *puttony* of dough (20–25 kg) added to a gönci cask (136 L barrel) of must. Today, the *puttony* number is based on the content of sugar and sugar-free extract in the mature wine. Aszú ranges from three *puttonyos* to six

Fig. 25.1 Aszú cellar with gönci casks

puttonyos, with a further category called Aszú-Eszencia representing wines above six *puttonyos*.

The wine is racked off into wooden casks where fermentation is completed, and the aszú wine is allowed to mature. Traditionally, the aszú wine maturation period was the same number of years as the puttony number it possessed. The fermentation is the key operation of the aszú wine production, but unfortunately the optimal process has not been documented fully. The yeast has to work against the very high sugar content, deficiency of nutrients caused by *Botrytis* and high initial alcohol (6–8 %). The fermentation itself starts very slow and finishes after years. Another problem is how to stop the fermentation when the required level of alcohol is reached. Pasteurisation, filtration, sulfuration and cooling are possible methods for terminating the fermentation, but this still needs to be optimised.

The casks are stored in a cool environment and are not tightly closed, so a slow fermentation process continues in the cask. The walls in the wine cellars are covered with *Cladosporium cellare* in this region which contributes to the final quality and is also a very important element in the secret of the aszú wine.

25.4 Other Types of "Aszú"

Dry wines: These wines, once referred to as common *ordinárium*, are now named after their respective grape varieties, Tokaji Furmint, Tokaji Hárslevelű, Tokaji Sárgamuskotály and Tokaji Kövérszőlő.

Szamorodni: This type of wine was initially known as *főbor* (prime wine), but from the 1820s Polish merchants popularised the name *samorodny* ("the way it was grown" or "made by itself"). What set Szamorodni apart from ordinary wines is that it is made from bunches of grapes which contain a high proportion of botrytised grapes. Szamorodni is typically higher in alcohol than ordinary wine. Szamorodni often contains up to 100–120 g of residual sugars. When the bunches contain less botrytised grapes, the residual sugar content is much lower, resulting in a dry wine. Its alcohol content is typically 14 %.

Eszencia: Also called nectar, this is often described as one of the most exclusive wines in the world, although technically it cannot even be called a wine because its enormous concentration of sugar means that its alcohol level never rises above 5–6 %. Eszencia is the juice of aszú berries which runs off naturally from the vats in which they are collected during harvesting. The sugar concentration of eszencia is typically between 500 and 700 g per litre, although the year 2000 vintage produced eszencia exceeding 900 g/L. Eszencia is traditionally added to aszú wines, but may be allowed to ferment, a process that takes at least 4 years to complete, and then bottled pure. The resulting wine has a concentration and intensity of flavour that is unequalled, but is so sweet that it can only be consumed in small quantities. Storage of eszencia is facilitated by the fact that, unlike virtually all other wines, it maintains its quality and drinkability for 200 years or more.

Fordítás (meaning "turning over" in Hungarian) is a wine made by pouring must on the aszú dough which has already been used to make aszú wine.

Máslás (derived from the word "copy" in Hungarian) is a wine made by pouring must on the lees of aszú.

Bibliography

Alkonyi L (2000) Tokaj—the wine of freedom. Spread BT, Budapest

Grossman HJ, Lembeck H (1997) Grossman's guide to wines, beers and spirits, 6th edn. Charles Scribner's Sons, New York, pp 172–174. ISBN 0-684-15033-6

Kállay M, Eperjesi I (1996) Borászati technológia. In: Kállay M (ed) Modern food microbiology, 5th edn. Chapman & Hall, New York, pp 38–66

Tradition and Innovation in the Tokaj Region PDF (328 kB) Tim Atkin, MW. masters-of-wine.org

Chapter 26
Mead: The Oldest Alcoholic Beverage

Rajko Vidrih and Janez Hribar

26.1 The Importance of Honey in the Past

The word honey conjures up images of warmth; we might address someone we love as 'honey', and a honeymoon is a time of hope and happiness. Although there are no exact figures for per capita consumption of honey through history, a general assumption is that it was a scarce commodity (Allsop and Miller 1996). However, according to a reappraisal of the evidence through history, honey was sold in large amounts, and mead, an alcoholic drink from the fermentation of honey, became a common alcoholic beverage (Allsop and Miller 1996). It is suggested that ordinary people consumed as much honey as the current consumption of refined sugar. As well as drawings in the Cave of the Spider (Spain), there are many other drawings that indicate the high value that honey had in the past. However, despite all of these indications of how well honey was appreciated, there appear to be no records of the amounts that were actually consumed. Instead, studies of present-day hunter–gatherers can provide approximations to the situation in the past, where honey was obviously plentiful. In Tanzania, for example, of all food eaten, meat and honey constituted 20 % by weight; in energy terms, this amounts to significantly more than 20 % (Woodburn 1963). In Congo, pygmies can cover up to 80 % of their energy needs with honey during the honey season (Crane 1983), while in Sri Lanka, honey was also used to preserve meat for times of scarcity (Crane 1983).

R. Vidrih (✉) • J. Hribar
Biotechnical Faculty, University of Ljubljana, Ljubljana, Slovenia
e-mail: rajko.vidrih@bf.uni-lj.si

© Springer Science+Business Media New York 2016
K. Kristbergsson, J. Oliveira (eds.), *Traditional Foods*, Integrating Food
Science and Engineering Knowledge Into the Food Chain 10,
DOI 10.1007/978-1-4899-7648-2_26

26.2   An Introduction to the History of Mead

While no one can pinpoint the exact time period when or location where mead was first discovered, there is evidence of it in nearly every ancient culture. Mead is an alcoholic drink produced by the fermentation of honey. After gathering the honey, the natural yeast and high moisture content cause the honey to ferment. Traditionally, mead is produced by diluting the honey with water and by addition of yeast. As in the past, mead can be produced in many variations, including as a plain honey wine (traditional, with honey only), flavoured with fruit (melomel), spices (metheglin), or vegetables, or with any combination of ingredients; here, the choice is only limited by the imagination of the brewer and the resources at hand. The alcoholic content of mead can range from about 8 to 18 % (Lichine 1987). It can be still, carbonated or sparkling, and it can be dry, semisweet or sweet (Rose 1977). In terms of the oldest alcoholic beverage on the earth beer, wine and mead remain in competition for the crown as the first fermented beverage.

No other sugar-containing foodstuff available in antiquity can surpass honey. Collecting honey was one of the first organised agriculture practices. Ancient cave paintings depict the earliest people stealing this sweet treasure from the bees. In the above-mentioned Cave of the Spider (Cueva de la Arana), a prehistoric cave discovered in Valencia, Spain, which is situated on the river Cazunta, drawings have been found that are estimated to be about 8000–15,000 years old. Here, two human figures are seen on a rope ladder, in the process of robbing the bees of their honey in a cave entrance (McGee 1984). Ancient people simply expelled the bees from their hives using smoke and then collected the honey. These bees, which would have been wild rather than domesticated, were then destined to die due to lack of food during the winter. Thus, archaeology acknowledges the existence of honey hunters from well before any evidence of beer or wine.

It is evident from written and pictorial records that beer making dates back to Mesopotamia, from about 4000 BC. Brewing and wine production have also been found in pictorial archaeological records dating back to 3000 BC, where both crafts had developed to the stage where the process had been standardised (Schramm 2003). Brewing and winemaking then evolved further, and they became sources of employment and trade for a considerable period. There is widely accepted evidence that cereal grains were cultivated from 9500 to 8000 BC, with evidence that wine was made at least in the Neolithic period; in 1996, a jar dating back to 5400–5000 BC was found at a Stone Age site in Iran that contained traces of wine residues (Riches 2009). The competition for the crown as the first fermented beverage continues to be fierce, however, with the end of the ice age bringing suitable conditions for viticulture and horticulture. Cereals and grapes were first used as food and then later on as fermentable substrates to produce alcoholic beverages.

Evidence of viticulture dates back to 8000–5000 BC, and this has been extrapolated, maybe debatably, to indicate winemaking. The first regions where grapes were grown included the previous Soviet republic of Georgia, Turkey, Syria and Jordan. During archaeological excavations in Armenia, a few dozen 6000-year-old karstic caves were discovered. In one cave, known as Areni-1 cave, evidence of the completion of the full cycle of winemaking was found in the Copper Age layer.

While there is substantial evidence regarding horticulture and viticulture, it is only colourful folklore that can enlighten about the primordial fermentation process. Although they lay claim to the crown for the first alcoholic beverage, mead makers have not been as well organised as brewers, who have even sponsored archaeological beer projects. Nevertheless, mead makers base their claim on the argument that honey was the first fermentable substrate, as is also well documented in the earliest pictorial evidence. Similarly, from archaeobotany, which is the discipline that deals with the archaeobotanical evidence of food and involves the search for plant microfossils that were present in food or beverages. Under certain conditions, food remains can be well preserved, preventing their decomposition. In the case of honey, pollen preserved by copper salts or drying might also indicate mead (Rosch 2005).

26.2.1   Mead in China

It appears that the earliest archaeological evidence for the production of mead itself dates to around 7000 BC. In the Neolithic village of Jiahu, in the Henan province of Northern China, remnants of an alcoholic beverage were found in 9000-year-old pottery jars (McGovern et al. 2004). These Neolithic people were brewing a type of mead, with an alcohol content of 10 % and which consisted of wild grapes, honey and rice. This thus produced the so-called wine–mead–sake, which indeed lays claim to the oldest record of any alcohol-containing beverage.

Compared to the grape and grain, the production of which is limited by pedoclimatic conditions; honey can be produced over a much wider geographical area. Schramm (2003) indicated that anthropologists are in good agreement that the beginnings of mead production should date back to around 8000 BC. Indeed, it appears that several cultures around the globe were producing mead at the same time, without having knowledge of each other.

26.2.2   Etymological Research of Mead

Etymological research has shown that mead is such an ancient beverage and that the linguistic root for mead, *medhu*, is the same in all of the Indo-European languages, where it took on a relatively wide range of meanings, including honey, sweet, intoxicating, drunk and drunkenness. This thus provides further evidence for the suggestion that fermented honey is the oldest form of alcohol known to man (Aasvad 1988) (Table 26.1).

Table 26.1 Examples from etymology dictionaries of the roots of mead in other languages

Language	"Mead" root	Meaning
Old English	*Medu*	Mead
Proto-Germanic	*Meduz*	Mead
Cf. O.N.	*Mjöðr*	Mead
Danish	*Mjød*	Mead
O. Fris. Middle Dutch	*Mede*	Mead
German	*Met*	Mead
From PIE base	*Medhu*	Honey, sweet drink
Sanskrit	*Madhu*	Sweet, sweet drink, wine, honey
Greek	*Methy*	Wine
O.C.S.	*Medu*	Mead
Lithuanian	*Medus*	Honey
Old Irish	*Mid*	Mead
Welsh	*Medd*	Mead
Breton	*Mez*	Mead
Slovenian	*Medica*	Mead
Polish	*Miód*	Mead
Baltic	*Midus*	Mead
Old Norse	*Mjöð*	Mead

26.2.3   Mead in Egypt

In Egypt, early evidence of beekeeping dates back to 2500 BC; at that time, honey was used as a household sweetener (Schramm 2003). Egyptians domesticated bees and they probably developed the first beehives, which were made of unbaked clay or of woven baskets covered with mud. The hives could be opened from the back, and the bees were forced out using smoke. The honey was harvested twice a year, with the combs placed on cow skin and crushed. The honey was then separated through small holes and filtered through blades of straw, to get rid of the impurities. All of these procedures have been depicted in different ancient Egyptian tombs. Honey was also mentioned as being fermentable in the Hymn to Ninkasi (Schramm 2003), and it became an important agricultural commodity in trading.

26.2.4   Mead in Ancient Greece

Honey was the first and only sweetener that the ancient Greeks had in their diet. It is worth mentioning that the Greeks develop an advanced hive system, with removable bars that enabled the harvest of the honey without disturbing the brood. Ancient Greek authors praised the medicinal and nutritional values of honey, and the Iliad refers to honey as the food of kings, while mead was considered a high beverage among ancient Greeks. The Greeks were very fond of using spices and herbs in their cuisine, including in their wine and mead.

26.2.5   Mead in the Roman Empire

In the ancient Roman Empire, honey was very popular, as it was in other Mediterranean states. It is speculated that the Phoenicians and early Greek settlers brought their apiculture knowledge to the Roman Empire, where it then flourished. Bees and honey were of great esteem for Virgil, who felt that honey was a gift from heaven. The famous Roman writer Pliny devoted many chapters to honey collected from the sweet juices of flowers and to the beneficial effects of honey on health. Beekeeping was also particularly advanced, as the Romans had at least nine types of beehives that were made of clay, cork, wicker and wood. Mead was omnipresent during the Roman Empire, and the Romans also produced *mulsum*, a traditional grape wine that was sweetened with honey. Pliny the Elder described how to make *hydromel*, a mead made of three parts water and one part honey, where this honey–water solution was left to ferment outside for 40 days. The Romans often used herbs infused in mead, where their extraction was helped by means of the alcohol. Schramm (2003) described in his book how the word *metheglin*, meaning 'mead with herbs', consisted of the Latin *medicus* (medicinal) and the Irish *ilyn* (liquor).

26.2.6   The Middle Ages

The oldest record of mead in Europe dates back to 2500 BC. That evidence was found in 1984, in an archaeological dig on Rum Island West of Scotland. Scientists proved that ingredients in a pot of dried mead appeared to be heather honey, oats, barley, royal fern and meadowsweet.

Lithuanian ancestors, Balts, were using mead thousands of years ago (since 1600 BC). Mead was drank out of drinking horns coated with metal in the period from the fourth to sixth century AD. The tradition of mead in Europe was pursued most in areas less suitable for wine grapes (England, Scandinavia, Eastern Europe), which explains the enthusiasm of the Vikings for mead. The monasteries were skilled in keeping bees and brewing, although mead making was a by-product of beekeeping.

The wooden bowls used by English mead drinkers during the Middle Ages were known as *mazers*.

The Book of Taliesin is one of the most famous of Middle Welsh manuscripts, where Kanu y med or song of mead was written probably around 550 AD. Mead is mentioned in many stanzas in Y Gododdin, a medieval Welsh poem dating to around 700 AD. In Scandinavian mythology, Valhalla is a majestic, enormous hall located in Asgard, ruled over by the god Odin. In Norse mythology, mead was the favourite drink of the Norse gods and heroes, e.g. in Valhalla as mentioned in Chap. 39. Snorri Sturlason (1179–1241) lived in Iceland, and he wrote a saga of all of the Viking kings, and most of it is believed to be at least near to being correct historically (Fig. 26.1). This saga has been printed in numerous editions over the last 200 years. The use of mead (and beer) is an element of many of these stories.

Fig. 26.1 Drawing by Erik Werenskiold showing King Fjolne falling into the mead barrel (Sturlason 1942). Some authors say that Fjolne became a King in Uppsala, or maybe he was just a mythical story

During reformation there was no need for luxury in Catholic churches and for their candles. Beekeepers lost an important part of their income due to the drop in the demand for wax. As a consequence, the prices of honey increased and mead was no longer affordable. In countries that remained Catholic, like Poland or Russia, mead remained a favoured drink among the nobles. As mentioned by Gayre and Papazian (1983) in their book, mead was for the great and grand occasions, while ale was for the masses and for all times. In Russia, mead remained popular as med-ovukha and sbiten long after its popularity declined in the West. Sbiten is often mentioned in the works by nineteenth-century Russian writers, including Gogol, Dostoevsky and Tolstoy.

After the introduction of wine from the Mediterranean countries, mead was also looked upon as more of a countryman's drink and lost its popularity among the nobility. However, it did not lose its popularity among the common folk because, like beer, it could be made from materials at hand, and unlike wine, it didn't require special temperatures for storage or ageing. However, beer was much less expensive to produce and gradually replaced mead as the daily drink of all classes.

26.2.7 Symbolism of the Honeymoon

According to legend, the word *honeymoon* derives from an ancient tradition of sending a newly married couple off to seclusion for a month with much mead. Drinking mead would have ensured their best chance to start a family quickly; indeed, mead has been, and still is, considered the drink of love. The term honeymoon is packed with symbolism, and mead, or honey wine, is sweet and symbolises the particular sweetness of the first month of marriage. It is a time free of the stresses and tensions that everyday life puts on a relationship as time goes by. The moon symbolises the phases or cycles of the couple's relationship as it extends from new moon to full moon. Like the moon, the couple's relationship would have its brighter moments and its darker ones. Being tied in with the moon cycle, the 1-month period of time was also considered to be associated with the woman's menstrual cycle and thus, her fertility. However, the real claim to fame for mead is in its origins during the wedding celebrations; it was believed that if the couple drank mead daily during the honeymoon, they would be assured of fertility and the birth of sons. Some customs sent the bride to bed and then filled the bridegroom with mead until he could no longer stand. He was then delivered to the bride's bedside to sire a son that very night. If the bride did indeed bear a son 9 months later, the maker of the mead would be complimented on its quality. This concept arose as it was widely believed that honey would promote the birth of a male child, and it was considered very lucky when the firstborn to a newly married couple was a boy. Therefore, on the day of their wedding, the father of the groom would give the couple enough honey to last a full month, or *moon*, thus hopefully assuring that the child conceived during the first days of the marriage would be a boy. Oddly enough, modern science appears to support this theory. Apparently, a diet rich in honey alters the female pH levels enough to provide a favourable environment for the conception of a male.

26.3   Mead Varieties

Many versions of mead are known today, and many of them are still being produced locally. Thus, many recipes include as a source of sugar, besides honey, also herbs, fruit or grain.

26.3.1   Meads from Caramelised Honey

Bochet mead or burnt mead is prepared in a way that honey is heavily caramelised or burnt. Afterwards, the procedure is similar to ordinary mead. Mead has a charred character with impressions of toffee, chocolate and marshmallow flavours.

26.3.2   Melomel Types of Mead

When honey is mixed with fruit juice, the resulting fermentation is a type of mead known as melomel. When the fruit juice used in making melomel is apple juice, the resulting fermentation is known as *cyser*. Mead, in Russia known as *medovukha* or *stavlenniy myod*, is made from honey, but a version with the addition of berries is also known. In Russia, mead was drunk to celebrate pagan god Perun. Lithuanian mead called midus is made similarly by fermenting honey and berry juice. Morat is considered a traditional drink of the Saxons, made with honey fermented with mulberries. Pyment is considered a type of fruit mead fermented with red or white varieties of grapes. It shares sensory properties of both traditional mead and wine. Grape varieties like Pinot noir and Barbera are mentioned. Pyment may be spiced by storing it in casks to enrich it with vanilla overtone.

Omphacomel is a style of mead mixed with an acidic liquid pressed from unripe grapes.

A Finnish type of mead called sima is another fruit type of mead flavoured by adding both the juice and zested rind of a lemon. When mead is brewed with black currant, it is called black mead.

Mead called acerglyn is produced from honey and maple syrup in the USA and Canada.

26.3.3 Mead with Spices

Meads seasoned with different spices, also called metheglin, have been produced throughout history.

In Mexico, Maya Indians are still producing mead called balche with the bark of a leguminous tree (*Lonchocarpus violaceus*), which is soaked in honey and water and fermented. Balche is an intoxicating beverage that was drunk during rituals and was considered to have magic power. Xtabentún is a similar beverage made of honey produced from the nectar of *Turbina corymbosa*. Pittarilla is a similar mead produced by Mayans using honey and balche tree bark.

Braggot is a mead made with malt and honey. It may be called honey bear because of the addition of hop. It is considered to be of Welsh origin.

Capsicumel or capsimel is a mead made with chillies. It is named for the spicy chemical capsaicin and the genus of chilli plants, *Capsicum*. The mead must be sweet to balance the heat, but not so sweet that the flavours of the base honey and chillies are masked.

Dandaghare is mead made in Nepal from honey and local herbs and spices. Croatian mead is called medovina or gverc and is made from honey and different spices.

Iqhilika is a traditional mead from South Africa. According to the traditional method, the ingredients are honey, water, roots of an indigenous succulent herb known as *imoela* and pollen.

Tej is a famous mead made in Ethiopia. It is made from local honey and flavoured with smoke and leaves and twigs of gesho (Rhamnus prinoides) that gives a hop-like taste.

Rhodomel is produced from honey, water and rose petals or distillate from rose petals.

Oxymel is made of honey and water, with vinegar added at the end of fermentation. It was used mainly for medicinal purposes.

26.3.4   Fortified Meads

Some meads are produced using more honey than traditionally used, and so the final product contains elevated levels of non-fermented sugars. This category includes sack mead and Polish meads called dwojniak and poltorak.

26.3.5   Plain Meads

Plain meads are produced from honey and water only. This category includes medicinal products produced in Slovenia and Croatia and show mead. In Slovenia, the oldest known recipe of mead originates in the seventeenth century. Czwórniak is a plain mead produced in Poland from nectar honey. Chouchen is also a plain mead produced in Brittany from honey and water, sometimes using seawater instead of freshwater.

26.3.6   Mead Distillates

Any type of mead may be distilled to produce brandy with higher alcohol content. Lithuanian mead midus enriched with the extracts from different herbs, flowers, or buds is distilled to produce a fortified beverage containing up to 75 % of alcohol. Medenovec is a Slovenian mead brandy that contains 32–35 % of alcohol. It can be made with any type of honey, but aromatic honeys are preferred.

26.4   Processes

26.4.1   Early Fermentation

Throughout history, the production of alcoholic beverages was a glorious habit of humans everywhere in the world. Indeed, many and various foodstuffs that contained sugar were used to produce alcohol. The most common foodstuffs were fruit, grain and vegetables, but the important advantage of honey was its good storability. Furthermore, much less care was needed to carry out the fermentation, compared to other foodstuffs. For this fermentation, the appropriate vessel seemed to have been of great importance. Although often representations in pottery are the most important historical proof, the first watertight vessels used by humans would probably have been animal skins and some animal organs (Schramm 2003). Crane (1983) supported this hypothesis further by indicating that African honey hunters used animal skins and gourd containers. In his book, Schramm (2003) hypothesised that the first spontaneous fermentation could have taken place in these skin containers. The hunters would fill their containers with water, and possibly later during their hunting they would have stored the honey together with this water. At the beginning, this honey–water beverage was probably esteemed by the hunters, due to the pleasant taste, and later on, once fermentation had taken place, this would have provoked great amazement.

26.4.2 Microorganisms in Mead Processing

Indigenous microorganisms naturally present either in the honey or on the equipment used usually promote the fermentation process (Ashenafi 2006). Yeast cannot grow in solutions where sugar concentrations exceed 22 %, while on the other hand, 14 % alcohol arrests fermentation (Stong 1972). Besides high osmotic pressure and increased alcohol content, some inhibitory compounds in honey can also make the fermentation process difficult to start (Pons and Schutze 1994). Due to these reasons, the fermentation of honey usually proceeds slowly and might get blocked before completion.

As well as these inhibitory agents, various nutrients are also present in honey. It is generally believed that dark-coloured honeys contain more nutrients compared to light-coloured honeys. To give an example, wort made from buckwheat honey is known to ferment much faster than one made from clover honey. Caridi et al. (1999) investigated honey fermentation with different yeast strains for each honey variety, because each honey source can contain specific factors that might hinder the fermentation process. They thus isolated four yeast strains that have suitable, but different, characteristics, such as high fermentation vigour, ability to produce mead with a low pH, good resistance to higher ethanol content and ability to produce low volatile acidity (Caridi et al. 1999). Among these characteristics that are related to the fermentation vigour, resistance to ethanol and the ability to produce a low content of volatile acidity are particularly important for mead production.

As indicated above, mead fermentation is a rather slow process, and it often takes from one to several months to accomplish. To increase the fermentation rate, Navratil et al. (2001) experimented with immobilised cells of *Saccharomyces cerevisiae*, using ionotropic gelation of pectate, which proved to be more stable when compared to alginate. This process was continuous and stable and took only a week to complete the fermentation (Navratil et al. 2001). Other authors have recommended the use of champagne yeasts (White 1994), while Vidrih and Hribar (2007) used *Saccharomyces bayanus* to carry out the fermentation.

26.4.3 The Importance of the Fermentative Microflora

It can happen that mead is not drinkable after fermentation, which would be due to the uncontrolled development of wild yeasts and bacteria (Mendes-Ferreira et al. 2010). Honey is considered an ideal medium for the growth of many undesirable microorganisms, which can multiply if they are not suppressed by the yeast (Dumont 1992). Thus, to prevent uncontrolled fermentation and to promote a faster start, selected strains of yeast or commercial yeast preparations have been introduced (Pereira et al. 2009). Using selected yeast significantly improves the overall quality of mead, but problems like a lack of uniformity across batches and an arrest of fermentation before completion are often found. Pereira et al. (2009) attributed these problems to the yeast strain. During mead processing, yeast encounters especially difficult conditions, due to high osmotic pressure, higher ethanol content, lack of nutrients and inhibitors present in some types of honey.

In old recipes, the addition of herbs, fruit and other additives was suggested, to enrich the flavour and taste of the mead. On the other hand, as speculated by Morse and Steinkraus (1975), the use of various additives could well have been an attempt to mask an incorrect aroma or flavour of a mead. More recent recipes can also recommend additives, such as grape juice, to produce red-coloured mead (White 1994).

26.4.4 Wort Pasteurisation Prior to Fermentation

Heat treatment of honey solutions is a traditional method that is still used by mead manufacturers today. As mentioned by Stong (1972), the boiling of diluted honey has the advantage of yielding a mead that clears more rapidly and has a better overall stability during its shelf life. The boiling of the diluted honey for 30 min is sufficient to prevent cloudiness, which might otherwise persist for several months. Changes like an altered colour and aroma profile were reported by Kime et al. (1991a). As a result of this heat treatment before fermentation, Wintersteen et al. (2005) found that the antioxidative potential of the mead was altered.

Another important advantage of this pasteurisation is the elimination of all of the microflora, so that the fermentation can start with the addition of the selected yeast. The goal of this procedure is thus also to produce a more uniform mead across different batches.

26.4.5 Application of Ultrafiltration

As mentioned above, heat treatment successfully prevents spontaneous microflora from carrying out the fermentation, and it promotes rapid clarification of the mead after fermentation. However, this also has its negative aspects, as heat treatment can lead to oxidation, browning and undesirable flavours. As already observed by Kime et al. (1991a), a further technique is to use pasteurisation at a high temperature, but for a short time (30 s), which has been shown to substantially improved the overall quality of the mead compared to traditional pasteurisation. The same authors studied the application of ultrafiltration of the wort before fermentation (Kime et al. 1991a, b). Ultrafiltration is widely used in the food industry to remove selected particles according to their molecular size. The solution to be filtered flows under hydrostatic pressure against a semipermeable membrane, which allows only the water and the low-molecular-weight solutes to pass through. The membrane surface is selective for a particular molecular weight, and so the solids and solutes that have higher molecular weights and that are responsible for haziness are retained and thus removed from the solution. The main advantage of applying ultrafiltration for mead production is the superior mead quality achieved, with a superior flavour, taste, colour and stability, and a mead that clarifies immediately after the fermentation. Such ultrafiltered mead is said to have a smoother and 'cleaner' flavour, compared to conventional processing (Kime et al. 1991a, b).

26.5   Consumer Aspects

26.5.1   Nutritionally Improved Mead

Nowadays, producers keep trying to improve the nutritional value of many traditional food products. A main route for these modifications is the addition of various ingredients, like vitamins or minerals. On the other hand, different plant extracts that are especially rich in polyphenols can be used to improve antioxidative potential, colour and/or flavour of food products. With regard to novel types of mead, producers have tried to enrich mead for attributes like a better antioxidative potential or improved sensory characteristics.

Koguchi et al. (2009) produced novel types of mead from honey that originated from Chinese milk vetch and from buckwheat, and they enriched these by adding grains of black rice (*Oryza sativa var. Indica cv. Shiun*). The milled black rice was added just before the start of the fermentation. The mead produced in this way was more red in colour and was characterised by a higher content of total polyphenols and a higher antioxidative potential, compared to the control mead.

26.5.2   Aroma and Flavour of Mead

Although mead is considered the oldest alcoholic beverage, scientific studies regarding all aspects of mead are very rare.

As is the case with other alcoholic beverages, the overall aroma of mead arises from the aroma compounds that originate from the floral source of the honey. These represent the so-called fermentative aroma compounds that are formed by microorganisms during fermentation and from the maturation aroma compounds formed during maturation. Each type of honey is characterised by its own typical aroma compounds, which originate from the primary floral source.

Having in mind the aroma compounds in mead, Vidrih and Hribar (2007) studied the synthesis of the higher alcohols, while Wintersteen et al. (2005) investigated the influence on the aroma volatiles of the heat treatment of the wort before fermentation. Higher alcohols are predominant aroma volatiles, and their synthesis is dependent on the honey source, the nitrogen content and the fermentation temperature (Vidrih and Hribar 2007; Mendes-Ferreira et al. 2010). As these higher alcohols can impart an unpleasant sensation to the mead (a grass-like taste and flavour), their synthesis should be limited. The widely accepted upper limit for the total content of higher alcohols is 300 mg/L, with higher content being detrimental to quality (Rapp and Mandery 1987). On the other hand, the higher alcohols are suitable substrates for the synthesis of their corresponding esters (butyl, isoamyl), which are formed during mead maturation. The practice of controlling the fermentation temperature and the use of supplementation of nitrogen at the start of the fermentation are widely accepted practices in wine processing, and they are also well accepted by mead producers.

As well as yeast, lactic acid bacteria also find a suitable medium during mead processing (Bahiru et al. 2006). Both yeast and lactic acid bacteria are known to synthesise numerous aroma volatiles that contribute to the flavour and aroma of mead.

References

Aasvad M (1988) History of mead. http://redstonemeadery.com/store/catalog/History-of-Mead-sp-17.html Accessed 2015

Allsop KA, Miller JB (1996) Honey revisited: a reappraisal of honey in pre-industrial diets. Br J Nutr 75(4):513–520

Ashenafi BM (2006) A view on the microbiology of indigenous fermented foods and beverages in Ethiopia. Ethiop J Biol Sci 5:189–245

Bahiru B, Mehari T, Ashenafi M (2006) Yeast and lactic acid flora of tej, an indigenous Ethiopian honey wine: variations within and between production units. Food Microbiol 23(3):277–282

Caridi A, Fuda S, Postorino S, Russo M, Sidari R (1999) Selection of Saccharomyces sensu stricto for mead production. Food Technol Biotechnol 37(3):203–207

Crane E (ed) (1983) The archaeology of beekeeping. Duckworth, London

Dumont DJ (1992) The ash tree in Indo-European culture. Mankind Q 32(4):323–336

Gayre R, Papazian C (eds) (1983) Brewing mead: Wassail! In mazers of mead. Brewers, Boulder

Kime RW, McLellan MR, Lee CY (1991a) An improved method of mead production. Am Bee J 131(6):394–395

Kime RW, McLellan MR, Lee CY (1991b) Ultra-filtration of honey for mead production. Am Bee J 131(8):517

Koguchi M, Saigusa N, Teramoto Y (2009) Production and antioxidative activity of mead made from honey and black rice (Oryza sativa var. Indica cv. Shiun). J Inst Brew 115(3):238–242

Lichine A (ed) (1987) Alexis Lichine's new encyclopedia of wines & spirits. Alfred A. Knopf, New York

McGee H (ed) (1984) On food and cooking: the science and lore of the kitchen. Harper Collins, London

McGovern PE, Zhang JH, Tang JG, Zhang ZQ, Hall GR, Moreau RA, Nunez A, Butrym ED, Richards MP, Wang CS, Cheng GS, Zhao ZJ (2004) Fermented beverages of pre- and proto-historic China. Proc Natl Acad Sci U S A 101(51):17593–17598

Mendes-Ferreira A, Cosme F, Barbosa C, Falco V, Ines A, Mendes-Faia A (2010) Optimization of honey-must preparation and alcoholic fermentation by Saccharomyces cerevisiae for mead production. Int J Food Microbiol 144(1):193–198

Morse RA, Steinkraus KH (eds) (1975) Wines from the fermentation of honey. Honey a comprehensive survey. Heinemann, London

Navratil M, Sturdik E, Gemeiner P (2001) Batch and continuous mead production with pectate immobilised, ethanol-tolerant yeast. Biotechnol Lett 23(12):977–982

Pereira AP, Dias T, Andrade J, Ramalhosa E, Estevinho LM (2009) Mead production: selection and characterization assays of Saccharomyces cerevisiae strains. Food Chem Toxicol 47(8):2057–2063

Pons MN, Schutze S (1994) Online monitoring of volatile compounds in honey fermentation. J Ferment Bioeng 78(6):450–454

Rapp A, Mandery H (1987) New progress in wine and wine research. Experientia 42:873–884

Riches HRC (2009) Mead: making, exhibiting & judging. Northerm Bee Books, Hebden Bridge

Rosch M (2005) Pollen analysis of the contents of excavated vessels—direct archaeobotanical evidence of beverages. Veg Hist Archaeobot 14(3):179–188

Rose A (ed) (1977) Alcoholic beverages. Academic, New York

Schramm K (2003) The complete meadmaker: home production of honey wine from your first batch to award-winning fruit and herb variations. Brewers, Boulder

Stong CL (1972) Mead, drink of vikings, can be made (legally) by fermenting honey in home. Sci Am 227(3):185

Sturlason S (1942) Snorres Konge Sogor. (The saga of Norwegian kings). Gyldendal Norsk Forlag, Oslo

Vidrih R, Hribar J (2007) Studies on the sensory properties of mead and the formation of aroma compounds related to the type of honey. Acta Aliment 36(2):151–162

White EC (1994) Honey concord wine—award-winning mead-made easy. Am Bee J 134(3):185–186

Wintersteen CL, Andrae LM, Engeseth NJ (2005) Effect of heat treatment on antioxidant capacity and flavor volatiles of mead. J Food Sci 70(2):C119–C126

Woodburn J (ed) (1963) An introduction to Hazda ecology. Man the hunter. Aldine de Gruyter, New York

Chapter 27
Midus: A Traditional Lithuanian Mead

Rimantas Kublickas

27.1 Introduction

Mead (also called honey wine) is almost certainly the oldest alcoholic beverage known to man. The earliest surviving description of mead is in the Hymns of the Rigveda, one of the sacred books of the historical Vedic religion dated around 1700–1100 BC (Rigveda, Book 5). The ancient Greeks honored Bacchus, who was widely regarded as the God of Mead long before he became accepted as the God of Wine. During the medieval times, it spread all over Europe, especially in the Scandinavian, Slavic, and Baltic lands.

A chance occurrence of mead was likely produced during the Stone Age when honey and non-concentrated liquid solution from rain and wild yeast in the air settled into the mixture. Historically, meads were fermented by wild yeasts and bacteria residing on the skins of the fruit or within the honey. Mead is known from many sources of ancient history throughout Europe, Africa, and Asia, although archaeological evidence of it is ambiguous (Hornsey 2003). Its origins are lost in prehistory.

Mead (according to modern understanding) is an alcoholic beverage that is produced by fermenting a solution of honey and water. The alcoholic content of mead may range from about 8 to 18 %. It may be dry, semisweet, or sweet, and it may be still, carbonated, or sparkling; it can be aged for short or very long (12–50 years) time. Honey comes from the nectar of flowers and is named according to the type of blossom from which the nectar is collected by the bees. Mead can have a wide range of flavors, depending on the source of the honey.

R. Kublickas (✉)
Department of Food Science and Technology, Kaunas University of Technology,
Kaunas, Lithuania
e-mail: rimantas.kublickas@ktu.lt

© Springer Science+Business Media New York 2016
K. Kristbergsson, J. Oliveira (eds.), *Traditional Foods*, Integrating Food
Science and Engineering Knowledge Into the Food Chain 10,
DOI 10.1007/978-1-4899-7648-2_27

Depending on local traditions and specific recipes, it may be flavored with spices, fruit, or hops; it may also be produced by fermenting a solution of water and honey with grain mash (Fitch 1990). The mead of honey, water, and yeast can be called basic mead. Mead can be also brewed in many forms and methods with different names. The following is a list of some of the numerous forms of mead:

- Melomel is mead made with fruit juices (can contain fruit).
- Pyment is mead made specifically with grape juice.
- Metheglin is mead made with herbs or spices or both.
- Hippocras is mead made with grape juice and with herbs and/or spices.
- Braggot (also called bracket) is a honey ale beverage (originally brewed with honey and hops, later with honey and malt—with or without hops added) made by fermenting honey and grains together.

Once the meadmakers in many countries begin adding fruits, spices, and herbs, it takes on an entirely different character and a new name. It can be made with many different ingredients: caramelized or burned honey, maple syrup, or even with wine vinegar or chili peppers.

Although the process of making mead is easier than brewing beer, the fermentation of mead takes much longer than the fermentation of beer, so mead lovers must be patient to reap the full benefits of their product. The equipment used is very similar to brewing equipment. Homemade mead can be of excellent quality—sometimes better than commercial.

27.2 Lithuanian Mead

Mead was a very important beverage in the life and religion of ancient inhabitants of Northern or Eastern part of Europe. Lithuanian ancestors Balts (Baltic region people, also called Aistians) were making mead thousands of years ago. Good conditions existed to make mead because Lithuanians took honey from wild bees in tree hollows since early times (Piccolomini 1477).

The Anglo-Saxon traveler Wulfstan of Hedeby (about 890 AD) while describing the Aistians referred to the fact that they have plenty of mead (Velius 1996). Mead was widely used by Lithuanian rulers (e.g., Vytautas the Great, the Lithuanian Grand Duke and the famous Lithuanian noble family, Radvila) for state representation. Mead ten or more years old was the landlord's pride, for mead's quality increases with age. The royal manor was keeping royal mead stock.

The biggest amount of mead was used by population in local inns, the network of which, in the fifteenth to eighteenth century, covered the entire territory of Lithuania.

The development of Lithuanian mead lasted till the eighteenth century. Since the seventeenth to eighteenth centuries, with the decline of wild beekeeping, mead production decreased and in that time the development of production of beer and vodka increased. Mead making in Lithuania was remaining only as homemade beverage production.

The exact recipes of the old Lithuanian mead that were made several hundred years ago have been lost. However, it is known (from remaining homemade mead traditions) that, in those times, a water solution of honey was simmered with various spices such as thyme, lemon, cinnamon, cherries, linden blossoms, juniper berries, and hops (Kriščiūnas 1955). The solution then was strained off and fermented with beer or wine yeast. Three types of sweet and mild mead were dominant in Lithuania—basic mead, some art of metheglin (very mild) and some art of melomel (with fruit and berry juice, predominant with wild berry juice) (Kriščiūnas 1926).

The first company that started producing mead in the twentieth century was the Prienai beer brewery. Four kinds of mead were produced there. Mead was left to mature for 5 years. In 1940, this brewery was nationalized and the production of mead was stopped.

In July 1960, the first 1200 bottles of new, very mild, racy, amber colored, 10 % alcohol by volume Lithuanian mead Dainava were produced in Stakliškės brewery of the Union of Lithuanian Cooperatives and commercialization began. Since then, Lithuanian mead is associated with sweet mead made of natural bee honey and fruit and berry juice, infused with carnation blossom, acorn, poplar buds, hops, juniper berries, and other herbs (Šimkevičius 1969). Since a successful start of the first industrial mead, several other mead varieties were created, and strong beverages were started to be produced using mead as their basis. The name "Lithuanian mead" sometimes associates with a mead nectar or mead balsam, some of the varieties having as much as 75 % of alcohol.

On March 25, 1969, the then invention committee of the Soviet Union registered the production technique of a Lietuviškas midus (Lithuanian mead) as an invention. The author of the invention, Aleksandras Sinkevičius, was awarded an author certificate No. V3114. On September 30, 1969, Elizabeth the Second, Queen of the United Kingdom of Great Britain and Northern Ireland, with patent No. 1280830, entitled Stakliškės factory Lietuviškas midus (Lithuanian mead) to solely use the invention of mead production and development of related activities in the territory of the United Kingdom for 16 years. Lithuanian mead has received a well-deserved evaluation at national exhibitions and abroad. Stakliškės mead played its role in preserving the Lithuanian identity in the Soviet era, and today it is an excellent representation of Lithuania abroad. Since the reviving production of mead, several production recipes have been selected and improved, and the same kind of mead is being produced up to date. Lithuanian mead of seven varieties is produced in Stakliškės: Gintaras, Dainava, Bočių, Trakai, Stakliškės, Prienų šilai, and Piemenėlių. Currently, UAB Lietuviškas midus (Lithuanian mead Ltd.) produces three varieties of mead and works according to the quality control system EN ISO 9001-2000. Today new technologies have been implemented in the company, but mead is matured in exactly the same way as several centuries ago, in the cellars of estates of castles. In 2002, Stakliškės mead was given the status of Culinary Heritage.

27.3 Technology of Lithuanian Mead

There are many small homemade mead producers in Lithuania. This mead is pro-duced using very simple equipment (very similar to homebrew beer equipment). Honey can be boiled or not. There is a traditional classification of such homemade mead that depends on the ratio of honey and water (or juice). For example, twofold mead is made from 1 part (liter) of honey and 1 part (liter) of water, threefold—from 1 part of honey and 2 parts of water and so on. One and half mead is the strongest mead in this classification and can be matured 5–10 years, and quadruple mead is relative weak mead and is drinkable after few months.

There is a mead industry in Lithuania too, and it has had a big influence on mead traditions in Lithuania. Lietuviškas midus Ltd. is the only (industrial) mead pro-ducer not only in Lithuania but also in the Baltic States. Lietuviškas midus Ltd. produces mead of three kinds: Bočių, Trakai, and Stakliškės. Each kind of mead differs in its taste and aroma, but all of them are a yellowish, aromatic, sweet bever-age with a low content of alcohol (12–15 % by volume) made from natural products with a natural fermentation using beer yeasts. This is an exceptional product that can be found nowhere else because this beverage is produced from local bee honey and from herbs grown in the same region.

Stakliškės mead, which contains 12 % alcohol by volume, is a sour vitamin-rich drink made from natural bee honey, flavored with hops, juniper berries, lime tree blossoms, and other extracts, which contain vitamin C. The taste of honey is strong in this mead. This drink is left to mature through natural fermentation for up to 12 months.

Bočių mead contains 14 % alcohol by volume and has the color of amber; its aroma is a wonderful mixture of bee honey, hops, and juniper berries.

Trakai mead has the color of amber and a unique aroma of honey enriched by hops, lime tree blossoms, juniper berries, and acorns.

Mead can be amber to a shade of brown. Its taste and aroma can be characterized as honey, with a sour and sweet flavor, and an aroma of hops. Honey mash for pro-duction of industrial mead can be obtained from a mixture of honey and water (a ratio of 1:1 to 1:4) or a mixture of honey, fruit juice, and water (a ratio of 1:1:3) (Notification 2008/443/LT). The Lithuanian mead technology (Fig. 27.1) realizes all three mead-making methods: basic mead, some art of metheglin, and some art of melomel.

Alongside these traditional meads, other mead beverages are made in Stakliškės too: four mead nectars and a mead balsam. Mead nectars are stronger than mead (which contains 25–50 % alcohol by volume) and are mixed with ethyl alcohol and mead distillate and flavored with natural berry (strawberries, blueberries, cranber-ries, cherries, and black and red currants), apple, or quince juices, flavoured with a bouquet of fragrant herbs. The production technology necessary to produce these mead nectars is extremely complicated, and the process may take up to 24 months. Later, the drink may be matured further. There is one extremely strong mead beverage in Lithuania—mead balsam (75 % alcohol by volume). It is bitter and is drinkable only in special occasions.

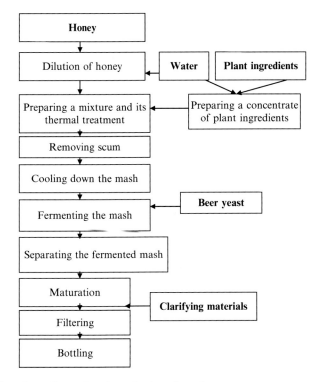

Fig. 27.1   Flow sheet of operations in production of mead

References

Fitch E (1990) Rites of Odin. Llewellyn Worldwide, St. Paul, p 290
Hornsey I (2003) A history of beer and brewing. Royal Society of Chemistry, Cambridge, p 7
Kriščiūnas J (1926) Kaip vaisvynis ir midus gaminti (How to make fruit wine and mead). Varpas Nr. 57, p 24
Kriščiūnas J (1955) Bitininkystė (Beekeeping). Valstybinė polit.ir moksl. lit. leidykla, p 540
Piccolomini E (1477) Cosmographia in Asiae et Europae eleganti description, p 445
Rigveda Book 5 v. 43:3–4
Šimkevičius J (1969) Senovės lietuvių midus (Old Lithuanian mead). Mūsų sodai. Nr. 11, p 20
Velius N (1996) Wulfstano pasakojimas (Wulftan's story) Baltų religijos ir mitologijos šaltiniai, t. 1, p 345

Chapter 28
Horchata

Idoia Codina, Antonio José Trujillo, and Victoria Ferragut

28.1 Introduction

"Horchata" is a vegetable beverage, similar in appearance to milk, directly obtained by mixing water with tigernuts. It is a traditional Spanish beverage, naturally sweet and nutritious, consumed specially in summer and served cold.

Tigernuts are the edible part of the plant *Cyperus esculentus* L. (Fig. 28.1). They are, in spite of its name, tubers of small dimensions, from 8 to 16 mm diameter, round to oval or elongated shaped, brown in color, and with typical concentric crowns. Its cultivation requires sandy soil and mild climate, such as that of the Mediterranean Basin, where this crop has historically been cultivated. These tubers are an important crop of Spain (in the area of Valencia), but they are also produced in other places such as in tropical and subtropical areas of Africa and Asia or in some American countries.

Composition data of tigernuts were reported by different authors (Codina-Torrella et al. 2014; Sánchez-Zapata et al. 2012). Lipid percentages ranged from 23 to 35 %; protein content varied from 3 to 9 %; extractable nitrogen-free material was the most abundant fraction which approximately ranged from 50 to 58 %, and it was composed of simple sugars (15–22 %) and starch (30–34 %); and other important components were fiber (8.5–10 %) and ashes (1.6–2 %), which corresponded to the mineral fraction. Starch granules, which are observable by optic microscope, are found to be round or oval, and they ranged between 5 and 9 μm in the case of tigernuts. The high starch content of this tuber is a technological factor to take into account in the processing of horchata, especially during the application of heat treatments, because the product experiments a high increase of viscosity since the starch granule is swollen when hydrated with temperature.

I. Codina • A.J. Trujillo • V. Ferragut (✉)
Veterinary Faculty, Autonomous University of Barcelona, Bellaterra, Spain
e-mail: Victoria.ferragut@uab.cat

© Springer Science+Business Media New York 2016
K. Kristbergsson, J. Oliveira (eds.), *Traditional Foods*, Integrating Food
Science and Engineering Knowledge Into the Food Chain 10,
DOI 10.1007/978-1-4899-7648-2_28

Fig. 28.1 Tigernuts (*left*) by Tamorlan (2009) and tigernut plant (*right*) by Joanbanjo (2014)

Sugars are the second carbohydrates in importance. Sucrose is the most abundant with 13–16 % (dry basis), and only small quantities of glucose, fructose, and galactose are found in their composition (less than 0.5 % dry basis for each one).

Lipids occur in foods as triglycerides, formed by fatty acids which are esterified with glycerin, which profile is important from a health point of view. Fatty acid composition of tigernuts is represented by oleic, palmitic, and linoleic acids, which represent 73 %, 14 %, and 9.8 %, respectively. Other fatty acids into a much lesser extent are stearic (2.4–3.2 %), myristic (0.2 %), and palmitoleic acids (0.3 %). Tigernut fatty acid profile shows a great similarity in composition to olive oil. Linssen et al. (1988) compared both oils and not only found strong similarities in their fatty acid profile (mainly represented by oleic) but also in the positional distribution structure of triglycerides. Since olive oil is considered a good quality product, due to its fatty acid profile, these results indicated that tigernuts can be used in the same manner as this oil. Nutritional and sensory characteristics of tigernut oil may increase the interest to produce value-added foods by means of the introduction of these tubers in other formulas.

In the protein fraction, which represents the most limiting component in tigernuts, the most abundant amino acid is the arginine, which is followed by aspartic and glutamic acids. On the contrary, the most limited amino acids are methionine, tryptophan, and valine. Mean values of tigernut amino acids showed that these tubers were not a good source of protein, neither in content nor in composition of amino acids. However, the protein fraction in tigernuts is mainly soluble in water, which means that most of this component is extracted in the horchata elaboration process. Codina-Torrella et al. (2015) showed that the protein fraction in tigernuts was represented by albumins (82.23–91.93 %), whereas the remaining ~3–7.5 % corresponded to globulins, prolamins, and glutelins.

Tigernuts are a good source of minerals. Potassium is found in high quantity (567 mg/100 g d.b.), followed by phosphorous (274 mg/100 g d.b.), while calcium, magnesium, and sodium are the least important components of this fraction (99, 94, and 41 mg/100 g d.b., respectively).

The most appreciated horchata is produced traditionally without applying any preservation treatment. In general, this beverage is sold at specialized shops, together with ice cream and other refreshing products. Traditional horchata is usually produced to be consumed during the same day of elaboration. The traditional recipe consists on blending hydrated tigernut with added water in a proportion of 1:5 (w/w) and with a further squeeze and filtration step to obtain a milky beverage to which sugar (up to 10 % of total weight in agreement to Spanish legislation) is finally added. The product is refrigerated at 2–4 °C. In general, horchata is a neutral beverage (pH 7) with soluble and suspended solids which has more than 10 % sugars, starch (2.5 %), and lipids (2.6 %). From a colloidal point of view, horchata is a system formed by a complex phase, composed of oil droplets, starch granules, and small solid particles, which are dispersed in an aqueous phase formed by water, sugars, and minerals. The tigernut compounds described above not only have a nutritional function but also an important role in colloidal stability and functionality. In this sense, in order to disperse the oil in the aqueous phase, an emulsifying agent is necessary. This function is carried out by proteins contained in tigernuts. Proteins have the property to be adsorbed to the oil–water interphase, in which the lipophilic groups of amino acids remain oriented to the oil phase, while the hydrophilic groups of proteins are oriented to the aqueous phase. This distribution of proteins, which form a cover around the oil droplets, makes possible for the oil droplets to remain "solubilized" in water for a long period of time. The emulsion stability, in terms of the capacity of oil droplets to be maintained in dispersion, depends on other factors such as oil droplet size and the viscosity of the aqueous phase. In the case of traditional horchata, emulsion stability is not usually a problem because it is consumed quickly, and it is enough to stir the horchata with a ladle previously to serving. However, the importance of obtaining a fine emulsion when horchata is stored during a long period of time will be discussed in a further section. Starch is another component which may have important implications in the stability of the system. As the traditional product is not heat treated, starch granules in native form are partially crystallized, and they maintain the original size and structure. In this situation, they behave as colloidal solid particles in suspension which tend to form a layer of sediment at resting. When horchata is stirred just before serving, the final product is enough homogeneous. However, when horchata is heat treated, starch granules lose the crystallized structure and adsorb water causing the swelling of granules, which provoke a big increase of viscosity. This modification is undesirable since horchata turns into a sticky product. For this reason, it is necessary the use of amylolytic enzymes before horchata processing by heat treatments.

28.2 Elaboration Process

The technological process for obtaining *horchata* is very simple; it consists in grinding tigernuts with water in order to obtain a paste, which is finally squeezed to obtain the milky extract. Then, this product is usually mixed with sugar. Due to the microbiological characteristics of tigernuts, it is necessary to disinfect the tubers before their use, in order to guarantee the microbiological quality of the final beverage. Although, nowadays, the *horchata* elaboration process has been improved by industry, the main steps of this process are still similar to those followed traditionally.

28.2.1 Selection of Tigernuts

Tigernuts are tubers, so they grow underground. When its vegetative cycle finishes, the tuft is cut and tubers are collected through the removal of the soil. Tigernuts are picked with different types of unwanted soil particles, such as little stones, sand, plant remains, metal objects, other grains, animal droppings, etc., as well as with parts of the tigernut plant (roots, small sticks, empty and broken nuts, and also with typical damaged nuts for the attack of parasites and insects). All these undesirable particles make increase the microbiological counts of collected tubers, which limit the beverage's shelf life. Owing to this, the removal of all these defective elements is essential before their packaging. Traditionally, this step was performed manually, but, nowadays, industry tends to use automatic systems, such as wet separating and sieving systems or the removal of heavy particles by gravity.

After this step, tigernuts are pre-cleaned with water showers, in order to eliminate the superficial dust. Consequently, their content of moisture increases to ~40–50 %, so it is therefore necessary to dry the tubers (to 10–11 % of moisture) for preserving their quality during the storage. Tigernuts are sundried (during 2–3 months, turning them daily to dry them up uniformly) or dried by forced convection. Once drying is complete, tigernuts are stored in ventilated areas, inside polypropylene bags in order to allow the product's transpiration during their storage. Tubers can also be fumigated in order to protect the tubers from the damage caused by insect plagues.

28.2.2 Production of Horchata

The beverage production starts with the rewashing of tigernuts, in order to eliminate the soil particles that still remain on their rough surface. This procedure is performed with specially designed washing machines, in which tigernuts are immersed with abundant water and cleaned in continuous agitation. Among others, this step contributes to reduce the high microbiological counts of collected tubers. Once clean, tigernuts are bathed into a salt solution to separate by floating the damage, broken, or excessively dry tubers from the rest. After this, tigernuts are rinsed and washed again with abundant water, in order to rinse off any salt residue.

Then, tubers are submerged in water to increase their moisture content. Soaking process is an important step because the water of tuber determines the extraction of nutrients during the grinding and also the nutritional quality of the beverage (Djomdi et al. 2007). Previously, different authors have studied the soaking behavior of tigernuts during this step, combining different parameters, such as the temperature of soaking, the time, and the tigernut size (Djomdi et al. 2007; Codina et al. 2009). These results showed that water absorption kinetics of tubers display an increase in accordance with the soaking temperature (higher temperatures allowing an easy penetration of water). The size of tubers also influences this kinetics, since the absorption of water in bigger size tubers is longer if soaking temperatures are not high. In this respect, industry tends to hydrate tigernuts at temperatures ≥ 60 °C, but it is very common to find who employs the traditional method, in which tigernuts were soaked at room temperature. Under these conditions, a long soaking time is required, which, in some cases, results in an increase of microbial counts of the medium affecting the quality of the final product (color, taste, shelf life, etc.). In this case, a good hygienic practice is to change the soaking water at intervals of 10–12 h (Djomdi et al. 2007; Monerris-Aparisi et al. 2001).

Once rehydrated, tigernuts are disinfected. This step is essential to improve the microbiological quality of this beverage and its shelf life. Some producers disinfect the tubers before the soaking step in order to avoid the possible microbiological growth during their rehydration and the consequent fermentation of tubers. However, some studies demonstrated that disinfecting the tubers after soaking is more effective, because their surface is smoothed and the disinfectant works better (Gallart 1999). There are different effective germicidal treatments, but, in general, *horchata* producers tend to immerse the tubers in chlorinated waters (1 % sodium hypochlorite). Once disinfected, the tubers can be peeled with scraped surface deposits, in which their skin is removed by scraping (Fig. 28.2).

After this step, tigernuts are triturated with added water in a hammer mill. According to the desirable composition of the final product, industry uses different ratios of tubers and water; in general, industry works with ratios of tigernuts: water corresponding to 1:8–1:10, but the best quality *horchata* is elaborated with a proportion 1:5–1:7. The paste obtained from the mill is finally squeezed with a special pressing system (form, in some cases, by a worm gear) and subsequently filtered

Fig. 28.2 Detail of a hammer mill (*left*) and a worm gear and its filter (*right*) used in the production of *horchata*

through different steel sieves (in a vibrating sieve systems), in order to remove the biggest size particles.

The final extract is mixed with the desired proportion of sugar, which, in general, corresponds to ~8–10 %, and then the bottled beverage is stored at cool temperature (≤ 4 °C).

28.2.3 Improving Yield of Horchata in Processing

Although *horchata* composition is mainly determined by tigernut characteristics, its final composition can be improved by optimizing the extraction yield. Traditionally, *horchata* producers improve the extraction of tigernuts' components through different consecutive washes of the solid paste, after obtaining the first liquid extract. This solid waste is still rich in nutritive components (protein, fat, fiber, etc.), which can be solubilized again in water and mixed with the initial extract (Morell and Barber 1983). Djomdi et al. (2007) reported that extraction yield improves with higher soaking temperatures of tigernuts, maybe due to the effect of temperature on the physical structure of tuber cells which facilitate their crumbling during the grinding and the pass of more solid particles to the final extract. However, these authors also observed that the final beverage decreases in the soluble extract, probably due to the transfer of these components from the tubers to the soaking solution. Codina et al. (2009) reported that no more than 12 h at ≥ 20 °C in tigernut soaking was enough for obtaining a good quality milky extract. These authors also observed that pre-warming the grinding water ($20 \leq T^a \leq 60$ °C) and adjusting its pH (6 or 9) neither improved the yield of extraction.

28.3 Preservation Treatments

Food preservation consists of the application of a series of available technologies and procedures to prevent deterioration and spoilage of food products and extend their shelf life, while assuring consumers a product free of pathogenic microorganisms. Preservation methods include both thermal and nonthermal technologies. Thermal or heat processing is widely applied to food, in spite of some problems such as the fact that heating, depending on its intensity, might reduce nutrients or deteriorate the sensory quality of foods. Nonthermal processing is considered an effective method that causes minimal deterioration of nutritional or sensorial qualities compared to thermal processing. Currently, the evaluation of nonthermal technologies is being widely pursued by researchers.

The most important alteration in *horchata* is mainly due to the high microbial load of tigernuts. It should be kept in mind that tigernut is a small root with a grooved surface; small particles of soil, which contain a high level of microorganisms, can be found in the grooves of the nuts. On the other hand, it should also be mentioned that *horchata* is a liquid product which is rich in sugars, fat, and other nutritional components, and it has a neutral pH; these conditions make *horchata* an ideal site for microorganisms to develop. This is why good manufacturing practices have established a series of compulsory steps for the production of raw *horchata* with the aim of reducing the microbial load of the nuts. These steps include nut rehydration, selection of nuts by flotation, washing, and germicidal treatment (1 % chlorine for at least 30 min).

Several studies about microbiological quality of raw *horchata* have been carried out (Monerris et al. 2001). The microbial results obtained in recent decades have shown that the microbiological quality of the product has improved noticeably thanks to the introduction of mandatory practices and to the principles of Hazard Analysis and Critical Control Points, or HACCP.

Fresh *horchata* obtained by the traditional procedure without using germicide reaches microbial counts of mesophilic microorganisms close to one million of colony-forming units per milliliter. When an identification study of microbial species from the naturally present microbiota of raw *horchata* is carried out, it is possible to observe a dominance of bacterial spores (*Bacillus* spp., *Clostridium* spp.) and bacteria of fecal origin (*Enterobacter aerogenes* and *Escherichia coli*), a fact that can be explained by considering that *horchata* contamination by these microorganisms is mainly due to the organic fertilization applied in tigernut cultivation, to the contaminated water used in irrigation and in the washing of roots, and to the microorganisms contained in the soil particles that are inevitably found on the tigernuts.

The sporulated and fecal microorganisms notably contribute to the *horchata* alteration due to their capacity to ferment different sugars, thereby producing acids and gas. On the other hand, these spores are very resistant to germicide action and even to the heat treatment, which can be applied to this product to increase its shelf life.

Raw *horchata* prepared from tigernuts with no disinfectant treatment may show signs of alteration 24 h after production. However, when a germicide (i.e., chlorine) is applied during tigernut washing, the *horchata* microbial load is considerably reduced, above all the presence of microorganisms of fecal origin and pathogens, thus increasing the shelf life of the product (2–3 days at 0–2 °C).

28.3.1   Preservation at Low Temperatures

Among the preservation treatments applied to food, physical treatments are the most appropriate, as long as its application does not substantially modify the physicochemical and sensory characteristics of the product. Preservation by using low temperatures is one of the best options for raw *horchata*, because the microbial, enzymatic, and chemical activities are slowed down. Refrigeration between 0 and 2 °C allows the preservation of raw *horchata* for a maximum of 2–3 days, depending on the initial microbial load, or for a few weeks if the product is previously pasteurized (see Sect. 28.3.2).

Freezing is a more effective process, depending on the temperature, and it allows a longer preservation for more than 1 week, in the case of slush *horchata* (it is obtained by cooling and stirring until achieving a slush at −2 to −3 °C, and it is then stored at these temperatures), and up to several months at −18 °C in frozen *horchata*. When this last freezing process is applied to unconcentrated *horchata*, it provokes phase separation and coagulation phenomena negatively affecting the colloidal characteristics and, hence, the general aspect. However, freezing is a suitable process for preserving concentrated *horchata* (see Sect. 28.3.2). Basically, concentrated *horchata* consists of a tigernut extract that is more concentrated than that which is usually used in the preparation of conventional *horchata*. In concentrated *horchata*, in order to avoid microbial alterations, sugar is added until soluble solids reach more than 60 °Brix. When this *horchata* is diluted with water (1 part *horchata* to 3–4 parts water), a *horchata* with similar characteristics to freshly prepared *horchata* is obtained. This product is called condensed frozen *horchata*.

28.3.2 *Preservation by Heat Treatments*

Preservation by heat treatments offers the possibility to destroy microorganisms and enzymes. However, as *horchata* is a product with a high concentration of microbial spores, the treatment conditions required can be very drastic and cause changes in the *horchata* constituents, mainly in starch, that modify the sensory characteristics of the product extensively. *Horchata* contains ~3 % of starch that gelatinizes at 75 °C and above. However, it is possible to eliminate easily the *horchata* starch by centrifugation or by total or partial hydrolysis of starch by the use of amylases. The elimination of *horchata* starch does not considerably reduce the sensory acceptance of *horchata*, allowing both sterilization and high pasteurization heat treatments.

Pasteurization is a heat treatment that prolongs the shelf life of this product which, while not sterilizing the product (spores are not destroyed), inactivates pathogenic microorganisms (killing the vegetative cells). Pasteurized products may still contain nonpathogenic bacteria, which eventually reduce shelf life. Furthermore, as spores are not destroyed and the beverage pH is neutral, the spores will eventually germinate and reestablish the vegetative population. Thus, pasteurized horchata still needs to be refrigerated to retard this process and give further shelf life. Early systems relied on heating products to about 63–65 °C for about 30 min in batch heating units, followed by rapid cooling; this process is referred to as low-temperature long-time (LTLT) pasteurization. This process is still used by small-scale *horchata* processors. However, when higher volumes are required, continuous flow plate or tubular heat exchanger-based pasteurizers, in which the product is heated to 72–74 °C for at least 15 s (process designated high-temperature short-time or HTST pasteurization), are largely used. In both types of processes, previous starch elimination is not necessary if temperature and time are controlled correctly, obtaining a shelf life of approximately 1 month at low temperatures (0–2 °C). This type of product is known as natural pasteurized *horchata*. However, when a more severe pasteurization (80–100 °C) or sterilization treatments are applied, starch must be eliminated in order to avoid gelatinization. On the other hand, it is also convenient to apply a homogenization treatment (i.e., 15 + 2 MPa at 65 °C) combined with heat treatment to stabilize the emulsion during storage, avoiding fat creaming. The product processed by high pasteurization with previous elimination of starch is known as pasteurized *horchata*.

The pasteurization of this product allows a noticeable reduction of microbial contamination (>99 %) without affecting its sensory quality.

The pasteurization process (90–95 °C) has also been successfully applied in the sterilization of concentrated *horchata*, with condensed pasteurized *horchata* being obtained as a product that is microbiologically stable at room temperature.

Sterilization consists in the complete destruction or elimination of all viable organisms (yeasts, molds, vegetative bacteria, and spores) and allows the food processor to store and distribute the products at ambient temperatures, thus extending the shelf life.

There exist two conventional types of heat treatments to sterilize products: in container sterilization or aseptic processing. Sterilization procedures for products in containers usually require longer times (120 °C for 15–20 min), due to the resistance to heat transfer of the product itself. Sterilization prior to filling in the container is usually accomplished by heating the product rapidly to 130–145 °C, holding for relatively short heating periods (2–4 s), then rapidly cooling the product, and it is known as ultra-high-temperature (UHT) treatment. In Spain, there are currently several companies that produce conventional sterilized and UHT *horchatas* using these treatments. However, on a sensory quality level, the sterilized and UHT *horchatas* are inferior compared to traditionally refrigerated *horchata* due to the drastic heat treatment applied and to the progressive loss of quality that occurs during storing at room temperature.

28.3.3 Preservation by Drying

Drying is defined as the removal of a liquid, usually water, from a product by evaporation under controlled conditions, leaving the solids in an essentially dry state. The main purpose of drying is to extend the shelf life of foods by reducing their water activity until reaching a level lower than the minimum required for the growth of microorganisms. This level is normally found in the range of 0.65–0.70, which corresponds approximately to a humidity of 5–10 % in the case of powdered *horchata* and similar products. In these drying conditions, a high quantity of microorganisms can be destroyed, although, generally, this effect is never complete, and, therefore, a dried product is not usually sterile. On the other hand, as the water activity decreases, so too will the reaction rates of some chemical reactions, such as oxidation of free fatty acids. However, for a number of chemical reactions, such as Maillard browning, the reaction rate increases in the water activity range 0.4–0.8.

A number of different drying processes are in use in the food, chemical, and pharmaceutical industries, such as spray drying, fluid bed drying, drum drying, batch drying in trays, freeze drying, microwave drying, and superheated steam drying (Refstrup 2002). There are two main techniques to obtain powdered *horchata*: freeze drying and spray drying.

Freeze drying, which is also known as lyophilization, is the process of removing water from a product by freezing it and lowering the pressure below the triple point of water and then sublimating the ice to vapor. Sublimation is a physical phenomenon by which solid ice is converted directly into vapor without passing through the liquid state. The absence of high temperatures protects the product against loss of important constituents and against chemical reactions that are associated with withdrawing or vaporizing liquid water (Mellor and Bell 2003).

The basic principle of spray drying is the exposure of a fine dispersion of droplets, created by means of atomization of preconcentrated liquid products, to a hot airstream. The small droplet size created, and, hence, large total surface area, results in very rapid evaporation of water at a relatively low temperature, whereby heat damage to the product is minimized obtaining a product of high quality.

When this process is applied in *horchata*, it produces a reduction in the microbiota counts (total counts ~90 %, coliforms ~90 %, molds and yeasts ~99 %, and spores ~50 %); it reduces the enzymatic activities (peroxidase 50 %, catalase 10 %), without affecting the *horchata* viscosity when the powdered product is reconstituted.

28.3.4 Preservation by Nonthermal Technologies

Minimally processed, fresh-like products have become usual in the food industry. This is partly the result of consumer demand for high-quality, minimally processed, additive-free, and microbiologically safe foods. In an effort to continue to meet this demand, the industry is developing alternatives to the use of heat preservation to eliminate or reduce levels of bacteria in foods. Heat treatment reduces nutritious components and can destroy the functionality and flavors of many foods depending on the conditions applied. Nonthermal processes can offer an alternative about it.

Among the nonthermal technologies, the application of pulsed electric fields (PEFs) and high hydrostatic pressures (HHPs) has proven capable of increasing the shelf life and to enhance the safety while preserving sensorial and nutritional characteristics of different food liquids.

PEF processing involves treating liquid foods with high-voltage pulses in the range of 20–80 kV for microseconds. The high-voltage pulses break the cell membrane of vegetative microorganisms by electroporation, thus increasing the permeability of membrane and eventually the disruption of cells (Sale and Hamilton 1967). Studies on the inactivation of microorganisms by PEF processing of orange, apple, tomato juices, and *horchata* have shown preservation of flavor, color, and nutrients in comparison to heat pasteurization (Evrendilk et al. 2000; Ayhan et al. 2001; Min and Zhang 2003; Cortés et al. 2005).

Cortés et al. (2005) studied the quality characteristics of *horchata* treated by pulsed electric fields (20 and 35 kV/cm, 100 and 475 µs) during refrigerated storage (2–4 °C). These authors observed faster decreases of pH during storage in untreated *horchata* in comparison to PEF-treated *horchata*, indicating a decrease of stability in the former. In addition, the longer the time over which pulses were applied (for a given field), the more effective was the treatment, allowing the stored *horchata* to have a longer shelf life. On the other hand, PEF treatment does not cause oxidation of fat, and *horchata*s treated by PEF can be kept under refrigeration for 5 days without oxidation of fat taking place, in comparison to other *horchata*s treated by severe heat treatments, which present higher levels of fat oxidation.

Selma et al. (2003, 2006) studied inactivation of *Enterobacter aerogenes* and *Listeria monocytogenes* in PEF-treated *horchata*, concluding that, to prevent growth of these microorganisms in this product, it is necessary to monitor contamination of the product in the production line guaranteeing a very low initial contamination in the raw ingredients and its refrigeration during distribution and storage.

Among the nonthermal food processes, ultra-high-pressure homogenization (UHPH) is a novel technology that improves the physicochemical and microbiological stability of liquid products. This technology is based on the same principle as the conventional homogenization, but works at pressures up to 400 MPa. Different studies have demonstrated the potential of UHPH technology to produce high-quality beverages, in comparison to the conventional heat treatments of sterilization or pasteurization (Pereda et al. 2007; Poliseli-Scopel et al. 2012; Valencia-Flores et al. 2013). In this sense, Codina-Torrella (2014) studied the effects of UHPH processing on *horchata* beverage, showing that it is possible to obtain a pasteurized beverage with improved physicochemical, microbial, and sensory characteristics, in comparison to the beverage treated by a conventional heat pasteurization.

References

Ayhan Z, Yeom HW, Zhang QH, Min DB (2001) Flavor, color, and vitamin C retention of pulsed electric field processed orange juice in different packaging materials. J Agric Food Chem 49:669–674

Codina I, Guamis B, Trujillo AJ (2009) Elaboración de licuado vegetal en base a chufa (*Cyperus esculentus* L.): influencia de los parámetros aplicados en las operaciones de hidratación y molturación del tubérculo sobre la composición final del licuado. Aliment. 404:62–66

Codina-Torrella I (2014) Optimización del proceso de elaboración y aplicación de la homogeneización a ultra alta presión como tecnología de conservación de licuado de chufa. Dissertation, Universitat Autònoma de Barcelona

Codina-Torrella I, Guamis B, Trujillo AJ (2015) Characterization and comparison of tiger nuts Cyperus esculentus L.) from different geographical origin: physico-chemical characteristics and protein fractionation. Ind Crop Prod 65:406–414

Cortés C, Esteve MJ, Frígola A, Torregrosa F (2005) Quality characteristics of *horchata* (a Spanish vegetable beverage) treated with pulsed electric fields during shelf-life. Food Chem 91:319–325

Djomdi, Ejoh R, Ndjouenkeu R (2007) Soaking behaviour and milky extraction performance of tiger nut (Cyperus esculentus) tubers. J Food Eng 78:546–550

Evrendilk GA, Jin ZT, Ruhlman KT, Qiu X, Zhang QH, Ritcher ER (2000) Microbial safety and shelf-life of apple juice and cider processed by bench and pilot scale PEF systems. Innov Food Sci Emerg Technol 1:77–86

Gallart J (1999) Guía para la aplicación del sistema de análisis de riesgos y control de puntos críticos en industrias elaboradoras de horchata de chufa natural. Generalitat Valenciana, Conselleria de Sanitat, Direcció General de Salut Pública, Valencia, España

Joanbanjo (2014) Camp de xufa al costat de l'ermita de Vera, Valencia. https://commons.wikimedia.org/wiki/Category:Cyperus_esculentus#/media/File:Camp_de_xufa_al_costat_de_l%27ermita_de_Vera,_Val%C3%A8ncia.JPG. Accessed 1 June 2015

Linssen JPH, Kielman GM, Cozijnsen JL, Pilnik W (1988) Comparison of chufa and olive oils. Food Chem 28:279–285

Mellor JD, Bell GA (2003) Freeze-drying. The basic process. In: Caballero B, Finglas P, Trugo L (eds) The encyclopedia of food sciences and nutrition, 2nd edn. Academic, London, pp 2697–2701

Min S, Zhang QH (2003) Effects of commercial-scale pulsed electric field processing on flavor and color of tomato juice. J Food Sci 68:1600–1606

Monerris C, Morcado G, Moreno P, Fagoaga F, Pérez N, Gisbert B, Murtó S, Blanquer C (2001) Sanitary training and microbiological quality of natural *horchata* de chufa. Alimentación, Equipos y Tecnología 20:87–93

Morell J, Barber S (1983) Chufa y *horchata*: características físicas, químicas y nutritivas. C.S.I.C. Instituto de Agroquímica y Tecnología de Alimentos, Valencia

Pereda J, Ferragut V, Quevedo JM, Guamis B, Trujillo AJ (2007) Effects of ultra-high pressure homogenization on microbial and physicochemical shelf-life of milk. J Dairy Sci 90:1081–1093

Poliseli-Scopel FH, Hérnández-Herrero M, Guamis B, Ferragut V (2012) Comparison of ultra high pressure homogenization and conventional thermal treatments on the microbiological, physical and chemical quality of soymilk. Food Sci Technol 46:42–48

Refstrup E (2002) Drying of milk. Drying principles. In: Roginski H, Fuquay JW, Fox PF (eds) The encyclopedia of dairy sciences. Academic, London, pp 860–871

Sale AJH, Hamilton WA (1967) Effects of high electric fields on microorganisms. I. Killing of bacteria and yeasts. Biochim Biophys Acta 148:781–788

Sánchez-Zapata E, Fernández-López J, Pérez-Álvarez JA (2012) Tiger Nut (Cyperus esculentus) commercialization: health aspects, composition, properties, and food applications. Compr Rev Food Sci Food Saf 11:366–377

Selma MV, Fernández PS, Valero M, Salmerón MC (2003) Control of *Enterobacter aerogenes* by high-intensity pulsed electric fields in *horchata*, a Spanish low-acid vegetable beverage. Food Microbiol 20:105–110

Selma MV, Salmerón MC, Valero M, Fernández PS (2006) Efficacy of pulsed electric fields for *Listeria monocytogenes* inactivation and control in *horchata*. J Food Saf 26:137–149

Tamorlan (2009) Chufas. https://commons.wikimedia.org/wiki/File:Chufas-2009.jpg. Accessed 1 June 2015

Valencia-Flores DC, Hernández-Herrero M, Guamis B, Ferragut V (2013) Comparing the effects of ultra-high-pressure homogenization and conventional thermal treatments on the microbiological, physical and chemical quality of almond beverages. J Food Sci 78:199–205

Part VI
Traditional Deserts, Side Dishes and Oil

Chapter 29
"Dobos": A Super-cake from Hungary

Elisabeth T. Kovács

29.1 Introduction

Hungary is situated in the middle of Central Europe in a very special position. Hungarian cuisine has influenced the history of the Hungarian people. The importance of livestock and nomadic lifestyle of the Hungarian people is apparent in the prominence of meat in Hungarian food and may be reflected in traditional meat dishes cooked over the fire like goulash. In the fifteenth century, King Matthias Corvinus and his Neapolitan wife Beatrice, influenced by the Renaissance culture, introduced new ingredients and spices like garlic and ginger. Later, considerable numbers of Saxons (a German ethnic group), Armenians, Italians, Jews, and Serbs settled in the Hungarian basin and in Transylvania. Elements of ancient Turkish cuisine were adopted during the Ottoman era, in the form of sweets. Hungarian cuisine was influenced by Austrian cuisine under the Austro-Hungarian Empire. Some cakes and sweets in Hungary show a strong German-Austrian influence. All told, modern Hungarian cuisine is a synthesis of ancient Asiatic components mixed with Germanic, Italian, and Slavic elements. The food of Hungary can be considered a melting pot of the continent, with its own original cuisine from its original Hungarian people.

E.T. Kovács (✉)
Faculty of Engineering, Institute of Process Engineering, University of Szeged,
Moszkvai krt 5-7, Szeged 6725, Hungary
e-mail: kovacs.elisabet@gmail.com

© Springer Science+Business Media New York 2016
K. Kristbergsson, J. Oliveira (eds.), *Traditional Foods*, Integrating Food
Science and Engineering Knowledge Into the Food Chain 10,
DOI 10.1007/978-1-4899-7648-2_29

29.2 Hungarian Confectionary, an Example: Dobos Cake

In the fourteenth and fifteenth centuries, honey was the only source of sugar/ sweetness in Europe, and only later was cane sugar used. The gingerbread makers were established in Hungary at the beginning of the seventeenth century. At this time, we can see the combination of the confectionary trade and gingerbread making. It was not until the second half of the nineteenth century at the advent of industrial sugar production and confectionary manufacturing. From the seventeenth century, Hungary became more closely connected to the West, not just politically but also economically. The Hungarian relationships changed to the Western values. Foreign specialists arrived from France, Switzerland, and Germany, and they were the pioneers of Hungarian confectionary. It was during this time that Hungarians began to prepare and consume desserts. On the basis of the decorative work, confectionary was viewed less as a type of industrial activity but more as a form of art, and for this reason the masters of the period never collaborated to form a guild.

The production of confectionary items for sale to the public began at the end of the eighteenth century and developed in parallel with confectioners working within the households of the nobility. The only baked floury product was the "biscuit." The range of baked floury products was widened by the "mandoletti bakers," who immigrated to Hungary from the Graubünden region of Switzerland and who gradually merged into the confectionary community.

By the seventeenth century, their influence had strengthened to such an extent that they took on the leading role in the confectionary industry. The Swiss confectioners were the first to work as professional confectioners. During the age of reform, shops that until then had only sold sweets, and confections became establishments where people could sit down and eat their products on-site. These quaint Biedermeier places that conformed to the bourgeois tastes were equipped with tables, chairs, musical clocks, sold gugelhupf (a traditional ring-shaped sponge cake), and coffee. The confectionary was especially popular among ladies, who could finally chat, gossip, and enjoy the sweets far from the ever-present eyes and ears that awaited them at home. The first shop of this kind in Pest was opened in 1814. Aside from the very popular coffeehouses, which were more favored by gentlemen, the confectionaries soon became the centers of city social life. Instead of the expressions "sugarwright" and "sugar cake maker," István Szechenyi used the now everyday Hungarian word for confectioner "cukrász" in 1930.

These years also brought other innovations for confectioners. Since the naval blockade ordered by Napoleon had obstructed the import of expensive cane sugar, they changed to using sugar beet as a source of sugar, a crop which could be grown in Hungary. By 1840, cane sugar had been almost totally replaced in the market. The 1867 Compromise with Austria led to swift changes in the economy, society, and culture alike. In the second half of the nineteenth century, the sphere of activities of confectionaries was much wider than it is today, but the choice within the various product types was much narrower. They produced their own boiled sweets, preserves, ice-cream flavorings, decorative items, wafers, etc. The buttercream desserts produced by Emil Gerbeaud and József C. Dobos were first experienced by the Pest public. The Dobos Confectionery Museum is now situated in Szentendre. József C. Dobos wrote several books, which are now very famous: in 1881, he published the *Hungarian French Cookery Book* and later in 1912 the book *Secrets for Women*.

Dobos torte utilizes some special processing techniques: the first is based on the sponge cake and the other on the process of caramelization. Sponge cake is a cake based on flour (usually wheat flour), sugar, and eggs, sometimes with baking powder which derives its structure from egg foam into which the other ingredients are folded. It may be soaked in rum. In addition to being eaten on its own, it lends itself to incorporation into a vast variety of recipes in which premade sponge cake serves as the base. The sponge cake is thought to be one of the first of the non-yeasted cakes, and the earliest recorded sponge cake recipe is to be found in a book of English poetry from 1615. Caramel is beige to dark-brown confection made by heating any variety of sugars. The process of caramelization consists of heating sugar slowly to around 170 °C (340 °F). When the sugar heats up, the molecules break down and reform into compounds with a characteristic flavor and appearance. Most linguists trace the origin of the word to Medieval Latin "cannamellis," or "honey cane" (i.e., sugar cane) or from Latin "callamellus" (little reed, is also presumed to refer to sugarcane). Some have suggested its origin in the Arabic phrase, which means "ball of sweet salt." Caramelization is the removal of water from a sugar, proceeding to isomerization and polymerization of the sugars into various high weight compounds. Compounds such as difructose anhydride may be created from monosaccharide after water loss. Fragmentation reactions result in low-molecular-weight compounds that may be volatile and may contribute to flavor. Polymerization reactions lead to larger-molecular-weight compounds that contribute to the dark-brown color. Caramel is a dark, rather bitter-tasting liquid, the highly concentrated product of near total caramelization that is bottled for commercial and industrial use.

The Dobos torte is a round torte, comprised of six thin layers of sponge cake filled with chocolate buttercream. The layers of pastry and chocolate filling give it a pleasant taste. The pastry layers of the original cake were made from a very thin sponge mixture, to which was poured melted butter to make it even crunchier. The mixture was poured onto a buttered and floured tin tray and spread into a circular cake shape. After baking, it was cut using a cake ring to ensure it was a regular circle. Confectioners today use a clever device known as the Dobos-layer drawer, to spread the mixture onto a baking tray (greaseproof paper). The Dobos sponge layer was a new creation in his recipe, but the real secret was the chocolate buttercream, made even smoother by using cocoa, which the master filled between the baked sponge layers. Dobos spread melted, golden-brown caramelized sugar onto the top-most thin layer of the sponge and immediately cut into equal segments, so that when it had cooled down, the cake could be easily cut into slices (Fig. 29.1).

29.3 History of the Dobos Cake

The torte bears the name of its inventor, József C. Dobos (1847–1924), one of the most interesting, versatile, and enigmatic figures of the Hungarian culinary world, a master chef, a tradesman, and a specialist. The offspring of a dynasty of cooks, he

Fig. 29.1 Dobos cake in the A Cappella Cafe and Confectionary of Szeged, Hungary (Photo by Elisabeth T Kovács, 26-06-2015)

learned the tricks of the trade from his father while working at the service of the noble family Andrassy. In 1879, he opened a delicatessen shop in the Kecskemet Street in Pest, where he sold foods and canned goods which he had prepared himself. In the laboratory behind the shop, he experimented ceaselessly (Gönzi 2001).

The cake which he became most famous for and which immortalized its inventor both at home and abroad was born as a result of continuous experimentation on the part of the ever-creative Dobos—and nobody managed to discover the recipe for decades. According to the more spiteful commentators, this sweet buttercream was not invented by Dobos at all but was the result of a mistake on the part of one of his apprentices, who instead of salt—which at the time was practically the only agent available with which to conserve butter—accidentally added sugar-loaf powder to it. He began to think about what he could do with the ruined expensive ingredients. "He realized how extraordinary it was, sweet butter. What a smooth and interesting taste." Dobos, who was hungry for innovations and forever experimenting, started to try to see how he could make the cream even more delicious.

The torte which was to become famous was first made in 1884 but did not receive public attention until 1885 at the Budapest General Exhibition. According to an exhibit in the newspaper "Sunday News," every person of importance stopped by his large establishment, a restaurant called Dobos C. Nagyvendéglő, measuring 832 square meters, in a pavilion in the Városliget. For those with less money at their disposal, the sensational cake could also be enjoyed at the refreshment stand in the building of the Dobos C. József Industrial Exhibition.

The torte was also highly successful at the Millennium Exhibition of 1896. The restaurant's owner personally offered his creation to the Austrian Emperor and King

Francis Joseph and his wife, Queen Elisabeth. (This event and the relevant period are quoted in the operetta entitled "Dobos Torte" by Bela Szakcsi Lakatos and Geza Csermer, which had its premier in 1996.)

In the 1890s, the Dobos torte was the highlight of every high-caliber evening event in Vienna, Budapest, and throughout the country. It also conquered Europe because it could be easily transported, unlike the other tortes of the time which had fillings or toppings that easily soured. They were transported in special cases and flat boxes which were lined with tinfoil and had paper cushions in the corners to protect the tortes. In Hungary, customers requested the new torte in sweet shops everywhere. Hence, the confectioners of Pest attempted to dissect the flavors of the Dobos torte but were unable to unlock the secret. The cake's secret remained protected until 1906. At that time, Dobos passed its recipe to the Budapest Confectioners' and Gingerbread Makers' Chamber of Industry for use by the general public before closing his shop. Soon thereafter, his torte circulated throughout the entire country. Its recipe appeared in practically every cookbook. In the Museum of the Hungarian Commerce and Catering Industry, approximately 130 variations of the recipe may be found.

Among cakes and tortes, the Dobos torte has a royal ranking. It is one of the most sought-after selections at sweet shops and a standard order for special occasions, such as engagement parties or wedding feasts as well as a regular part of the weekend package of cakes.

29.4 The Original Recipe of Dobos Torte

The original recipe is as follows (Csapó and Éliás 2010):
One cake 22 cm (9 in.) in diameter requires six thinly spread sponge layers.

The recipe:
Mix 6 eggs yolks thoroughly with 3 lat* (about 5 dkg**) of icing sugar***, beat the whites of 6 eggs with 3 lat (5 dkg) of icing sugar until hard, then combine with the egg-yolk mixture, 6 lat (5 dkg) of flour and 2 lat (about 3.5 dkg) of melted butter.
To prepare the cream for one cake:
4 whole eggs, 12 lat (20 dkg) icing sugar, 14 lat (23.5 dkg) unsalted butter, 2 lat (3.5 dkg) solid cocoa mass, 1 lat (1.7 dkg) vanilla sugar, 2 lat (3.5 dkg) cocoa butter, 1 bar of genuine, dark chocolate (20 dkg).
*lat: old unit of Hungary ~ 1 lat = 17.5 g
**1 dkg = 10 g
***Icing sugar: very fine sugar or confectioners' sugar 1 dkg = 10 g
Take the 4 eggs and 12 lat icing sugar and beat them above a gas flame until hot, then remove from gas and continue beating until it cools. Beat 14 lat butter thoroughly, then add 1 lat vanilla sugar, 2 lat melted cocoa and 2 lat melted cocoa butter and the 12 lat of warmed, softened chocolate. Combine with the cooled egg mixture and when well mixed, fill between six layers of sponge. Coat the sixth with golden-brown Dobos-sugar and cut into 20 slices. (According to a later recipe, the master also spread a thin layer of apricot or raspberry jam under the sugar coating)

29.5   Introduction to Dobos Cake Making at Home

The preparation of Dobos torte is not a simple process, and several complex steps must be taken in order to produce this very special cake at home (Borbás 2009).

29.5.1   The Cake

Prepare six 9-in. cake tins for baking: cut six circles of waxed paper, brown paper, or baking parchment to fit the bottom of the pans, grease the bottom of each one with butter, place the paper in, and grease as well. Set the pans aside until ready to use.

Preheat the oven to 400 °F (204 °C). Beat the egg whites with a pinch of salt until foamy; continue beating until stiff peaks are formed. Set aside. Using an electric mixer, beat the egg yolks and the sugar together until lemon-colored and very thick. About 1/4 cup at a time, sift the flour on top of the egg yolk and sugar mixture, and fold it in. Mix a tablespoonful of beaten egg whites into the batter to lighten it, and then gently fold in the rest of the whites. Keep a light touch throughout, handling the batter just enough to make sure it is evenly blended.

Take a prepared pan, spread one-sixth of the batter on the bottom as evenly as possible, and let the batter touch the sides of the pan at several points. Place it in the middle of the preheated oven. Bake for 10–12 min or until the cake hardens and begins to change color. Remove from the pan with a spatula, invert, and quickly but carefully tear off all the paper. Cool on a cake rack. Continue in this fashion until all the layers are baked. During the baking time, prepare the filling as follows.

29.5.2   The Filling

Melt the chocolate with the coffee in a double boiler or over very low heat. Cream the butter with the sugar and beat until fluffy. Add the melted chocolate and beat until it is well blended. Beat in the eggs, one at a time, and continue beating until the cream is light and fluffy. Keep the filling in the refrigerator until ready for use.

29.5.3   The Glaze

When all the layers are ready, pick the best one for the top. Place it on a piece of waxed paper and set it aside. Spread filling on the first four layers and stack them; put the fifth layer on top. Saving sufficient filling for the fifth layer plus a little extra, frost the outside of the cake and then the fifth layer.

Meanwhile, melt the sugar in a light-colored heavy skillet over low heat. Continue cooking until the caramel is smooth and quite brown. Do not touch or taste the caramel: it is very hot! When ready, pour it quickly over the sixth layer, and spread it evenly with a spatula.

With an oiled or buttered knife, quickly cut the caramel-topped layer into 12 or 16 wedges before the caramel hardens. As soon as it dries, place the wedges on top of the cake and use the rest of the filling to frost the outside of the fifth and sixth layers. If there is enough filling left, put it in a pastry bag, and pipe a design along the top edge of the torte. Leave the cake in a cool place until ready to serve. Keep any leftovers in the refrigerator.

References

Borbás M (2009) Values saving in other mode. DVD. Produced by Television HIR TV ZRT (Chapter 1)

Csapó K, Éliás T (2010) Dobos and 19th century confectionery in Hungary. Hungarian Museum of Trade and Tourism, Budapest, pp 10–94. ISBN 978-615-5021-02-2

Gönzi T (2001) Traditions-flavors'-regions: collection of Hungary's traditional and local agricultural products, vol II. Published Ministry of Agriculture and Rural Developing, AMC Kht, Budapest, pp 98–101. ISBN 963 00 90228, 963 009023 6

www.europeancuisines.com/Hungary-Dobos-Dobosh-Torte-Torta-Chocolate-Buttercream-Layer-Cake—40k

Chapter 30
Traditional Green Table Olives from the South of Portugal

Maria Alves and Célia Quintas

30.1 Introduction

The olive tree (*Olea europaea* L.) is one of the most important trees in Mediterranean countries. Its products, olive oil and table olives, are relevant components of the Mediterranean diet and are also consumed worldwide. Olives are known sources of monounsaturated lipids, phenolic compounds, and triterpenic acids, which have been associated to numerous biological activities (Kountouri et al. 2007; Romero et al. 2010).

The International Olive Oil Council (IOOC) estimates that the world production of table olives reached around 1,823,000 t in the 2009/2010 crop year. In the European Union, Spain was the leading producer with 492,600 t, followed by Greece with 107,000 t, Italy with 60,000 t, Portugal with 15,000 t, and France with 1100 t. In Europe, table olives are, probably, the most common fermented food of plant origin consumed (IOOC 2010).

Table olives are prepared from "the sound fruit of varieties of the cultivated olive trees that are chosen for their production of olives whose volume, shape, flesh-to-stone ratio, fine flesh taste, firmness and ease of detachment from the stone make them particularly suitable for processing; treated to remove its bitterness and preserved by natural fermentation; or by heat treatment, with or without the addition of preservatives; packed with or without covering liquid." The most common trade preparations are classified as "treated olives," "natural olives," and "olives darkened by oxidation." "Treated olives" are green or black fruits that have undergone alkaline

M. Alves • C. Quintas (✉)
Departamento de Engenharia Alimentar, Campus de Gambelas, Instituto Superior de Engenharia, Universidade do Algarve, Faro 8005-139, Portugal

Center for Mediterranean Bioresorces and Food (Meditbio), Campus de Gambelas, Faro 8005-139, Portugal
e-mail: cquintas@ualg.pt

© Springer Science+Business Media New York 2016
K. Kristbergsson, J. Oliveira (eds.), *Traditional Foods*, Integrating Food Science and Engineering Knowledge Into the Food Chain 10,
DOI 10.1007/978-1-4899-7648-2_30

treatment, followed by complete or partial fermentation. "Natural olives" are green, changing color to black, placed directly in brines where they undergo a complete or partial fermentation. "Olives darkened by oxidation" undergo an alkaline treatment where the oxidation occurs due to the aeration and may be fermented or not (IOOC 2004).

The industrial processes of greatest importance are the Spanish-type green olives, the Greek-type natural black olives, and the olives produced by alkaline oxidation called Californian style (Garrido-Fernández et al. 1997). Spanish-style olives are an example of "treated olives" and are harvested when the skin color is green to straw yellow. They are then treated with a sodium hydroxide solution (2–5 %, w/v), during a period of time that depends on the variety. This treatment aims to debitter the olives through the degradation of polyphenols (with the chemical hydrolysis of oleuropein) and to increase the permeability of the olive skin, resulting in the efflux of nutrients into the brines. After various washings with water to remove the lye, the olives are brined in 10–11 % (w/v) NaCl solutions in which they undergo lactic acid fermentation (Garrido-Fernández et al. 1997; Sánchez-Gómez et al. 2006).

Unlike Spanish-style green olives, Greek-style black olives are harvested mature, when the surface color is black, and are not submitted to lye treatments, being classified as "natural olives." After harvesting, olives are washed and brined in 8–10 % (w/v) NaCl solutions, where fermentation takes place and oleuropein removal is slow and partial (Tassou et al. 2002).

Californian black olives are lye treated, washed in water, and aerated between lye applications, in order to promote oxidation processes dependent on the air introduction throughout the suspension of olives in the liquid. Oxygen promotes chemical oxidation reactions in which phenolic compounds are polymerized to form dark pigments. In the case of California-style green olives, the fruits are lye treated through several applications, followed by steps of washing, brining, and canning. To finalize, olives are thermally processed. Unlike the other olive preparations described, California-style olives do not undergo a fermentation process (Hutkins 2006).

Before commercialization, green olives may undergo a series of complementary operations as pitting, stuffing, and seasoning.

The primary purposes of table olive processing and fermentation are to remove the natural bitterness, to achieve a preservation effect, and to enhance the organoleptic and nutritional attributes of the final product.

In the Mediterranean region, there are many other traditional/industrial ways of processing table olives according to the olive variety, maturation of fruits (green, turning color to black), fermentation conditions (temperature, salt content), type of microorganisms, type of compounds diffusing from drupes, strategies to debitter, and season. In some of them, lye-untreated olives are directly brined after harvesting. Once in brine, olives undergo fermentation and are maintained in these solutions until they lose their natural bitterness at least partially. The organoleptic properties of these untreated olives classified as "natural olives" are different from the lye-treated ones mainly due to the residual bitterness they retain. However, they are very much appreciated by consumers due to their flavor and to the willingness for traditional and natural foodstuffs (Alves et al., 2012; Garrido-Fernández et al. 1997; Oliveira et al. 2004; Panagou et al. 2008; Hurtado et al. 2008; Aponte et al. 2010; Bautista-Gallego et al. 2010).

The major microbial groups involved in olive fermentation are lactic acid bacteria and yeasts. When lactic acid bacteria outgrow yeasts, lactic acid fermentation is favored, and a product with a lower pH is obtained. On the other hand, when yeasts become the dominant microbial group, the olives produced have a milder taste and a higher pH value. The fermentation of Spanish-style olives is mainly due to lactic acid bacteria (starter cultures); however, yeasts can be present. In the Greek processed olives, the main microorganisms responsible for the fermentation are both yeast and lactic acid bacteria (Nychas et al. 2002; Tassou et al. 2002; Chammem et al. 2005; Panagou et al. 2008). Microorganisms use the nutrients from the olives, especially sugars, and produce acids leading to a pH drop. Consequently, the brine acidity increases and the pH decreases, which are determining factors for the success of the fermentation and safety of table olives (Sánchez-Gómez et al. 2006).

In the southern part of Portugal (Algarve), cracked green table olives (Manzanilla variety) not treated with NaOH solutions to debitter is the most popular method of processing fermenting olives. The processing includes harvesting, transporting, sorting, and washing with water to remove superficial dirt. Then, olives are broken and brined and a spontaneous fermentation takes place. The process remains empirical, where only the salt content and the overall sensory characteristics are verified. In general, these olives are consumed seasoned with aromatic herbs.

30.2   Harvesting and Production

Olives from the Manzanilla variety are harvested by hand, from September to November, when the fruits have reached their maximum development/size and when the surface color is green. They are then transported in plastic containers to the processing units (commercial or homemade productions). There, fruits are washed in water without chlorine to eliminate dirt and are manually selected to remove fruits showing blemishes, cuts, and insect damage. After selection, they are cracked by passing the olives between two stainless steel plates. The distance between the plates is calibrated according to olive size to ensure that the force applied cracks the olive flesh, while the pit remains intact. In homemade preparations, machetes are used to perform the cracking. Cracked olives are transferred to fermentation vessels and covered with brine (8–12 %). A pretreatment with water, to enhance the dilution of the bitter compound oleuropein, before the immersion of the fruits in the brines, may be applied. Once in brines, a fermentation process begins, based on the growth of wild microbiota. After the fermentation period, olives are seasoned with garlic, lemon, thyme, and other aromatic herbs and prepared to be consumed/commercialized. Figure 30.1 shows a diagram of the production of cracked green table olives in comparison to Spanish-style green olives and California-style green olives.

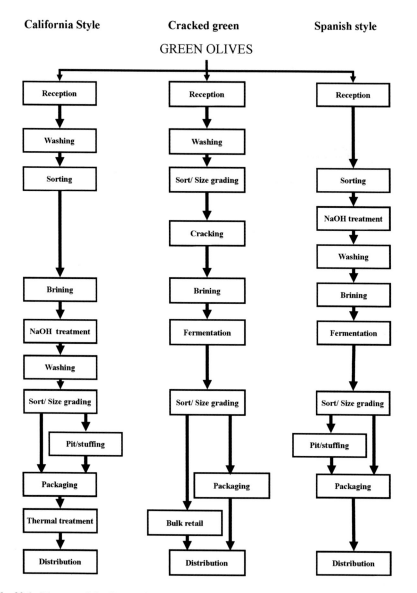

Fig. 30.1 Diagram of the Green table olives:harvesting and production:fermentationproduction of cracked green table olives, California-style green olives, and Spanish-style green olives

30.2.1 Fermentation: Microbiota Evolution

The fermentation of the cracked green table olives is carried out at room temperature due to the activities of the microorganisms present in the fruits and the environment. At the same time, the cracking of the olives facilitates the solubilization of

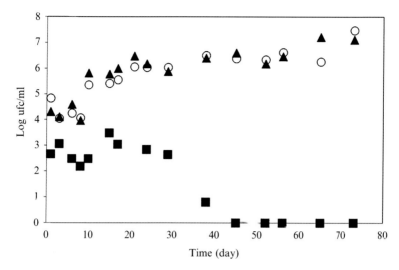

Fig. 30.2 Evolution of microbiota in the brines, during the fermentation of cracked green olives. Total microbiota (*circle*), yeasts (*filled triangle*), and *Enterobacteriaceae* (*filled square*)

sugars which will be used by the microorganisms resulting in the production of acids, alcohols, and other compounds that will contribute to the organoleptic, nutritional, and preservation characteristics of the final product. During the first weeks (3–4), Gram-negative bacteria (*Enterobacteriaceae* and *Pseudomonas* sp.) are present. However, a decline in the *Enterobacteriaceae* and *Pseudomonas* sp. populations occurs thereafter, and no viable counts are found at the end of fermentations. The evolution of total viable counts, yeasts, and *Enterobacteriaceae* in a typical fermentation process is shown in Fig. 30.2. Yeasts were present all along the fermentation period and seem to be the group of organisms with a relevant activity in the processing of this traditional product. On the other hand, lactic acid bacteria are not detected during the fermentation period. The growth inhibition of these bacteria has been attributed to the presence of antimicrobial phenolic compounds in the brines (Medina et al. 2008, 2009). Consequently, the development of a lactic fermentation is very difficult in this type of olives.

30.2.2 Fermentation: pH and Total Acidity

At the beginning of the fermentation processes, the pH of the brines depends on the olives and the water, but values of 4.7–5.0 are common. The development of microorganisms produces different compounds including acids resulting in a reduction of the pH value and an increase of the total acidity. As an example, the evolution of the pH and titratable acidity during a cracked green table olive production process is shown in Fig. 30.3. In general, the pH values of brines of Manzanilla olives showed

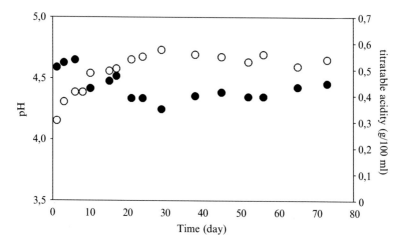

Fig. 30.3 Evolution of pH values and total acidity in the brines, during a fermentation of cracked green olives. Total acidity (*circle*) and pH (*filled circle*)

a slow drop during the first weeks of fermentations. After this decrease, the pH reached a plateau of about 4.1–4.5. The acidity levels attained values of 0.30–0.60 g lactic acid/100 mL brine. The main acids found in the fermentation brines were lactic, acetic, citric, and succinic. According to IOOC (2004), natural fermented olive brines should have a maximum pH value of 4.3 and a minimum acidity of 0.3 g lactic acid/100 mL brine.

30.2.3   Fermentation: Reducing Sugars and Phenolic Compounds

The initial cracking of the olives causes the rupture of the fruits facilitating the diffusion of various compounds from the olives' cells into the brines, namely, phenolic compounds as well as reducing sugars. Figure 30.4 shows the evolution of reducing sugars and phenolic compounds in cracked green table olive fermentation.

Reducing sugars represent a fermentable material that diffuses from the olives' tissues into the brines and are metabolized by the microorganisms. The levels of those fermenting molecules increased during the first 5–10 days and constantly decreased thereafter. At the end of the fermentation, values of reducing sugars from 2.0 to 6 g/L in the brine have been measured. The high levels of reducing sugars present in the brines at the final stages of fermentation could compromise the storage of this type of olives due to the growth of microbial populations (Alves et al. 2015).

The level of phenolic compounds increased during the first 20 days, remaining at 3–4 g/L of brine until the end of fermentation. During this type of biological processes, the enzymatic transformation of oleuropein in its hydrolysis compounds

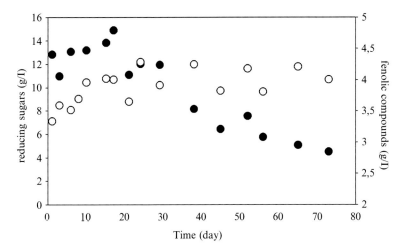

Fig. 30.4 Evolution of reducing sugars and phenolic compounds in the brines, during the fermentation of cracked green olives. Reducing sugars (*filled circle*) and phenolic compounds (*circle*)

(aglycon, elenolic acid, and hydroxytyrosol) with less bitter taste certainly occurs, justifying the levels of phenolics throughout the whole fermentation process. Phenolic substances are implicated in the inhibition of bacteria (lactic acid bacteria and *Enterobacteriaceae*) and are responsible for the bitterness and color of olives (Ruiz-Barba et al. 1993; Rodríguez et al. 2009).

The fermentation is ended after 1–4 months depending on the development of the flavor characteristics and on the overall eating quality.

The main differences between cracked green table olive processing and other fermented preparations to obtain green olives, namely, the Spanish-type olives, reside in the following aspects:

1. Cracked green table olives are not lye treated giving them a more bitter flavor than the Spanish-type green olives or the California-type green olives.
2. The fermentation is mediated by yeasts as lactic acid bacteria are not found, probably due to the high levels of phenols which are bacterial growth inhibitors.
3. This type of fermentation gives origin to less acidic final products with total acidity levels lower than 0.60 % (g lactic acid/100 mL of brine).
4. The processing allows for final products with higher contents of compounds with biological activity.

30.2.4   Color

The fresh appearance of cracked green table olives is related to the green color and is one of the characteristics of this type of fermented food. An example of the evolution of the surface color of the drupes, in terms of **L**, **a**, and **b** in the Hunter system,

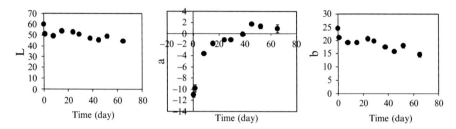

Fig. 30.5 Evolution of olives' surface color parameters **L**, **a**, and **b**, during the fermentation of cracked green olives

during the fermentation is shown in Fig. 30.5. The fruits showed a slight decrease in the lightness parameter (**L**), from initial values of 50–60 in fresh olives to final values of about 44–47 at the end of the fermentation (Fig. 30.5). The **b** (yellow to blue) value changed from initial values of 25 in fresh olives to final values of 14–17 (Fig. 30.5). The gradual decrease of these parameters indicates that the surface of the drupes was becoming darker. The parameter **a** (red to green) changed from the green region (negative values) to red region (positive values) (Fig. 30.5) reaching values of 0.4–0.9. The increase of the **a** values is related to a loss in the green color and a gain in the brown color. The change of the magnesium atom by hydrogen atom in the tetrapyrrole ring of chlorophylls results in the loss of green color due to the production of pheophytins, which are brown. This phenomenon is known as pheophytization (Martins and Silva 2002). However, the retention of the natural green color in cracked green table olives, which is highly appreciated by consumers, enhances the fresh attribute of the final product.

30.2.5 Packing and Storage

By the end of the fermentation, olives must have reached the proper characteristics to be consumed and commercialized. Cracked green table olives are, in general, sold in bulk. Nowadays, there is an increasing demand for packed traditional olives, and plastic containers have been introduced by some industries. In these cases, the final packing is achieved with fresh brine containing garlic, lemon, thyme, oregano, and local aromatic herbs (as *Calamintha baetica*).

References

Alves, M, Esteves, E, Quintas, C (2015) Effect of preservatives and acidifying agents on the shelf life of packed cracked green table olives from Maçanilha cultivar. Food Packaging and Shelf live 5:32–40

Alves, M, Gonçalves, T, Quintas, C (2012) Microbial quality and yeast population dynamics in cracked green table olives' fermentations. Food Control 23(2):363–368

Aponte M, Ventorino V, Blaiotta G, Volpe G, Farina V, Avellone G, Lanza CM, Moschetti G (2010) Study of green Sicilian table olive fermentations through microbiological, chemical and sensory analyses. Food Microbiol 27(1):162–170

Bautista-Gallego J, Arroyo-López FN, Durán-Quintana MC, Garrido-Fernández A (2010) Fermentation profiles of Manzanilla-Aloreña cracked green table olives in different chloride salt mixtures. Food Microbiol 27(3):403–412

Chammem N, Kachouri M, Mejri M, Peres C, Boudabous A, Hamdi M (2005) Combined effect of alkali pretreatment and sodium chloride addition on the olive fermentation process. Bioresour Technol 96(11):1311–1316

Garrido-Fernández A, Fernández-Díaz MJ, Adams RM (1997) Table olives: production and processing. Chapman & Hall, London

Hurtado A, Reguant C, Esteve-Zarzoso B, Bordons A, Rozès N (2008) Microbial population dynamics during the processing of *Arbequina* table olives. Food Res Int 41(7):738–744

Hutkins RW (2006) Fermented vegetables. In: Microbiology and technology of fermented foods, 1st cdn. IFT Press Blackwell, London, pp 233–259

International Olive Oil Council (IOOC) (2004) COI/OT/NC no. 1-Trade standard applying to table olives. International Olive Oil Council, Madrid, p 17

International Olive Oil Council (IOOC) (2010) Production tables. http://www.internationaloliveoil.org/estaticos/view/132-world-table-olive-figures. Accessed 10 Feb 2011

Kountouri AM, Mylona A, Kaliora AC, Andrikopoulos NK (2007) Bioavailability of the phenolic compounds of the fruits (drupes) of *Olea europaea* (olives): impact on plasma antioxidant status in humans. Phytomedicine 14(10):659–667

Martins RC, Silva CLM (2002) Modelling colour and chlorophyll losses of frozen green beans (Phaseolus vulgaris, L.). Int J Refrig 25(7):966–974

Medina E, Romero C, de Castro A, Brenes M, García A (2008) Inhibitors of lactic acid fermentation in Spanish-style green olive brines of the Manzanilla variety. Food Chem 110(4):932–937

Medina E, García A, Romero C, de Castro A, Brenes M (2009) Study of anti-lactic acid bacteria compounds in table olives. Int J Food Sci Technol 44(7):1286–1291

Nychas GJE, Panagou EZ, Parker ML, Waldron KW, Tassou CC (2002) Microbial colonization of naturally black olives during fermentation and associated biochemical activities in the cover brine. Lett Appl Microbiol 34(3):173–177

Oliveira M, Brito D, Catulo L, Leitão F, Gomes L, Silva S, Vilas-Boas L, Peito A, Fernandes I, Gordo F, Peres C (2004) Biotechnology of olive fermentation of Galega portuguese variety. Grasas y Aceites 55(3):219–226

Panagou EZ, Schillinger U, Franz C, Nychas GJE (2008) Microbiological and biochemical profile of cv. Conservolea naturally black olives during controlled fermentation with selected strains of lactic acid bacteria. Food Microbiol 25(2):348–358

Rodríguez H, Curiel JA, Landete JM, de las Rivas B, de Felipe FL, Gómez-Cordovés C, Mancheño JM, Muñoz R (2009) Food phenolics and lactic acid bacteria. Int J Food Microbiol 132(2–3):79–90

Romero C, García A, Medina E, Ruíz-Méndez MV, de Castro A, Brenes M (2010) Triterpenic acids in table olives. Food Chem 118(3):670–674

Ruiz-Barba JL, Brenes M, Jiménez R, García P, Garrido A (1993) Inhibition of *Lactobacillus plantarum* by polyphenols extracted from two different kinds of brine. J Appl Bacteriol 74:15–19

Sánchez-Gómez, A, Garcia, P, Navarro L (2006) Trends in table olive production. Grasas y Aceites 57(1): 86–94

Tassou CC, Panagou EZ, Katsaboxakis KZ (2002) Microbiological and physicochemical changes of naturally black olives fermented at different temperatures and NaCl levels in the brines. Food Microbiol 19(6):605–615

Chapter 31
Extra Virgin Olive Oil and Table Olives from Slovenian Istria

Milena Bučar-Miklavčič, Bojan Butinar, Vasilij Valenčič, Erika Bešter, Mojca Korošec, Terezija Golob, and Sonja Smole Možina

31.1 Introduction

Olive oil production has a long tradition in Slovenian Istria: it was mentioned for the first time by the Greek historian Pausanias. There are many established sources on olive oil production in the area dating from the thirteenth to the seventeenth century. Until the second half of the nineteenth century, olive oil production was the main industry of the area, witnessing a decline after the hard frost of 1929.

Olive growing and olive oil production are an important complementary economic activity and at the same time they are important for the local tradition, culture and lifestyle. Olive growing was revitalised after 1985, and since then the surface area of olive groves has increased from 500 to 1800 ha. Extra virgin olive oil from Slovenian Istria with protected designation of origin is the first Slovenian agricultural product to be registered as such at the European level, while table olives from Slovenian Istria have this protected designation of origin at the national level.

M. Bučar-Miklavčič (✉)
LABS LLC, Institute for Ecology, Olive Oil and Control, Zelena ulica 8, Izola 6310, Slovenia

Science and Research Centre of Koper, University of Primorska,
Garibaldijeva 1, Koper 6000, Slovenia

Faculty of Health Science, University of Primorska, Polje 42, Izola 6310, Slovenia
e-mail: Milena.Miklavcic@guest.arnes.si

B. Butinar • V. Valenčič • E. Bešter
Science and Research Centre of Koper, University of Primorska,
Garibaldijeva 1, Koper 6000, Slovenia

M. Korošec • T. Golob • S.S. Možina
Department of Food Science and Technology, Biotechnical Faculty, University of Ljubljana,
Jamnikarjeva 101, Ljubljana 1000, Slovenia

© Springer Science+Business Media New York 2016
K. Kristbergsson, J. Oliveira (eds.), *Traditional Foods*, Integrating Food
Science and Engineering Knowledge Into the Food Chain 10,
DOI 10.1007/978-1-4899-7648-2_31

31.2   History and Olive Varieties in Slovenia

In addition to certain places in Lombardy (around Lake Garda), Treviso and Veneto, the Slovenian coastal belt along the northern Adriatic is the northernmost geographic area where olives are still grown in vast cultivated areas. In Slovenia, the olive tree prospers mainly in the coastal area extending up to the Kras edge (Slovenian Istria), though it can also be found in the Gorica/Gorizia district (including Goriška Brda) and in some places in the Vipava Valley. Olive plantations are rather small because of specific relief (terraces) and climate conditions (low temperatures in southern locations, altitude), with their size ranging from 0.2 to 3 ha and with plantations exceeding 10 ha being quite a rare sight.

The olive tree is a typical Mediterranean plant traditionally cultivated in Slovenian Istria. The beginnings of olive growing in Istria are related to those of olive growing along the northern Mediterranean coast where olive trees were introduced to the region by the Phoenicians in 600 BC. Two hundred years later the Greek colonisers brought it to Istria (Hugues 1999). The Greek historian Pausanias (180–115 BC) mentions Istrian olive oil in Paragraph 10.32.19 of his *Description of Greece* (Vesel et al. 2009). In Roman times—the Romans conquered Istria in 177 BC—olive growing was a popular agricultural activity in Slovenian Istria, with Istrian olive oil having already earned a good reputation according to Pliny the Elder (23–79 AD) (Darovec 1998).

Olive growing also prospered under the Venetian and French rule. The first written documents related to olive oil date to 1281 when *La Serenissima* issued a decree stipulating that no ship was allowed to transport olive oil without a certificate issued by the Office of *Ternaria*. The authorities in the Istrian town of Piran issued a similar decree in 1307. Olive oil accounted for not only the most important but also the highest income from taxes levied by the Venetian authorities in Istria (once it had replaced salt as Istria's most valuable commodity). Oil was mostly used for the illumination of towns and soap making, while olive skins, which remained the property of the olive grower, were used as fuel. Oil was produced in special presses, with the bigger ones being locally called *torkl(j)a* and the smaller ones *torkolas*. The data available reveal that in those times there were approximately 800,000 olive trees growing in the area covering 2600 km². Towards the end of the nineteenth century, Slovenian Istria was home to approximately 320,000 olive trees.

The first data on olive varietal diversity were recorded between 1882 and 1902 and published in the book *Studio Tecnico comparato ed atlante delle principali varietà d'olivo coltivato nell'Istria* (*Comparative Technical Study and Atlas of Major Olive Varieties Grown in Istria*) written by Prof. Carlo Hugues and illustrated in oils and watercolours by the formally trained painter Giulio De Franceschi. The book describes the following Istrian varieties: 'Piranska Štorta', 'Piranska Rosulja', 'Piranska Žižula', 'Piranska Buga', 'Piranski Zmartel', 'Piranska Mata' and 'Piranska Črnica'. Classification and naming was based on the know-how available at the time and on the knowledge of olive-growing areas. As a result, the olives were classified into subvarieties of individual Italian varieties and named after the place where they had been discovered (Vesel 2006). According to Hugues, the following varieties could be used for pickling and table olive production: 'Piranska kriva' or 'Štorta', 'Piranska Mata' and 'Piranska Žižula'. Today's appearance and presence of olives in Slovenia is a result of poor crops and frosts in 1604, 1763, 1782, 1789, 1887, 1914, 1929, 1956, 1985 and 1996, as well as of different economic and political conditions. The most fatal winter was that of 1929; it struck the majority of olive groves and together with the general economic crisis of the late 1920s forced the majority of olive growers to abandon the business.

In 1955, Prof. Stanko Kovačič published a survey on the actual state of affairs in olive growing in Slovenian Istria in the journal *Kmetijski vestnik*. On the basis of statistical data, he reported that before the frost of 1929, there were 300,000 olive trees in Slovenian Istria, while in 1955 their number dropped to 120,000. The average number of trees per hectare (i.e., 150) helped him to calculate the average number of trees per hectare before and after the great frost, i.e., 2000 and 800, respectively. The most common varieties were 'Navadna belica' ('Drobnica' or 'Comuna'), 'Črnica' ('Karbona'), 'Buga' and 'Žlahtna belica' ('Istrska belica'). The variety 'Istrska belica' was most popular in the Milje/Muggia hillocks and in the villages of Boljunec/Bagnoli della Rosandra, Dolina/San Dorligo della Valle, Osp and Dekani (Bandelj Mavsar et al. 2008).

Following the frost of 1956, the varietal distribution selection changed in favour of 'Istrska belica', especially due to grafting of the old local varieties 'Črnica' and 'Drobnica'. Lack of plant material resulting from the frost spurred the import of saplings from Italy and, consequently, the introduction of Italian sorts into Slovenian olive groves. After the frost of 1985, the following varieties have been grown in new olive tree orchards: 'Istrska belica', Italian 'Leccino', 'Frantoio' and 'Pendolino' and the table olive variety 'Ascolana tenera', with 'Istrska belica' being the prevalent one. In 2010/2011, approximately 2000 olive oil producers produced around 600–700 t of olive oil in 1800 ha.

31.3   Extra Virgin Olive Oil with Designation of Origin

31.3.1   Quality and Traceability Guaranteed

The extra virgin olive oil 'Ekstra deviško oljčno olje Slovenske Istre' with designation of origin (EDOOSI ZOP) has been produced and bottled (packaged) in the area of Slovenian Istria. The rules of olive growing and oil production, internal control and the certification body, but particularly the knowledge and efforts of producers, are guarantees of quality and genuineness of the product. Land surface, the proportion of olive varieties, integrated production, pressing procedure, oil storage and chemical and sensory quality parameters are under permanent control, which ensures product traceability. The buyer of the extra virgin olive oil with designation of origin of Slovenian Istria thus purchases a genuine olive oil of the topmost quality.

During the last 5 years, the number of producers of EDOOSI ZOP increased from 37 to 72 and the production increased from 11 to 45 t of olive oil.

The designation 'EDOOSI ZOP' has been applied for and held by the Society of Olive Growers of Slovenian Istria (DOSI) established in 1992.

The certification body is BUREAU VERITAS, d.o.o., Ljubljana, and the organisation authorised for internal control is LABS d.o.o., Institute for Ecology, Olive Oil and Control, Izola.

31.3.2   Sort Selection

EDOOSI ZOP is produced from different olive varieties or from a single olive variety grown in the area of Slovenian Istria. Blended olive oils have to meet the following requirements:

- They must contain at least 80 % of Istrska belica, Leccino, Buga, Črnica, Maurino, Frantoio and/or Pendolino.
- They may contain a maximum of 20 % of other varieties not mentioned above.
- There must be at least 30 % of Istrska belica olives in total volume of olives processed.

The blended olive oil is characterised by a special bitter and piquant taste resulting from the fact that it contains at least 30 % of Istrska belica olives. Processed at optimal ripeness, its fruits are characterised by a high content of biophenols (natural antioxidants), which protect oils from spoiling. Such oils keep their freshness longer (even after a year) and are stable, which makes them highly valued for their quality.

The oil made from a single variety of autochthonous or domesticated (local) olive varieties has to contain at least 80 % of the declared variety (Bučar-Miklavčič et al. 2004).

31.3.3   Quality Parameters

EDOOSI ZOP has to meet higher criteria than those stipulated by the Commission Regulation (EEC) No. 2568/91.

The quality parameters, shown in Table 31.1, are taken from the Study on Accreditation Procedure for the Designation of Geographical Origin 'EDOOSI' No. 324-01-7/2002/27 (Bučar-Miklavčič et al. 2004).

Cold extraction or cold pressing of EDOOSI ZOP can take place only in accredited olive oil mills in Slovenian Istria. During processing, the temperature should

Table 31.1 Limit values in accordance with the Study on Accreditation Procedure for the Designation of Geographical Origin 'EDOOSI' No. 324-01-7/2002/27 (Bučar-Miklavčič et al. 2004)

Parameter		Limit value
Acidity (wt%, expressed as oleic acid)		≤0.5
Peroxide number (mmol O_2/kg)		≤7
Spectrophotometric investigation in the UV	K_{232}	≤2.5
	K_{270}	≤0.2
	ΔK	≤0.01
Content of total biophenols (mg/kg)		≥100
Content of oleic acid, C18:1 (wt%)		≥72
Content of linoleic acid, C18:2 (wt%)		≤8.0
Organoleptic assessment	Median for 'fruity' (Mf)	≥2.0
	Median of defects (Md)	=0.0

not rise above 27 °C, and no additives, except water, can be used. Olives have to be processed within 48 h after picking.

EDOOSI ZOP has been labelled in accordance with the Commission Regulation (EC) No. 1019/2002 on marketing standards for olive oil. In order to ensure traceability, the Society of Olive Growers of Slovenian Istria (DOSI) issues labels with serial numbers in accordance with the rules on the procedure for the acquisition and annual validation of EDOOSI ZOP.

As stipulated by Article 6(2) of the Council Regulation (EC) No. 510/2006 on the protection of geographical indications and designations of origin for agricultural products and foodstuffs, the application for registration of EDOOSI ZOP, including the product specification, was published in the Official Journal of the European Union of 31 May 2006 (C 127/16).

The designation 'EDOOSI ZOP' has been entered in the register of protected designations of origin and protected geographical indications in accordance with the Commission Regulation No. 148/2007 of 15 February 2007, which entered into force on 26 February 2007. Figure 31.1 shows an example of the sensory profile of a typical olive oil from Slovenian Istria.

31.4   Table Olives

Table olives from Slovenian Istria are traditionally present in the region. It is well known that olive fruits could not be directly edible at harvest. The olive's bitter taste and the astringent sensation must be reduced using an appropriate production technology, which first involves a debittering phase and subsequently a fermentation process that is carried out by spontaneous microorganisms. Slovenian table olives represent less than 1 % of the annual Slovenian olive production. 'Štorta', 'Istrska belica', 'Buga', 'Mata', 'Žižula' and other local olive varieties are traditionally

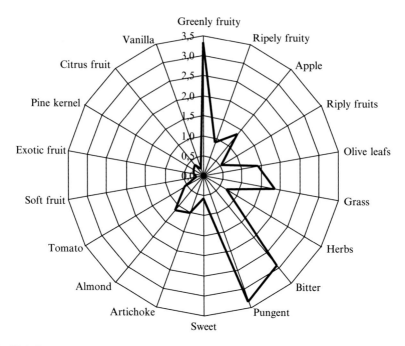

Fig. 31.1 Sensory characteristics of a typical sample of extra virgin olive oil from Slovenian Istria

processed for table olives. With the exception of 'Istrska belica', the other olive varieties are present in very small quantities, and also for that reason, the processing of table olives is still limited to home preparation.

In Slovenian Istria, table olives are usually processed using the traditional production technology, which involves initial debittering in water for 10–20 days, which is replaced every 2 days, and fermentation in brine solution. The fermentation takes place for the first 2 days in a 3.5 % brine solution, then 5–7 days in a 4.2 % brine solution and after that for 6–8 months, until the end of the process, in a 6.2 % brine solution, part of which could be periodically replaced with a new one. Usually, black olives are processed using the processing technology described, but also green and turning colour olives can be processed in this way. Besides the traditional process, a modification of the Spanish-style production technology is used especially for green Slovenian table olives. The olives are treated with a lye solution (1.3–2.5 % sodium hydroxide solution) for 8–12 h, until the lye has penetrated the three quarters of the fruits. In the next step, the olives are washed 15 min under running tap water. Subsequently the olives are placed in water for 30 h, which is replaced five or six times. The process is followed by a spontaneous fermentation in 6–10 % brine solution for 2 months.

Lactic acid bacteria and yeasts are part of autochthonous microbiota of raw vegetables, adapted to their intrinsic characteristics. When lactic acid bacteria outgrows yeasts, lactic acid fermentation is favoured rendering a more acidic product with lower pH, which is greatly desirable in the fermentation of naturally black olives.

The opposite happens when yeasts become the predominant group, resulting in a product with milder taste and fewer self-preservation characteristics. Yeast metabolites contribute to the formation of aroma compounds such as glycerol; higher alcohols; esters; volatile compounds; acetic, succinic and formic acid; ethanol; methanol; and acetaldehyde, which affect the sensory properties of table olives (Arroyo López et al. 2008). In a recent study it was ascertained that yeasts are responsible for the fermentation process of Slovenian table olives with high biophenol content, which presumably inhibited the growth of lactic acid bacteria (Fig. 31.2). The species of *Aureobasidium pullulans*, *Cryptococcus adeliensis*, *Metschnikowia pulcherrima*, *Rhodotorula mucilaginosa*, *Pichia anomala* and *Candida oleophila* represented the leading microflora of Slovenian table olive fermentations (Valenčič et al. 2010).

The traditional table olives preparation is closely related to the traditional production of Piran salt. Table olives from Slovenian Istria, with protected designation

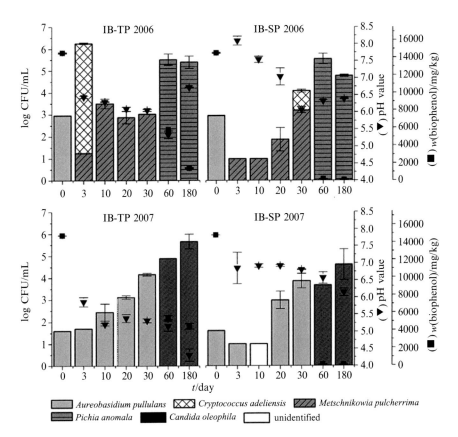

Fig. 31.2 Yeast population dynamics, pH values and total biophenol content during processing of Istrska belica (IB) olive fruits before debittering (0 days) and table olives, processed with traditional regional (TP) and modified Spanish-style (SP) technology, crop years 2006 and 2007. The results are averaged from four replicates and are expressed as means ± standard deviations (*error bars*) (Valenčič et al. 2010)

of origin, are currently registered at the national level and produced and packed for the market in the geographical area of Slovenian Istria. The production is restricted to registred olive orchards of the region. The characteristics of table olives are influenced by local climatic conditions, olive varieties, storage time of olives before processing, production technology using the brine from Piran salt with a protected designation of origin and an appropriate pest control of the fruits according to integrated or biological management (Deržek et al. 2008).

The colour of table olives is strongly related to the olive variety. It can vary from green to straw yellow or from dark pink to brown through deep dark violet. The sensory characteristics of Slovenian table olives are due to the fermentation process. The table olives produced are lightly sour, bitter and salty and are characterised by a good ratio flesh to stone (Fig. 31.3). Especially the flesh of 'Štorta' table olives, that is very popular among consumers, it easily detaches from the stone.

The name 'Namizne oljke Slovenske Istre' (table olives from Slovenian Istria) with protected designation of origin can be used only for the certified product that is processed and prepared for the market in the area of Slovenian Istria and that complies with the specification for the mentioned product, available at the Ministry of Agriculture, Forestry and Food of Slovenia.

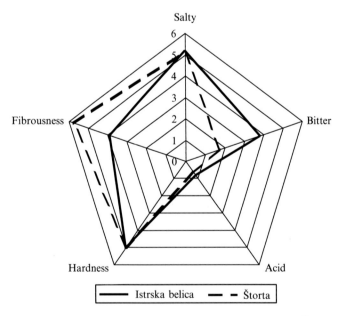

Fig. 31.3 Sensory characteristics of a typical sample of 'Istrska belica' and 'Štorta' table olives from Slovenian Istria

References

Arroyo López FN, Querol A, Bautista-Gallego J, Garrido- Fernández A (2008) Role of yeasts in table olive production. Int J Food Microbiol 128:189–196

Bandelj Mavsar D, Bučar-Miklavčič M, Podgornik M, Valenčič V, Raffin G, Mihelič R, Režek Donev N (2008) Pregled stanja oljkarstva in možnosti širjenja oljčnikov v Slovenski Istri. In: Bandelj Mavsar D, Bučar-Miklavčič M, Obid A (eds) Sonaravno ravnanje z ostanki predelave oljk, Univerza na Primorskem, Znanstveno-raziskovalno središče, Založba Annales. Zgodovinsko društvo za južno Primorsko, Koper, pp 11–26

Bučar-Miklavčič M Dujc V, Hlaj A (2004) Specifikacija za 'Ekstra deviško oljčno olje Slovenske Istre' z zaščiteno označbo porekla. http://www.mkgp.gov.si. Accessed 2015

Commission of the European Communities (2002) Commission Regulation (EC) No. 1019/2002 of 13 June 2002 on marketing standards for olive oil. Official J Eur Communities L155:27–31

Commission of the European Communities (2007) Commission Regulation No. 148/2007 of 15 February 2007 registering certain names in the Register of protected designation of origin and protected geographical indications …—Ekstra deviško oljčno olje Slovenske istre (PDO). Off J Eur Union L46:14–17

Council of the European Union (2006) Council Regulation (EC) No. 510/2006 of 20 March 2006 on the protection of geographical indications and designations of origin for agricultural products and foodstuffs. Off J Eur Union L93:12–25

Darovec D (1998) Proizvodnja oljčnega olja kot osrednja gospodarska panoga slovenske Istre v preteklosti. In: Glasnik ZRS Koper, vol 3 (5), pp 36–49

Deržek P, Dujc V, Valenčič V, Bučar-Miklavčič M, Butinar B (2008) Specifikacija za 'Namizne oljke Slovenske Istre' z zaščiteno označbo porekla. http://www.mkgp.gov.si. Accessed 2015

Hugues C (1999) Maslinarstvo Istre/Elaiografia Istriana. Miljkovič I. (ur.). Zagreb, Ceres

Valenčič V, Bandelj Mavsar D, Bučar-Miklavčič M, Butinar B, Čadež N, Golob T, Raspor P, Smole Možina S (2010) The impact of production technology on the growth of indigenous microflora and quality of table olives from Slovenian Istria. Food Technol Biotechnol 48(3):404–410

Vesel V (2006) "Sorte oljk v Slovenski Istri". 1. mednarodni simpozij o avtohtonih oljkah, Portorož

Vesel V, Valenčič V, Jančar M, Čalija D, Butinar B, Bučar-Miklavčič M (2009) Oljka—živilo, zdravilo, lepotilo. Kmečki glas, Ljubljana, p 141

Chapter 32
A Perspective on Production and Quality of Argentinian Nut Oils

Marcela Lilian Martínez and Damián Modesto Maestri

32.1 Introduction

Nuts are recommended constituents of the daily diet, although their real intake differs remarkably. In particular, they are part of healthy diets such as the Mediterranean diet. In the traditional Mediterranean population, mortality rates from coronary heart disease and cancer are low (Simopoulos 2001). Emerging proofs indicate that nuts may be a source of health-promoting compounds that elicit cardioprotective effects (Kris-Etherton et al. 1999). Nuts are crops of a high economic interest for the food industry. Nut fruits contain high levels of oil (50–70 %), and they are consumed fresh, toasted or salted, or alone or in other edible products. Among the byproducts of the nuts industry, in general, the oil has not yet obtained popularity, although it has been demonstrated that its consumption has many nutritional benefits. The composition of nut oils from different geographic origins has been reported extensively. Their major constituents are triglycerides, of which unsaturated fatty acids are present in high amounts. The presence of other bioactive minor components, such as tocopherols and phytosterols, has been also documented.

A major goal in nut oil production is to find an appropriate method to recover oil from the seeds. To get high-quality virgin oils, the process includes the selection of the fruits, the conditions of their transport and storage, the choice of the steps of the mechanical processing, and finally the control of the storage and packaging conditions of the virgin oils.

M.L. Martínez (✉) • D.M. Maestri
Facultad de Ciencias Exactas, Físicas y Naturales, Instituto Multidisciplinario de Biología Vegetal (IMBIV. CONICET-UNC)—Instituto de Ciencia y Tecnología de los Alimentos (ICTA), Universidad Nacional de Córdoba, Av. Vélez Sarsfield 1611, Córdoba X5016GCA, Argentina
e-mail: mmartinez@efn.uncor.edu

© Springer Science+Business Media New York 2016
K. Kristbergsson, J. Oliveira (eds.), *Traditional Foods*, Integrating Food Science and Engineering Knowledge Into the Food Chain 10, DOI 10.1007/978-1-4899-7648-2_32

In recent years, there has been a resurgence of the interest of continuous mechanical screw presses to recover oil from oilseeds. Screw pressing will not replace solvent extraction in commodity oilseeds extensively because it recovers a lower proportion of oil; in the case of new specialty edible oils, however, screw pressing provides a simple and reliable method of processing small batches of seed. Moreover, mechanical pressing oil extraction is technically less extensive and less labor intensive than solvent extraction. The safety and simplicity of the whole process is advantageous over the more efficient solvent extraction equipment. Furthermore, materials pressed out generally have better preserved native properties, end products are free of chemicals, and it is a safer process.

32.2   Oil Production

The purpose of an oil recovery process is to obtain oil in high yield and purity and to produce coproducts of maximum value.

Screw press performance depends on the method of preparing the raw material, which consists of unit operations such as cleaning, cracking, cooking, drying, or moistening to optimal moisture content. The application of a thermal treatment before or during pressing generally improves oil recovery, but it may adversely influence the oil quality by increasing oxidative parameters. The seed moisture content at the time of pressing is another key process variable. It is known that moisture increases plasticity of seed materials and contributes to press feeding due to its effects as a barrel lubricant. However, high moisture contents may result in poor oil recovery because of insufficient friction during pressing and may affect negatively the chemical quality of the oils, for example, by hydrolysis of glycerides with the consequent increase of oil acidity (Khan and Hanna 1983; Singh and Bargale 1990; Fils 2000; Singh and Bargale 2000; Wiesenborn et al. 2001; Singh et al. 2002; Martínez et al. 2008). An improvement in oil extraction yield may be achieved by an enzymic breakdown of the cell wall coats and by a partial decomposition of matter fiber content, diminishing its resistance to pressing (Dominguez et al. 1996; Rosenthal et al. 1996; Moure et al. 2002; Zúñiga et al. 2003; Soto et al. 2007; Szydlowska-Czerniak et al. 2010). A bigger extraction yield can be attained as a result of a cell membrane rupture obtained using microwave radiation, and permanent pores can be generated, enabling the oil to move through the permeable cell walls during pressing (Uquiche et al. 2008; Azadmard-Damirchi et al. 2010).

There is a consensus on the importance of low temperature in cold pressing, but many different limits are found in literature. In order to be labeled "cold pressed" in the UK, the oil temperature when exiting the screw press should be less than 50 °C (Russo 2009). Singh and Bargale (2000) reported exiting oil temperatures of less than or equal to 70 °C. According to Kirk and Barchmann (2006), truly cold-pressed oil (less than 32.2 °C) has very special quality. Ferchau (2000) stated that temperatures above 20 °C have no additional benefit for oil quality and must be avoided.

Mechanical screw presses typically recover 86–92 % of the oil from oilseeds (Singh and Bargale 2000; Martínez and Maestri 2008). Cold-pressed seed oils may retain more natural beneficial components of the seeds, including natural antioxidants, and are free of chemical contamination. This procedure involves neither heat nor chemical treatments with solvents, and no further processing other than filtering is applied. It is becoming an interesting substitute for conventional practices because of consumers' desire for natural and safe food products (Yu et al. 2005).

In cold-pressed products, minor components affecting the color and flavor and keeping qualities of the oil are thus preserved. These minor constituents can have either prooxidative (free fatty acids, hydroperoxides, chlorophylls, carotenoids) or antioxidative (tocopherols, phenols, phospholipids) effects (Russo 2009).

The quality of all edible oils is the sum of their hygienic, safety for consumption, organoleptic, and nutritional properties; this is determined in turn by the plant source from which oil is produced, the nature of the production process, and last but not least the practices applied in storage and packing (Russo 2009).

Packaging of oils must be considered as carefully as the processing, if acceptable quality is to be maintained. The challenge is to understand the manner in which the package can modify the quality of fats and oils and the measures necessary to maintain quality. The container can only influence the accessibility of light, oxygen, heat, and moisture to the product. Regulations for packaged food, including oils, have a long history dating back to Roman times. Nowadays, great care is taken because of the potential toxicity of some of the monomers and other substances used to produce packaging materials. Many studies have compared different materials suitable for oil packaging. The most important parameters, as mentioned before, are the permeability of package to oxygen and the permeability to light. In order to determine which of these parameters is more important, most studies were performed in the presence and absence of light. It is commonly recognized that glass package is the best material for packaging of oil, due to its impermeability to gases (Martínez 2010; Russo 2009).

32.3   Nontraditional Nut Oils

32.3.1   *Walnut* (**Juglans regia** *L.*)

Walnut has been used in human nutrition since ancient times. At present, walnut is cultivated commercially throughout southern Europe, northern Africa, eastern Asia, the USA, and western South America. World production of whole walnut (with shell) was around 1.5×10^6 t in 2008 (FAOSTAT 2008). China is the leading world producer, followed by the USA, Iran, Turkey, Ukraine, Romania, France, and India, but production in other countries such as Chile and Argentina has increased rapidly in recent years.

Table 32.1 Fatty acid composition (% of total) of oil extracted from nuts

Oil sample	Fatty acid					
	Palmitic (16:0)	Palmitoleic (16:1)	Stearic (18:0)	Oleic (18:1)	Linoleic (18:2)	Linolenic (18:3)
Walnut	6.1–8.1	nd–0.23	0.9–2.9	16–28	50–62	9.6–19
Almond	5.0–9	nd–0.63	1.5–4	50–70	10–26	nd–0.16
Pistachio	9.2–13	0.5–1.4	0.5–2.0	52–72	12–31	0.1–0.4
Hazelnut	4.1–6.0	nd–0.29	1.6–2.7	74–80	10–16	nd–0.46

nd not detected

Walnut presents a high phenotypic variation in the shape and size of fruits, color, thickness, and hardness of shell and kernel (Sharma and Sharma 2001; Yarilgac et al. 2001; Solar et al. 2002; Koyuncu et al. 2004; Zeneli et al. 2005).

The walnut seed (kernel) represents between 40 and 60 % of the nut weight. The seed contains high levels (67 % in average) of oil; the range varies from 60 to 74 % depending on the cultivar, crop year, geographic location, and irrigation rate (Amaral et al. 2003; Prasad 2003; Martínez et al. 2006). The major components of walnut oil (WO) are TAG (up to 98 % of the oil), in which monounsaturated (mainly oleic acid) and polyunsaturated (linoleic and α-linolenic acids) are present in high amounts (Prasad 2003; Crews et al. 2005; Martínez et al. 2006; Martínez and Maestri 2008; Martínez 2010) (Table 32.1).

Fatty acid composition of WO is strongly affected by the genotype and, in a lesser extent, by the crop year and geographical origin (Crews et al. 2005; Martínez et al. 2006). Interestingly, it has been demonstrated that the fatty acid profile of walnut genotypes may influence flavor stability during storage and, therefore, it may affect walnut palatability (Sinesio and Moneta 1997; Savage et al. 1999).

The slightly astringent flavor of walnut fruit has been associated to the presence of phenolic compounds (Prasad 2003). Walnut phenolics are found at higher concentration in the seed coat, the skin that lines the pulp of the nut. They are principally polyphenolics of the non-flavonoid type and fall into the category of ellagitannins (Fukuda et al. 2003; Colaric et al. 2005). Polyphenolics from walnut are reported to have favorable effects on human health due to their anti-atherogenic, anti-inflammatory, and antimutagenic properties (Anderson et al. 2001). Moreover, they could be the most effective antioxidant compounds in preserving WO stability (Fukuda et al. 2003). In effect, among nuts, walnut was shown to be one of the richest sources of total antioxidants (Halvorsen et al. 2002). Nevertheless, due to their low oil solubility, walnut polyphenols are recovered in minor amounts in the extracted oils. Besides these compounds, tocopherols occur in WOs, and they are also important in providing some protection against oxidation (Demir and Çetin 1999; Savage et al. 1999). The measurement of tocopherol isomers in WOs is well documented. Significant differences related to genotype, crop year, and geographical origin were observed (Lavedrine et al. 1997; Savage et al. 1999; Crews et al. 2005; Martínez et al. 2006). Total tocopherol content in WO ranges from 200 to 400 mg/kg. γ-Tocopherol is the main component, followed by δ- and α-tocopherol.

Table 32.2 Tocopherol content of oil (mg/kg) extracted from nuts

Oil sample	α-Tocopherol	β-Tocopherol	γ-Tocopherol	δ-Tocopherol
Walnut	8.7–40.5	nd	172–265	8.2–42
Almond	240–440	nd	12–100	nd
Pistachio	nd	nd	100–434	nd–23
Hazelnut	200–409	6.2–17	19–149	1.0–6.6

nd not detected

Table 32.3 Sterol content of oil (mg/kg) extracted from nuts

	Oil sample			
Sterol	Walnut	Almond	Pistachio	Hazelnut
Cholesterol	nd–38	nd	nd	7.0–12
Campesterol	44–121	55–66	30–45	78–114
Stigmasterol	nd–16	52–56	6.4–9.5	nd–38
Clerosterol	11–50	nd	nd	nd
β-Sitosterol	772–2520	2072–2097	855–880	1416–1693
Δ5-Avenasterol	25–153	nd	57–92	110–170
Δ5-24-Stigmasterol	nd–19	nd	nd	nd
Δ7-Avenasterol	nd–28	nd	nd	37–60
Δ7-Stigmasterol	nd	nd	nd	21–58

nd not detected

No β-tocopherol is identified at most cases (Table 32.2). The presence of other minor unsaponifiable lipid constituents, such as sterols, hydrocarbons, and volatile compounds, is less documented. It appears that variations in sterol composition of WO from different countries are higher than variations among genotypes (Amaral et al. 2003; Crews et al. 2005; Martínez et al. 2006). Beyond this aspect, there is no doubt that β-sitosterol is the predominant compound of the sterol fraction, followed by campesterol and Δ5-avenasterol in similar amounts (Table 32.3). In addition to sterols, small amounts of methylsterols may be detected in WOs (Martínez et al. 2006).

A hydrocarbon fraction containing *n*-alkanes and minor quantities of branched and unsaturated hydrocarbons was isolated from press-extracted WO (Martínez et al. 2006). The *n*-alkane profile is characterized by even carbon-numbered components among which C_{14}–C_{20} compounds predominate. Discrepancies were observed among these data and those from McGill et al. (1993) who found that odd-numbered *n*-alkanes were the major components.

WO volatile compounds include low-molecular weight (C_5–C_9) *n*-alkanes, aliphatic alcohols and aldehydes, and furan derivatives (Clark and Nursten 1976; Caja et al. 2000; Torres et al. 2005). Saturated and unsaturated aldehydes are the most abundant flavor components. The evidence indicates that short- and medium-chain alkanes and aldehydes in WO are produced from unsaturated fatty acids (linoleic acid mainly) through oxidative pathways (Torres et al. 2005).

Although WO is not described by the current Committee on Fats and Oils of the CODEX Alimentarius, it is produced at a small scale in many countries such as France, Spain, Chile, and Argentina. The oil is used directly (without refining) for edible purposes, mainly as a salad dressing. It is also used in the cosmetic industry, as a component of dry skin creams and antiwrinkle and antiaging products.

Due to the high unsaturation level of WO, extreme care needs to be taken to prevent oxidative degradation reactions during storage and processing of fruits and extraction of oils. Typically, oil extraction is carried out by pressing, either by using a screw press or a hydraulic press.

In summary, walnut is an important source of edible oil with high content of essential polyunsaturated fatty acids, pleasant flavor, and good consumption prospects. Although the former aspect is important from a nutritional point of view, it may result in a poor oxidative stability and a shorter shelf life of the oil. The screening of new cultivars devoted to improve the fatty acid and tocopherol profiles, the assessment of oil extraction methods that enhance and preserve the antioxidant substances present naturally in the nut, and the addition of natural antioxidant additives are some aspects that should be considered for a successful commercial production of WO.

32.3.2   Almond *[Prunus dulcis (Mill.) D.A. Webb, Syn. Prunus amygdalus Batsch, P. communis L., Rosaceae]*

Almond is a major tree nut crop cultivated extensively in the Mediterranean region and the USA, which is the world's major producer. The fruit (nut) is utilized as shelled and peeled kernels and packed nuts, or as an ingredient of many food products such as bakery products and confectionary, and as flavoring agent in beverages and ice creams.

Almond quality is primarily genotype dependent (Romojaro et al. 1988). The almond kernel contains mainly lipids (48–67 %).

Almond oil (AO) contains a substantial quantity of triacylglycerols (TAG), which were identified as OOO, OLO, OLL, POO, and PLO (where O = oleic, L = linoleic, P = palmitic) in a decreasing order of abundance (Prats-Moya et al. 1999; Cherif et al. 2004). Differences in TAG composition may be useful to distinguish almond cultivars (Martín-Carratalá et al. 1999). The total lipid accumulation in developing almond kernels follows a sigmoidal curve similar to other seeds and kernels, and significant differences in TAG composition are observed among the different stages of kernel development (Soler et al. 1988; Cherif et al. 2004).

The fatty acid composition of AO has been extensively reported for several cultivars from different geographical origins including the USA, Spain, Italy, Tunisia, and Turkey (García-López et al. 1996; Abdallah et al. 1998; Martín-Carratalá et al. 1998; Askin et al. 2007). The major fatty acid in AO is oleic, representing 50–70 % of the total fatty acid content. Values as high as 80 % have been reported in some genotypes from Turkey (Askin et al. 2007) and Greece (Nanos et al. 2002). Linoleic, palmitic, and stearic acids are present in smaller amounts (Table 32.1). Both kernel weight and storage conditions were found to influence the fatty acid composition of the oil (Nanos et al. 2002; Askin et al. 2007).

Minor components characterized in AO include: tocopherols (about 450 mg/kg oil), sterols (2170 mg/kg), and squalene (95 mg/kg) (Maguire et al. 2004; Kornsteiner et al. 2006). The major tocopherol is α-tocopherol, together with minor amounts of γ-tocopherol (Table 32.3). The sterols are mainly β-sitosterol (95 % of the total sterols) and campesterol and stigmasterol in similar amounts (2.4 %) (Table 32.3).

Freshly extracted AO has a low peroxide value (lesser than 0.2 meq O_2/kg oil). This means that AO is readily preserved from oxidation. Besides tocopherols, which exert a protective role against lipid oxidation, recent studies have shown that almonds also contain a diverse array of phenolic and polyphenolic compounds. These chemical components are mainly present in the skin (seed coat), and they are effective inhibitors of oil oxidative deterioration (Sang et al. 2002; Amarowicz et al. 2005; Wijeratne et al. 2006).

Extended storage of almonds with retention of quality is one of the major needs to the food industry. Long-term storage leads to slow but perceptible deterioration in almond quality. The most common defect of this type is the off-flavor development owing to oxidative rancidity. Relationships among the shelf life of almonds, fatty acid composition, and lipoxygenase (LOX) activity have been reported by Zacheo et al. (1998, 2000). Short-time heat treatments can reduce (LOX) activity, retard the oxidative rancidity development, and extend the shelf life of almonds during storage (Buranasompob et al. 2007).

AO can be obtained by solvent extraction, but it is more frequently by pressing. Good quality AO is light amber in color, with free fatty acid content lesser than 0.5 %. It is consumed unrefined, mainly for edible purposes. Only the low-grade oils are used in cosmetics, for soap manufacturing or other industrial applications.

32.3.3 Pistachio (Pistacia vera L.)

Pistachio is one of the most important tree nuts. It can be cultivated in dry and hot areas and under saline conditions (Metheney et al. 1998). Iran, the USA, Turkey, Syria, Italy, and Greece are the main producers of pistachio nut, and the global production increases steadily (FAOSTAT 2008).

Pistachio is a nut with peculiar organoleptic characteristics. It is widely consumed as a raw or toasted ingredient of many desserts, ice cream, cake, and pastry and for the production of some sausages. The pistachio seed (kernel) contains mainly lipids (50–60 %). The major components are monounsaturated oleic acid, the polyunsaturated linoleic acid, and the saturated palmitic acid (Tsantili et al. 2010) (Table 32.1). The ratio between unsaturated and saturated fatty acids (about 6.5) and the ratio between oleic and linoleic acid (about 4.5), as well as the distribution of fatty acids, show that pistachio oil (PO) is very similar to olive oil (Arena et al. 2007).

The fatty acid composition is influenced mainly by the ecological conditions (Seferoglu et al. 2006). High temperature seems to reduce the production of saturated fatty acid; in fact, pistachios grown in hot regions (over 25 °C) have a smaller content than those grown in more temperate regions (about 22 °C) (Satil et al. 2003). Differences in ecological conditions between different countries are expected to be greater than differences between different locations within the same country. Recently, it was suggested that the geographical origin of pistachios could be discriminated from their fatty acid distribution (Arena et al. 2007). However, these findings are valid only provided that the same varieties are studied, as differences among varieties may exceed differences between countries (Tsantili et al. 2010).

Total tocopherol content in PO ranges from 100 to 500 mg/kg. The major tocopherol is γ-tocopherol, together with minor amounts of δ-tocopherol (Table 32.2). The sterols are mainly β-sitosterol, Δ5-avenasterol, campesterol, and stigmasterol (Kornsteiner et al. 2006; Arranz et al. 2008; Miraliakbari and Shahidi 2008) (Table 32.3).

Extended storage of pistachio with retention of quality is one of the major needs of the food industry. Long-term storage leads to oil oxidation which negatively influences flavor. Moreover, the typical green color and flavor of pistachio might be reduced or lost during the storage period. Some studies were conducted under modified atmosphere in order to improve the storage stability of pistachio, in particular on the potential protective effect of carbon dioxide (Maskan and Karatas 1999) and pure nitrogen (Andreini et al. 2000). Modified atmosphere with high CO_2 improves the storage stability, especially at low temperature.

32.3.4 Hazelnut (Corylus avellana L.)

Turkey is the most important producer of hazelnuts in the world, contributing with 70 % of the total hazelnut world production (Kirbaslar and Erkem 2003).

Hazelnuts are consumed all over the world in dairy, bakery, candy, and chocolate products. Besides being highly nutritious, hazelnuts have other properties that might be useful in the development of products other than foods, for example, cosmetics and pharmaceuticals.

Hazelnuts have a high fat content (55–69 %); monounsaturated (mainly oleic acid) and polyunsaturated fatty acids (mainly linoleic acid) are the major compounds in hazelnut oil (HO) (Savage et al. 1997; Delgado et al. 2010) (Table 32.1). The fatty acid composition is affected by variety and geological, climate, and growing conditions. It is important because it correlates very closely with the stability and degradation kinetics of hazelnuts, especially during storage. Linoleic acid content plays an important role in stability because it is easily oxidized, which causes flavor degradation and rancidity in the nut (Savage et al. 1997; Kirbaslar and Erkem 2003). Other technical conditions, lipase and lipoxygenase activity, thermal processes, storage temperatures, and humidity are also very important factors. If storage conditions are not suitable, it is well known that, in a short period, the nutrient value of hazelnuts decreases and unpleasant odors appear. This odor and flavor are still an important problem, especially in roasted hazelnuts, which lower the economic value of the product (Savage et al. 1997; Seyhan et al. 2002; Kirbaslar and Erkem 2003).

Total tocopherol content in HO ranges from 225 to 550 mg/kg. The major tocopherol is α-tocopherol with minor amounts of γ-tocopherol, β-tocopherol, and δ-tocopherol (Table 32.2) (Savage et al. 1997). The major sterols are β-sitosterol, campesterol, and Δ5-avenasterol. Other minor components are Δ7-avenasterol, Δ7-stigmasterol, and a small amount of cholesterol (Savage et al. 1997) (Table 32.3).

HO can be obtained by solvent extraction, but preferably it is by pressing (Zúñiga et al. 2003; Uquiche et al. 2008). HO can be used for edible purposes and also has cosmetic and pharmacology properties, being used in the formulation of creams, soaps, shampoos, balsams, suntan oils, and solar protectors (Zúñiga et al. 2003).

References

Abdallah A, Ahumada MH, Gradziel TM (1998) Oil content and fatty acid composition of almond kernels from different genotypes and California production regions. J Am Soc Hort Sci 123:1029–1034

Amaral JS, Casal S, Pereira JA, Seabra RM, Oliveira BPP (2003) Determination of sterol and fatty acid compositions, oxidative stability, and nutritional value of six walnut (*Juglans regia* L.) cultivars grown in Portugal. J Agric Food Chem 51:7698–7703

Amarowicz R, Troszynska A, Shahidi F (2005) Antioxidant activities of almond seed extract and its fractions. J Food Lipids 12:344–349

Anderson KJ, Teuber SS, Gobeille A, Cremin P, Waterhouse AL, Steinberg FM (2001) Walnut polyphenolics inhibit in vitro human plasma and LDL oxidation. J Nutr 131:2837–2842

Andreini T, Piergiovanni L, Beninato S (2000) Conservazione in atmosfere protettive di pistacchi (*Pistacia vera*): modificazioni chimiche e sensoriali in un test accelerato di conservazione. Ricerche e innovazioni nell'industria alimentare, vol IV, Pinerolo, Chiriotti Ed., pp 261–267

Arena E, Campisi S, Fallico B, Maccarone E (2007) Distribution of fatty acids and phytosterols as a criterion to discriminate geographic origin of pistachio seeds. Food Chem 104:403–408

Arranz S, Cert R, Pérez-Jiménez J, Cert A, Saura-Calixto F (2008) Comparison between free radical scavenging capacity and oxidative stability of nut oils. Food Chem 110:985–990

Askin MA, Balta MF, Tekintas FE, Kazankaya A, Balta F (2007) Fatty acid composition affected by kernel weight in almond [*Prunus dulcis* (Mill.) D.A. Webb.] genetic resources. J Food Comp Anal 20:7–12

Azadmard-Damirchi S, Habibi-Nodeh F, Hesari J, Nemati M, Achachlouei BF (2010) Effect of pretreatment with microwaves on oxidative stability and nutraceuticals content of oil from rapeseed. Food Chem 121:1211–1215

Buranasompob A, Tang J, Powers JR, Reyes J, Clark S, Swanson BG (2007) Lipoxygenase activity in walnuts and almonds. Food Sci Technol Int 40:893–898

Caja MM, Ruiz del Castillo ML, Martínez Alvarez R, Herraiz M, Blanch GP (2000) Analysis of volatile compounds in edible oils using simultaneous distillation-solvent extraction and direct coupling of liquid chromatography with gas chromatography. Eur Food Res Technol 211:45–50

Cherif A, Sebei K, Boukhchina S, Kallel H, Belkacemi K, Azul J (2004) Kernel fatty acid and tria-cylglycerol composition for three almond cultivars during maturation. J Am Oil Chem Soc 81(10):901–905

Clark RG, Nursten HE (1976) Volatile flavour components of walnuts (*Juglans regia* L.). J Sci Food Agric 27:902–907

Colaric M, Veberic R, Solar A, Hudina M, Stampar F (2005) Phenolic acids, syringaldehyde, and juglone in fruits of different cultivars of *Juglans regia* L. J Agric Food Chem 53:6390–6396

Crews C, Hough P, Godward J, Brereton P, Lees M, Guiet S, Winkelmann W (2005) Study of the main constituents of some authentic walnut oils. J Agric Food Chem 53:4853–4860

Delgado T, Malheiro R, Pereira JA, Ramalhosa E (2010) Hazelnut (*Corylus avellana* L.) kernels as a source of antioxidants and their potential in relation to other nuts. Ind Crops Prod 32:621–626

Demir C, Çetin M (1999) Determination of tocopherols, fatty acids and oxidative stability of pecan, walnut and sunflower oils. Dtsch Lebensmitt Rundsch 95:278–283

Dominguez H, Sineiro J, Núñez MJ, Lema JM (1996) Enzymatic treatment of sunflower kernels before oil extraction. Food Res Int 28(6):531–545

FAOSTAT (2008) Food and Agriculture Organization. FAOSTAT data, Rome

Ferchau E (2000) Equipment for decentralized cold pressing for oil seeds. Folkecenter for renewable energy. www.folkecenter.dk. Accessed July 2008

Fils JM (2000) The production of oils. In: Hamm W, Hamilton RJ (eds) Edible oil processing. Sheffield Academic, Sheffield, pp 47–78

Fukuda T, Ito H, Yoshida T (2003) Antioxidative polyphenols from walnut (*Juglans regia* L.). Phytochemistry 63:795–800

García-López C, Grané-Teruel N, Berenguer-Navarro V, García-García JE, Martín-Carratalá ML (1996) Major fatty acid composition of 19 almond cultivars of different origins. A chemometric approach. J Agric Food Chem 44:1751–1756

Halvorsen BL, Holte K, Myhrstad MCW, Barikmo I, Hvattum E, Remberg SF, Wold AB, Haffner K, Baugerod H, Andersen LF, Moskang JO, Jacobs DR, Blomhoff RA (2002) A systematic screening of total antioxidants in dietary plants. J Nutr 132:461–466

Khan LM, Hanna MA (1983) Expression of oil from oilseed: a review. J Agr Eng Res 28:495–503

Kirbaslar FG, Erkem G (2003) Investigation of the effect of roasting temperature on the nutritive value of hazelnuts. Plant Foods Hum Nutr 58:1–10

Kirk A, Barchmann J (2006) Oilseed processing for small-scale producers. National Sustainable Agriculture Information Service. http://attra.ncat.org/attra-pub/oilseed.html. Accessed 2015

Kornsteiner M, Wagner K-H, Elmadfa I (2006) Tocopherols and total phenolics in 10 different nut types. Food Chem 98:381–387

Koyuncu MA, Ekinci K, Savran E (2004) Compositional changes of fatty acids during the development of kernel of Yalova-1 and Yalova-4 walnut cultivars. Biosyst Eng 87:305–310

Kris-Etherton PM, Yu-Poth S, Sabaté J, Ratcliffe HE, Zhao G, Etherton TD (1999) Nuts and their bioactive constituents: effects on serum lipids and other factors that affect disease risk. Am J Clin Nutr 70(3):504–511

Lavedrine F, Ravel A, Poupard A, Alary J (1997) Effect of geographic origin, variety and storage on tocopherol concentrations in walnuts by HPLC. Food Chem 58:135–140

Maguire LS, O'Sullivan SM, Galvin K, O'Connor TP, O'Brien NM (2004) Fatty acid profile, tocopherol, squalene and phytosterol content of walnuts, almonds, peanuts, hazelnuts and the macadamia nut. Int J Food Sci Nutr 55:171–176

Martín-Carratalá ML, García-López C, Berenguer-Navarro V, Grané-Teruel N (1998) New contribution to the chemometric characterization of almond cultivars on the basis of their fatty acid profiles. J Agric Food Chem 46:963–968

Martín-Carratalá ML, Llorens-Jordá C, Berenguer-Navarro V, Grané-Teruel N (1999) Comparative study on the triglyceride composition of almond kernel oil. A new basis for cultivar chemometric characterization. J Agric Food Chem 47:3688–3693

Martínez ML (2010) Doctoral thesis: Extracción y caracterización de aceite de nuez (*Juglans regia* L.): influencia del cultivar y de factores tecnológicos sobre su composición y estabilidad oxidativa. National University of Cordoba, Argentina

Martínez ML, Maestri DM (2008) Oil chemical variation in walnut (*Juglans regia* L.) genotypes grown in Argentina. Eur J Lipid Sci Technol 110:1183–1189

Martínez ML, Mattea M, Maestri DM (2006) Varietal and crop year effects on lipid composition of walnut (*Juglans regia*) genotypes. J Am Oil Chem Soc 83:791–796

Martínez ML, Mattea M, Maestri DM (2008) Pressing and supercritical carbon dioxide extraction of walnut oil. J Food Eng 88:399–404

Maskan M, Karatas S (1999) Storage stability of whole-split pistachio nuts (*Pistacia vera* L.) at various conditions. Food Chem 66:227–233

McGill AS, Moffat CF, Mackie PR, Cruickshank P (1993) The composition and concentration of n-alkanes in retail samples of edible oils. J Sci Food Agric 61:357–362

Metheney PD, Reyes HC, Ferguson L (1998) Blended drainage water irrigation of pistachios, cv 'Kerman', on four rootstocks in the southern San Joaquin Valley of California. Acta Hortic 470:493–501

Miraliakbari H, Shahidi F (2008) Antioxidant activity of minor components of tree nut oils. Food Chem 111:421–427

Moure A, Domínguez H, Zúñiga ME, Soto C, Chamy R (2002) Characterisation of protein concentrates from pressed cakes of *Guevina avellana* (Chilean hazelnut). Food Chem 78:179–186

Nanos GD, Kazantzis I, Kefalas P, Petrakis C, Stavroulakis GG (2002) Irrigation and harvest time affect almond kernel quality and composition. Sci Hortic 96:249–254

Prasad RBN (2003) Walnuts and pecans. Encyclopedia of food science and nutrition. Academic, London, pp 6071–6079

Prats-Moya MS, Grané-Teruel N, Berenguer-Navarro V, Martín-Carratalá ML (1999) A chemometric study of genotypic variation in triacylglycerol composition among selected almond cultivars. J Am Oil Chem Soc 76:267–272

Romojaro F, Riquelme F, Giménez JL, Llorente S (1988) Fat content and oil characteristics of some almond varieties. Int J Fruit Sci 15:53–58

Rosenthal A, Pyle DL, Niranjan K (1996) Aqueous and enzymatic processes for edible oil extraction. Enzyme Microb Technol 19:402–420

Russo GL (ed) (2009) Final report and scientific handbook of the specific support action—MAC-Oils—mapping and comparing oils, sixth framework programme—priority 5—food quality and safety. Institute of Food Sciences—National Research Council, Avellino

Sang S, Lapsley K, Jeong WS, Lachance PA, Ho CT, Rosen RT (2002) Antioxidative phenolic compounds isolated from almond skins (*Prunus amygdalus* Batsch.). J Agric Food Chem 50:2459–2464

Satil F, Azcan N, Baser KHC (2003) Fatty acid composition of pistachio nuts in Turkey. Chem Nat Compd 39:322–324

Savage GP, McNeil DL, Dutta PC (1997) Lipid composition and oxidative stability of oils in Hazelnuts (*Corylus avellana* L.) grown in New Zealand. J Am Oil Chem Soc 74(6):755–759

Savage GP, Dutta PC, McNeil DL (1999) Fatty acid and tocopherol contents and oxidative stability of walnut oils. J Am Oil Chem Soc 76:1059–1064

Seferoglu S, Seferoglu SG, Tekintas FE, Balta F (2006) Biochemical composition influenced by different locations in Uzun pistachio cv. (*Pistacia vera* L.) grown in Turkey. J Food Comp Anal 19:461–465

Seyhan F, Tijskens LMM, Evranuz O (2002) Modelling temperature and pH dependence of lipase and peroxidase activity in Turkish hazelnuts. J Food Eng 52:387–395

Sharma OC, Sharma SD (2001) Genetic divergence in seedling trees of Persian walnut (*Juglans regia* L.) for various metric nut and kernel characters in Himachal Pradesh. Sci Hortic 88:163–168

Simopoulos AP (2001) The Mediterranean diets: what is so special about the diet of Greece? The scientific evidence. J Nutr 131(11):3065–3073

Sinesio F, Moneta E (1997) Sensory evaluation of walnut fruit. Food Qual Prefer 8:35–40

Singh J, Bargale PC (1990) Mechanical expression of oil from linseed. J Oilseeds Res 7:106–110

Singh J, Bargale PC (2000) Development of a small capacity double stage compression screw press for oil expression. J Food Eng 43(2):75–82

Singh KK, Wiesenborn DP, Tostenson K, Kangas N (2002) Influence of moisture content and cooking on screw pressing of crambe seed. J Am Oil Chem Soc 79(2):165–170

Solar A, Ivancic A, Stampar F, Hudina M (2002) Genetic resources for walnut (Juglans regia L.) improvement in Slovenia. Evaluation of the largest collection of local genotypes. Genet Resour Crop Evol 49:491–496

Soler L, Canellas J, Saura-Calixto F (1988) Changes in carbohydrate and protein content and composition of developing almond seeds. J Agric Food Chem 36:695–700

Soto C, Chamy RL, Zuniga ME (2007) Enzymatic hydrolysis and pressing conditions effect on borage oil extraction by cold pressing. Food Chem 102(3):834–840

Szydlowska-Czerniak A, Karlovits G, Hellner G, Szlyk E (2010) Effect of enzymatic and hydrothermal treatments of rapeseeds on quality of the pressed rapeseed oils: Part II. Oil yield and oxidative stability. Process Biochem 45:247–258

Torres MM, Martínez ML, Maestri DM (2005) A multivariate study of the relationship between fatty acids and volatile flavor components in olive and walnut oils. J Am Oil Chem Soc 82:105–110

Tsantili E, Takidelli C, Christopoulos MV, Lambrinea E, Rouskas D, Roussos PA (2010) Physical, compositional and sensory differences in nuts among pistachio (Pistachia vera L.) varieties. Sci Hortic 125:562–568

Uquiche E, Jeréz M, Ortíz J (2008) Effect of pretreatment with microwaves on mechanical extraction yield and quality of vegetable oil from Chilean hazelnuts (Gevuina avellana Mol). Innov Food Sci Emerg Technol 9:495–500

Wiesenborn D, Doddapaneni R, Tostenson K, Kangas N (2001) Cooking indices to predict screw-press performance for crambe seed. J Am Oil Chem Soc 78(5):467–471

Wijeratne SS, Abou-Zaid MM, Shahidi F (2006) Antioxidant polyphenols in almond and its coproducts. J Agric Food Chem 54:312–317

Yarilgac T, Koyuncu F, Koyuncu MA, Kazankaya A, Sen SM (2001) Some promising walnut selections (Juglans regia L.). Acta Hortic 544:93–98

Yu J, Ahmedna M, Goktepe I (2005) Effects of processing methods and extraction solvents on concentration and antioxidant activity of peanut skin phenolics. Food Chem 90:199–206

Zacheo G, Cappello AR, Perrone LM, Gnoni GV (1998) Analysis of factors influencing lipid oxidation of almond seeds during accelerated ageing. Lebensm Wiss Technol 31:6–11

Zacheo G, Cappello MS, Gallo A, Santino A, Cappello AR (2000) Changes associated with postharvest ageing in almond seeds. Lebensm Wiss Technol 33:415–420

Zeneli G, Kola H, Dida M (2005) Phenotypic variation in native walnut populations of northern Albania. Hortic Sci 105:91–96

Zúñiga ME, Soto C, Mora A, Chamy R, Lema JM (2003) Enzymic pre-treatment of Guevina avellana mol oil extraction by pressing. Process Biochem 39:51–57

Chapter 33
Pupunha (*Bactris gasipaes*): General and Consumption Aspects

Carolina Vieira Bezerra and Luiza Helena Meller da Silva

The *Pupunha* tree is a tropical plant native to the Americas, with a wide variety of breeds and ecotypes. According to studies, there are approximately 73 species deployed around the south Mexico, Caribe, Brazil and Paraguay. It was one of the first plants domesticated by the indigenous people in pre-Columbian times. The roots were used as vermifuge; its trunk was used as wood to make hunting, fishing, and housing tools; and its "core" and fruits were used as food (Shanley and Medina 2005; Ferreira 2005).

Brazil is considered one of the largest producers and consumers of the heart of palm in the world. The extraction techniques employed put in danger some producer species of heart of palm, opening a potential market for the *Pupunha* plantation. Nowadays, the cultivation of this species has been going through an intense process of dissemination outside the Amazon, especially in the states of Bahia, Espírito Santo, Rio de Janeiro, São Paulo, Paraná, and Santa Catarina (Neves et al. 2007).

The heart of palm is removed from the top of the stalk corresponding to the central part. It has a pleasant taste and it is smooth, having low calorific value. Several studies report the advantages of the *Pupunha* tree as the heart of palm producer, such as precocity, with the first cut from 18 months after planting; tillering of the mother plant in a profuse way; quality; and profitability of the heart of palm, where 1 ha can generate 5000–12,000 hearts of palm per year (Resende et al. 2004; Bogantes 1998; Vargas 1994; Marques and Coelho 2003).

Another feature that stands out in *Pupunha*'s heart of palm is the low activity of oxidative enzymes (peroxidase and polyphenol oxidases), which reduces the risk of color changes in the final product (Clement and Pastore 1998; Clement and Galdino 2008; Kapp 2003).

C.V. Bezerra (✉) • L.H.M. da Silva
Food Engineering Faculty, Physical Measurements Laboratory,
Federal University of Para, Para, Brazil
e-mail: cvb@ufpa.br

© Springer Science+Business Media New York 2016
K. Kristbergsson, J. Oliveira (eds.), *Traditional Foods*, Integrating Food
Science and Engineering Knowledge Into the Food Chain 10,
DOI 10.1007/978-1-4899-7648-2_33

Fig. 33.1 Variations in color of the epicarp, varying texture of the mesocarp: oily and starch, with seed and without seed

The most commonly used species for the production of the heart of palm include those that do not grow thorns on their stalk, which facilitates their handling, cultivation, and processing. Studies also show that *Pupunha* trees without thorns have higher yield compared to *Pupunha* trees with thorns, because more sheaths should be removed from the latter in order to eliminate the insertion of thorns (Moro 1993; Morera 1988).

In addition to the consumption in the form of the heart of palm, the *Pupunha* tree stands out for the exploration of its fruits, popularly known as *Pupunha*.

In the Amazon region, the main use of the *Pupunha* tree is in the form of fruit, which is consumed after being cooked in water and salt. Such a practice is required to inhibit anti-nutritional factors (trypsin inhibitors) present in the mesocarp. When ripe, it has a fibrous epicarp ranging in color with shades of red, orange, green, or yellow and a mesocarp ranging from starchy to oily (Fig. 33.1). The fruits may also vary in weight (20–200 g) and size (2–7 cm), presenting themselves with or without seeds, according to the species (Bergo et al. 2004; Borém et al. 2009; Pesce 2009; Leterme et al. 2005).

Nutritionally, the *Pupunha* may also differ mainly in lipids and carbohydrates. Studies involving proximate characterization show lipid values ranging from 3.7 to 15.7 % and carbohydrates varying between 50 and 80 % mainly represented in the form of starch (Clement 1991; Piedra et al. 1995; Silvestre 2005).

On the street markets of the Amazon, the fruits with oily characteristic are the most wanted by the people. To know if the fruit has this characteristic, traditionally, consumers grind a piece of raw *Pupunha* between their fingers and with rubbing movements they identify the oily aspect (Clement 1999). Studies that correlate data of physical characterization and chemical composition show that generally smaller fruits tend to be more oily and fibrous, while bigger fruits tend to be starchy, with these two main components (starch and lipid) in an inversely proportional relationship (Leakey 1999; Leterme et al. 2005; Garcia et al. 1998).

In addition to direct consumption, the fruit can be used as feedstock for production of various products, especially the *Pupunha* flour, the oils, and a fermented beverage, for example. The *Pupunha* in the form of flour is an alternative to avoid the market saturation of fresh fruit and also to diversify the demand for *Pupunha*. The species with mesocarp with high starch content are the most appropriate, because starch, when heated, tends to generate technological properties of interest for certain types of products such as breads, cakes, soups, creams, sauces, and porridge's bases. The incorporation of amounts of *Pupunha* flour in products has been successfully realized. Several studies show that this incorporation adds nutritional value, especially in lipids, beta-carotene, and fiber, with good acceptance by consumers, when this flour is incorporated in proportions of 10–25 % (Oliveira and Marinho 2010; Carvalho et al. 2009; Ferreira and Pena 2003).

The *Pupunha* oil can be extracted from the mesocarp or the seed, presenting in its composition a prevalence of unsaturated fatty acids, ranging from 51.8 to 69.9 %, specially oleic (C18:1) (Gomez and Amelotti 1983; Hammond et al. 1982; Zapata 1972; Zumbado and Murillo 1983).

Even in the lipid fraction, it is possible to find various forms of sterols, with emphasis on sitosterol (Bereau et al. 2003). Its use covers the area of food and cosmetics, due to its emulsifying properties and emollients. Although there are excellent crops for oil in the Amazon region, such as coconut and *dendë* oils, the need for greater diversity is well demonstrated by diseases and pests that limit the productivity of crops, boosting the use of the palm tree for these purposes.

Another byproduct of *Pupunha* that is quite common, especially in rural Amazonia, is a fermented beverage. The indigenous peoples of the Amazon region were the first to use the *Pupunha* as raw material for fermentation. After fermentation, they obtained a dense turbid liquid, with fruit traces, called *caiçuma*. In the literature it is possible to find some studies of improvement of the fermented beverage obtained from *Pupunha*, especially the use of microbial strains to improve sensory characteristics and performance (Andrade et al. 2003; Pantoja et al. 2001).

The *Pupunha* is a typical fruit of Northern Brazil, and it is incorporated in the food habits by being present in the northern breakfast, in snacks in the afternoon, or even in culinary preparations of sophisticated cuisine. The fruit is energetic, with considerable amounts of carbohydrates and lipids, as well as being an important source of vitamin A, which has justified its use in school lunches in the northern region, where it is widely grown, and in several other cities where it is widespread. Some researchers advocate its use as a substitute for corn because of its nutritional value (Table Table 33.1) (Salas and Blanco 1990; Clement and Mora 1987; Kerr et al. 1997).

In the Amazon region, there is a wide variety of fruits with great technological, economic, and nutritional potential, and among these is the *Pupunha* (*Bactris gasipaes*).

Several studies show that *Pupunha* has a lipid profile with a prevalence of unsaturated fatty acids (over 50 %), highlighting the oleic one (Gomez and Amelotti 1983; Hammond et al. 1982; Zapata 1972; Zumbado and Murillo 1983).

Unsaturated fatty acids have been the target of numerous studies, valued for their nutritional properties that may play an important role in lowering levels of LDL-C which reduces the risk of cardiovascular disease (Brandão et al. 2005; Innis 1991; Lima et al. 2000; Galli et al. 1994; Esrey et al. 1996). The lipid content also contributes to increasing the percentage of calories from the fruit, which can be used as a

Table 33.1 Pupunha's composition

Composition (%)	Garcia et al. (1998)[a]	Ferreira and Pena (2003)[a]
Fat	4.29	3.40
Protein	7.64	6.11
Fiber	1.37	1.10
Carbohydrate	40.64	32.50
Ash	1.04	0.89
Moisture	45.50	56.00
Minerals	Leterme et al. (2005)[a]	
Ca (g)	0.10	
P (g)	0.08	
Mg (g)	0.06	
Na (g)	0.03	
Fe (mg)	4.40	
Cu (mg)	0.40	
Zn (mg)	1.00	

[a]Average composition

nutritional strategy for weight gain, showing itself as an interesting tool for malnourished patients (Leterme et al. 2005).

Appreciable levels of carotenoids are found inside the *Pupunha*. According to the authors, these levels can vary from 17 to 150 µg (Garbanzo et al. 2011; Aguiar et al. 1980). Carotenoids are important elements to the body because they are precursors of vitamin A, which is essential for growth, development, maintenance of epithelial tissues, reproduction, and immune system and, in particular, for the operation of the visual cycle in the photoreceptors' regeneration. Deficiency in vitamin A is a serious public health problem and affects mainly children, causing clinical manifestations known as night blindness. Works developed with *Pupunha's* flour showed that this seems to be a good strategy to prevent or remedy the deficiencies of vitamin A, since, after being inserted in the diet, it increases the hepatic content (Yuyama et al. 1991; Yuyama and Cozzolino 1996).

Tocopherols can also be identified in *Pupunha*, popularly known as vitamin E, as well as sitosterols. The former has a chemical structure similar to cholesterol, with bactericidal, fungicidal, and anti-inflammatory action. Studies have shown that it plays an important role in reducing the blood cholesterol by blocking intestinal absorption (Tango et al. 2004; Ye et al. 2010).

The Pupunha, due to the absence of gluten and richness of starch (depending on the species), can be an interesting strategy for development of products for patients with celiac disease (Zumbado and Murillo 1983).

Fiber is another component present in the *Pupunha* tree, with concentration ranging from 2 to 10 g/100 g of fruit. Due to the functional properties of the fibers, the interest in manufacturing products that have a higher percentage of nondigestible fractions is growing. By not being digested by the organisms initially, they end up being fermented by microorganisms in the intestines, producing compounds (butyrate, ethyl, and propionate) that can exert a positive action in the individual health.

References

Aguiar JPL, Marinho HA, Rebêlo YS, Shrimpton R (1980) Aspectos nutritivos de alguns frutos da Amazônia. Acta Amaz 10(4):755–758

Andrade JS, Pantoja L, Maeda RN (2003) Improvement on beverage volume yield and on process of alcoholic beverage production from pejibaye (Bactris gasipaes Kunth). Food Sci Technol 23(Suppl):34–38

Bereau D, Mlayah BB, Banoub J, Bravo R (2003) FA and unsaponifiable composition of five Amazonian palms kernel oils. J Am Oils Chem Soc 80:49–53

Bergo CL, Mendonça HA, Ledo FJS (2004) Estimation of genetic and phenotypic parameters in different half-brothers of peach palm (Bactris gasipaes Kunth Palmae) in the Western Amazon. Ciênc Agrár 42:127–142

Bogantes A (1998) Recomendaciones para El manejo de pejiabaye (Bactris gasipaes) para palmito. Boletín Tecnico, 9p

Borém A, Lopes MTG, Clement CR (2009) Domesticação e melhoramento: espécies amazônicas. Universidade Federal de Viçosa, 486p

Brandão PA, Costa FGP, Barros LR, Nascimento GAJ (2005) Fatty acids and cholesterol in food. Agropec Técn 26(1):5–14

Carvalho AV, Vasconcelos MAM, Silva PA, Aschere JLR (2009) Production of third generation snacks by extrusion-cooking of pupunha and cassava flour mixtures. Brazilian J Food Technol 12(4):277–284

Clement CR (1991) Pupunha: uma árvore domesticada. Revista Ciência Hoje. Volume especial, pp 43–47

Clement CR (1999) Revista da pupunha. Disponível em. http://www.inpa.gov.br/pupunha/revista/clement

Clement E, Galdino NO (2008) Heart of palm (Bactris gasipaes Kunth): mineral composition and kinetics of oxidative enzymes. J Sci Technol 28(3):540–544

Clement CR, Mora JU (1987) The pejibaye (Bactris gasipaes H.B.K.): multi-use potential for the lowland humid tropics. J Econ Bot 41(2):302–311

Clement E, Pastore GM (1998) Peroxidase and polyphenoloxidase, the importance for food technology. Food Sci Technol 32(2):167–171

Esrey KL, Joseph L, Grover SA (1996) Relationship between dietary intake and coronary heart disease mortality: lipid research clinics prevalence follow-up study. J Clin Epidemiol 49(2):211–216

Ferreira CD, Pena RS (2003) Hygroscopic behavior of the pupunha flour (Bactris gasipaes). Food Sci Technol 23(2):251–255

Ferreira SAN (2005) Pupunha, Bactris Gasipae Kunth in: Ferraz IDK; Camargo JLC. Manual de sementes da Amazônia. Fascículo 5. 12p. INPA Manaus Brazil.

Galli C, Simopoulos AP, Tremoli E (1994) Effects of fatty acids and lipids health and disease. World Rev Nutr Diet 27:152

Garbanzo CR, Pérez AM, Carmona JB, Vaillant F (2011) Identification and quantification of carotenoids by HPLC-DAD during the process of peach palm (Bactris gasipaes H.B.K.) flour. Food Res Int 44:2377–2384

Garcia DE, Sotero VE, Lessi E (1998) Caracterización de la fraccion lipídica de tres razas de Pijuayo (Bactris gasipaes H.B.K.). Folia Amaz 9(1–2):29–43

Gomez WS, Amelotti G (1983) Composiziones della sostanza grassa Del fmtto di Guilielma speciosa (Pupunha). Riv Ital Sostanze Gr 60:767–770

Hammond EG, Pan WP, Mora UJ (1982) Fatty acid composition and glyceride structure of the pejibaye palm (Bactris gasipaes H.B.K.) mesocarp and oil. Rev Biol Trop 30(1):91–93

Innis SM (1991) Essential fatty acids in growth and development. Prog Lipid Res 30:39–103

Kapp EA (2003) Tempo de preservação de tolete de palmito pupunha (Bactris gasipaes) minimamente processado e armazenado sob refrigeração. Publicatio 9(3):51–57

Kerr LS, Clement CR, Clement RNS, Kerr WE (1997) Cozinhando com a pupunha. Instituto Nacional de Pesquisa da Amazônia, 95p

Leakey RRB (1999) Potential for novel food products from agroforestry trees: a review. Food Chem 66:1–14

Leterme P, García MF, Londõno AM, Rojas MG, Buldgen A, Souffrant WB (2005) Chemical composition and nutritive value of peach palm (Bactris gasipaes Kunth) in rats. J Sci Food Agric 85:1505–1512

Lima FEL, Menezes TN, Szarfarc SC, Fisberg RM (2000) Acidos graxos e doenças cardiovasculares: uma revisão. Rev Nutr 13(2):73–80

Marques PAA, Coelho RD (2003) Estudo da viabilidade econômica da irrigação da pupunheira (Bactris gasipaes H.B.K) para Ilha Solteira Cienc Rural 33(2):291–297

Morera JA (1988) Caracterizacion de los estipetes de pejibaye (Bactris gasipaes) en base a las espinas. Agron Costarric 13(1):111–114

Moro JR (1993) A breeding program for Bactris gasipaes (Pejibaye Palm). Acta Hort 360:135–139

Neves EJM, Santos AF, Rodhigheri HR, Junior CC, Bellettini S, Tessmann DJ (2007) Cultivo da pupunheira para palmito nas regiões sudeste e sul do Brasil. Embrapa Floresta 143, 9p

Oliveira AMMM, Marinho HA (2010) Development of the basis of panettone peach palm flour (Bractis gasipaes Kunth). Aliment Nutr 21(4):595–605

Pantoja LO, Maeda RN, Andrade JS, Pereira N, Carvalho SMS, Astolfi-Filho S (2001) Bebida alcoólica fermentada a partir de pupunha (Bactris Gasipaes Kunth). Biotecnologia Ciência e Desenvolvimento 43(19):50–54

Pesce C (2009) Oleoaginosas da Amazônia. Museu Paraense Emílio Goeldi. Nucro de Estudos Agrários e Desenvolvimento Rural, 334pPiedra MF, Metzler AB, Mora JU (1995) Contenido de ácidos grasos em cuatro poblaciones de pejibaye (Bactris gasipaes). Rev Biol Trop 43:61–63

Resende JM, Fiori JE, Junior OJS, Silva EMR, Botrel N (2004) Processamento do palmito de pupunheira em agroindústria artesanal: uma atividade rentável e ecológica. Embrapa Agrobiologia, Sistema de produção

Salas GG, Blanco A (1990) Un alimento infantil com base em pejibaye: su desarollo y evaluación. Bol Inf 2(2):12–14

Shanley P, Medina G (2005) Frutíferas e plantas úteis na vida Amazônica. CIFOR, Imazon, Belém, 304p

Silvestre S (2005) Frutas Brasil Fruta. Empresa das artes. São Paulo, 321p

Tango JS, Carvalho CRL, Soares NB (2004) Caracterização física e química de frutos de abacate visando a seu potencial para extração de óleo. Rev Bras Frutic 26(1):17–23

Vargas A (1994) Evaluacion de brotos de pejibaye para palmito em relacion com su posición em La cepa y bajo 2 formas de colocación Del fertilizante. Corbana 19(41):15–17

Ye JC, Chang WC, Hsieh DJU, Hsiao MW (2010) Extraction and analysis of B-sitosterol in herbal medicines. J Med Plants Res 47:522–527

Yuyama LKO, Cozzolino SMF (1996) Efeito da suplementação com pupunha como fonte de vitamina A em dieta: estudo em ratos. Rev Saúde Pública 30(1)

Yuyama LKO, Aguiar JPL, Yuyama K, Macedo SHM, Fávaro DIT, Afonso C, Vasconcelos MBA (1991) Determinação de elementos essenciais e não essenciais de pupunheira. Hort Bras 17(2):91–95

Zapata A (1972) Pejibaye palm from the Pacific Coast of Colombia, a detailed chemical analysis. Econ Bot 26(2):156–159

Zumbado ME, Murillo MG (1983) Composition and nutritive value of pejibaye (Bactris gasipaes) and pejibaye meals for animals. Rev Biol Trop 32:51–56

Index

A
Aging, Pálinka, 319
Alcoholic products
    classification, 49
    Khao-Maak, 49, 50
    Loog-Paeng, 50, 51
    Nam-Dtaan-Mao, 51, 54
    Ou, 51, 53
    Sato, 50, 52
Almond
    extended storage, 395
    fatty acid composition, 394
    fruit (nut), 394
    genotype dependent, 394
    minor components, 395
    oxidation, 395
    solvent extraction, 395
    triacylglycerols (TAG), 394
Amaranth crop, America
    current challenges, 229–230
    description, 218
    functional properties (*see* Functional
        properties)
    history, 217, 218
    leaves, 223, 224
    potential industrial applications, 229
    seed, 218–223
Amylopectin, 219
Amylose/amylopectin ratio, 227
Ancient Greece, Mead, 330

Animal products
    cooked/roasted rice, 42
    Naem, 43, 44
Argentine Food Code, 123, 124, 132
Argentine quality protocol, 133–135
Austrian dumplings
    beef soup, 149, 150
    boiler cultures, 140
    bread, 142, 147
    cooking and parboiling methods, 140
    equipment, 140
    garnish/main dish, 150–152
    historical development, 139
    Klöße, 139
    knife, 141
    potato, 147, 148
    prototypes, 140
    qualification, 141
    Romans, 140
    soup, 148–150
    sweet, 152–155
Ayran, 121

B
Backhendl (fried chicken)
    history, 258, 259
    ingredients and preparation, 259, 260
Baking chamber, 161
Biopeptides, 220

© Springer Science+Business Media New York 2016   407
K. Kristbergsson, J. Oliveira (eds.), *Traditional Foods*, Integrating Food
Science and Engineering Knowledge Into the Food Chain 10,
DOI 10.1007/978-1-4899-7648-2

BIS. *See* Bureau of Indian Standards (BIS)
Boiler cultures, 140
Boknafisk, 266
Brazilian charqui meats
    carne seca, 293
    classes, 293
    feijoada, 294
    financial incentives, 296
    jerked beef processing, 294, 295
    manufacture processes, 295
    preparation, 294
    Quechuan language, 293
    salting and drying, 294
Bread dumpling, 142, 147
Bread production
    banking, 161
    kneading, 159
    mechanical and heat energy, 158
    shaping, 160
    tank fermentation, 159, 160
    types, 158
    weighing-dividing shaping and resting, 160
Brewing, German beer
    boiling process, 308
    cold liquor, 308
    compounds, 306
    enzymatic processes, 306
    flower, 307
    mashing process, 307
    withering and curing, 306
    wort and forms, 306
Brewing liquor, 312
Bulgarian Yogurt
    acid milk, 116
    development, sheep farming, 116
    fair, 118, 119
    fresh milk, 118
    home preparation, 118
    homeland, 118
    katuk, 116
    kuthuk, 116
    lactic acid bacteria, 117
    microflora, 116
    microorganism, 117
    prokish, 116
Bureau of Indian Standards (BIS), 103

C
"*Cabrito da Gralheira*" (PGI)
    animal health, 27
    demographic date, kids meat, 23–24
    kid's meat, 21
    KMO value, 25

    mean value, 24, 25
    perceived quality, 27
    quality assurance and tradition, 27
    Serrana breed goats (*Capra hircus*), 21
    sociodemographic consumer, 22
    Spearman's correlation, 25
    statistical analysis
        Cronbach's alpha coefficient, 22
        IBM SPSS, 22
Carne seca, 293
Cement, 87
China, Mead production, 329, 330
Cod, salting and drying
    black membrane, 283
    brine, 284
    chemical changes, 286
    convection, 285
    curing methods, 283
    dehydration and protein leaching, 284
    desalting, 288, 289
    factors, 278
    fish
        chemical composition, 280
        components, 282
        freshness, 281
        granulometry, 282
        marine, 282
        parasites, 281, 282
        quality parameters, 280, 287
        sodium chloride, 282
        species, 279
    halophilic bacteria, 286
    operations, 283
    pickling, 284
    Portuguese industry, 285
    preservation techniques, 278
    protein denaturation and reduction, 286
    seafood consumption, 278
    sensory properties, 277
    sodium chloride crystals, 286
    split fish thickness, 286
    storage, 287, 288
    structural and mechanical properties, 284
    water content, 283
Common agricultural policy (CAP)
    EU quality/financial schemes, 18
    geographical environment, 18
    legal protection, 18
    PDO/PGI kid meat, 19
    traditional and regional food products, 17
    wine and spirits, 18
Consumer aspects, Turkish traditional foods
    area and region, 87, 88
    black tea, 86

childhood, 88
coffee, 86
    production, 87
    purposes, 86, 87
Coronary heart disease, 222
Cured lamb thigh (Fenalår)
    coastal regions, 270
    salting and drying process, 270, 271
    smoking, 270

D

Dahi
    Aryans, 103
    attributes, 104
    Ayurveda, 103
    body and texture, 104
    Charaka Samhita, 103
    colour and appearance, 104
    curd, 102
    flavour, 104
    Leuconostocs, 103
    local confectionery, 103
    manufacture, 103, 104
    milk, 103
    panchamrut, 102
    Sushruta Samhita, 103
    technological developments, 105, 106
DDL. *See* Dulce de leche (DDL)
Desalting, 288, 289
Distillation, Pálinka, 319
Dobos cake
    history, 363–365
    home, 366, 367
    Hungarian confectionary, 362–364
    Hungary, 361
    recipe, 365
Dried fish
    Boknafisk, 266
    history, 261
    Klippfisk/*Bacalao* (cliff fish), 264
    Lutfisk, 264, 265
    raw material, 262
    stockfish (Bacalao), 262, 263, 265, 266
    traditional and modern processing, 262
Dry wines, 324
Dulce de leche (DDL)
    Argentine Food Code, 123, 124
    Argentine production, 125
    Argentine regulations, 125
    breakfast, 123, 124
    formulation and implication, 130–132
    gastronomic and cultural heritage, 124
    homemade, 124

Latin American country, 125
    manufacturing process
        evaporation, 125
        food industry, 129
        heat concentration process, 126
        homogenization stage, 129
        industrial manufacturing stages, 126
        internal view, 126, 129
        lactose crystals cause, 129
        packing equipment and operations, 129
        provision and production matrix,
            126, 127
        requirements, 125
        safety and quality requirements, 126
        stainless steel jam pail, 126, 128
        sucrose, 126
        traditional jam pail, 126, 128
        ultra filtration technology, 130
    milk caramel, 123
    quality and consumer aspects, 132, 133, 135
    SAGPyA, 124

E

Egypt, Mead production, 330
English/French bread production, 158
Erdepfel, potato, 147
Eszencia, 325
European consumers
    cross-sectional consumer survey, 6–7
    focus group discussions, 6
    Free Word Associations, 6
    gastronomic traditions and eating habits,
        14
    perceive TFP (*see* Traditional food
        products (TFPs))
    truefood (EU FP6), 14
Extra virgin olive oil, Slovenian Istra
    coastal area, 380
    complementary economic activity, 379
    property, 380
    quality and traceability guaranteed, 381, 382
    quality parameters, 382, 383
    sensory characteristics, 384
    sort selection, 382

F

Fermentation, Green table olives
    microbiota evolution, 373
    pH and total acidity, 373, 374
    sugars and phenolic compounds, 374, 375
Fermentation, Mead, 335
Fermentation, Pálinka, 318

Fermentation, Thailand
  development, 53, 57
  microorganisms, 51, 54–57
  technology/equipments, 57
Fermentation–maturation–storage
  alcohol and carbon dioxide, 308
  aluminum/stainless steel, 309
  brewing process, 308
  categories, 308
  compounds, 308
  cylindro-conical tanks, 309
  oxygen, 308
  uniform beer quality, 309
Fishery products
  Budu, 36
  categorization, 32
  classification, 32, 35
  fermentation processes, 32
  fruits and salt, 41, 43
  Kapi, 37, 38
  Nam-Pla, 32–37
  Pla-Paeng-Daeng, 39, 40
  Pla-Raa, 38, 39
  Pla-Som, 39, 40
  salt and carbohydrates, 38
  Som-Fak, 41, 42
Fluffy curd cheese dumplings, 153, 155
Fordítás, 325
Free Word Associations
  stereotype response behavior, 6
  XLSTAT 2006 v.4 software, 6
French bread baking
  consumption, 157
  Egyptians and Babylonians, 157
  fermentation stage, 160
  kneading process, 159
  materials, 157
  ovens, 161–165
  production (see Bread production)
  types, 157, 158
Functional properties
  emulsification, 226
  foaming, 225, 226
  gelling
    frequency sweeps, 227, 228
    proteins, 227
    starches, 227, 228
  hydration, 225

G
Garnish dumplings
  bread crumb/potato, 150
  diced bacon and bread rolls, 151

Klosterneuburger, 151
salad and mushroom sauce, 152
German beer
  biotechnology, 300
  brewing, 306–308
  brewing liquor, 312
  cereal grains, 304
  civilization, 299
  coke, 303
  communal drinking, 300, 301
  cooler and wetter climate, 301
  doppelbock, 303
  fermentation–maturation–storage, 308, 309
  filling and packaging
    barrels, 311
    bottling, 311
  filtration, 310
  glucose, 301
  hops, 312, 313
  malt, 304–306, 312
  microbes, 300
  Neolithic revolution, 299
  sulfate-rich water, 303
  types, 303
  weight and density, alcohol, 302
  yeast, 313, 314
Goa bean, 194
Green salted cod, 285
Green table olives
  applications, 370
  aromatic herbs, 371
  classification, 369
  harvesting and production
    color, 376
    fermentation, 372, 374, 375
    packing and storage, 376
  industrial processes, 370
  lactic acid bacteria, 371
  organoleptic properties, 370
  sources, 369

H
Hazelnut
  fatty acid composition, 396
  properties, 396
  solvent extraction, 397
  tocopherol content, 397
Honey, Mead production, 327
Hops, 312, 313
Horchata
  carbohydrates, 348
  components, 348
  composition data, 347

elaboration process
  characteristics, 352
  production, 350–352
  tigernuts, 350
lipids, 348
mineral fraction, 347
preservation treatments, 352–357
production, 351
protein fraction, 348
tigernuts, 348, 349
tropical and subtropical areas, 347
vegetable beverage, 347
Hungarian confectionary, Dobos cake, 362, 363
Hypocholesterolaemic effect, 220

I

Indian traditional fermented dairy products
  characteristics, 102
  classification, 102
  cultures, 101
  dahi, 102–106
  fresh bovine milk, 102
  lactic acid bacteria, 102
  lassi, 106, 107
  milk, 102
  misti dahi, 109, 111
  principles, 101
  requirement, 101
  shrikhand, 107–109

K

Kaiser-Meyer-Olkin (KMO), 22, 25
Kashkaval
  cheddar types, 120
  homemade, 121
  recipe, 121
Katyk
  contemporary recipe, 120
  lactic acid bacteria, 119
  milk, 120
  national dairy product, 119
  preparation, 119, 120
Klippfisk/*Bacalao* (cliff fish), 264
Klöße, 139
Klosterneuburger dumpling, 151, 154
KMO. *See* Kaiser-Meyer-Olkin (KMO)
Knife dumpling, 141
Knödelhenker, 142
Knödelwürger, 142
Knöderln, 148

L

LAB. *See* Lactic acid bacteria (LAB)
Lactic acid bacteria (LAB)
  clusters, 177, 179
  genetic characterization, 177, 179
  leavening action, 174
  rye flour and water, 173
  16S rDNA gene sequence, 179–181
Lactobacillus bulgaricus, 116
L-ascorbic acid (L-ascorbate), 200
Lassi
  advantages, 106
  consistency, 106
  flavour, 106
  salted version, 106
  technological developments, 106, 107
Leaves, Amaranth
  amino acid profiles, 224
  anti nutrient components, 224
  antioxidant activity, 223
  mineral ratio, 223
  toxic factors, 224
Legume grains
  ammonia, 190
  animal feed, 202, 203
  dry mature seeds, 191
  factors, 192
  food source, 190
  human food and animal feed, 190
  Mediterranean diet, 192, 193
  nomenclature and classification, 189, 190
  nutrients, 190
  nutrition (*see* Nutrition, Legume grains)
  soybean, 191
Lithuanian Mead
  alcohol, 343
  development, 342
  honey, 342, 344
  natural berry, 344
  production technique, 343
  spices, 343
  technology, 344, 345
  traditional classification, 344
Lithuanian spontaneous rye
  characterization, 175
  factors, 175
  genetic characterization, 181, 182
  LAB, 175, 177–179
  lactic microflora, 176
  organoleptic and structural properties, 175
  phenotypic characteristics, isolates, 177, 178
  physicochemical and microbiological characteristics, 176

Lutfisk
    cooking, 265
    mellow treatment, 264
    NaOH, 264
    serving, 265
Lye treatment, 370

M

MADR. *See* Ministry of Agriculture and Rural
    Development (MADR)
Main dish dumplings. *See* Garnish dumplings
Malt, German beer
    cylindroconical vessels, 305
    enzymes, 304
    factor, 304
    ingredients, 304, 312
    natural germination process, 304, 305
    withering, 305
Máslás, 325
Mead
    Ancient Greece, 330
    caramelised honey, 333
    China, 329, 330
    consumer aspects
        aroma and flavour, 338, 339
        nutrition, 338
    distillates, 335
    Egypt, 330
    evidence, 328
    fermentation process, 329, 335, 336
    fortified, 335
    honey, 327
    honeymoon, 332, 333
    melomel types, 333, 334
    microflora, 336, 337
    microorganisms, 336
    pictorial archaeological records, 328
    plain, 335
    reformation, 332
    Roman Empire, 331
    spices, 334
    storage/ageing, 332
    sugar-containing foods, 328
    ultrafiltration, 337, 338
    variations, 328
    wine grapes, 331
    Wort Pasteurisation prior, 337
Mediterranean diet
    components, 192
    gastric adenocarcinoma, 193
    healthy lifestyle, 192
    insulin resistance, 193
    lipid consumption, 192

Melomel types, Mead, 334
Microscopic and macroscopic approach,
    bread-baking
    3D tomography image, 167, 168
    factors, 170
    internal pressure and thickness, 167, 168
    organoleptic characteristics, 166
    product mass, 169
    SEM image, 163, 166
    standard conditions, 165, 166
    steam injection, 167, 170
    thick and crunchy brown crust, 166
    types, 166, 167
Midus
    alcoholic content, 341
    Lithuanian Mead, 342–344
    meadmakers, 342
    numerous forms, 342
Milk caramel, South America. *See* Dulce de
    leche (DDL)
Ministry of Agriculture and Rural
    Development (MADR), 64
Misti Dahi
    firm curd, 110
    industrial manufacture, 109
    manufacture, 111
    preparation, 109
    technological developments, 111

N

Napkin dumpling, 148
Natural resources
    forest fruits, 72
    fungi, 73–74
    herbs, 71–72
    medicinal plants, 74
    spices and seasonings, 72–73
Neolithic revolution, 299
Nockerln, 150
Non alcoholic products
    Kanom-Jeen, 45, 47
    Kanom-Thuai-Fu, 46, 48
    Khao-Daeng, 46, 48
    Thua-Nao, 46, 49
Non-salted fermented foods
    alcoholic products, 49–51
    category, 45
    non alcoholic products, 45–49
Non thermal technology, 356, 357
Nut oils
    coronary heart disease and cancer, 389
    non ttradional
        Almond, 394, 395

Hazelnut, 396, 397
Pistachio, 395, 396
Walnut, 391–394
production, 390, 391
Nutrition
bread
kvasok, 80
podplamenník, 81
vahan, 80
fruit dishes
lizak, 83
ripe and unstoned plums, 82
fungal dishes
Boletus reticulatus, 83
Cantharellus cibarius, 83
paprikash, 83
legume dishes, 81
legume grains
carbohydrates, 195, 196
Goa bean, 194
lipids, 195, 197–199
minerals, 199–201
non-native environments, 194
nutraceuticals, 201, 202
phytonutrients, 194
proteins, 196, 197
sources, 193
split pea and fava bean, 193, 194
vitamins, 198–200
mash, 78
pasta dishes
halushky, 78
mrvenica, 78
Pirohy, 79
trhance, 78
soups, 77
demikat, 77
kisel, 77
vegetable dishes
kapustnica, 82
kapustníky, 82
meadow plants, 81

O
Oil production
cold-pressed seed oils, 391
minor constituents, 391
packaging, 391
raw material, 390
Omphacomel, 334
Ovens, baking
categories, 161
crumb, 164, 165

crust, 164
description, 161
expansion, volume, 165
loading, 162
microscopic and macroscopic approach,
165–167, 169, 170
steam, 162
temperature, 163
time, 164
types, 162

P
Pálinka, Hungarian distilled fruit
aging and storage, 319
alcoholic fermentation and distillation, 315
distillation, 319
fermentation, 318
mashing, 317, 318
raw materials
apricot, 316
berries, 316
cherry, 316
pear, 316
plum, 316
quince, 316
wild fruits, 316
specialities, 319
PGI. *See* Protected geographical indication
(PGI)
Physicochemical and microbial changes,
Xuanwei ham
color
metmyoglobin, 243
nitrate/nitrite, 243
oxymyoglobin, 243
lipids
glycerides, 244
intramuscular lipids, 244
oxidation, 244
phospholipids and free fatty acids, 245
triglycerides and phospholipids, 244
microflora
dry-cured ham, 246
"green growth", 247
humidity and temperature, 247
mycotoxins, 247
Streptomyces bacteria, 246
proteins
DPP, 245
endogenous enzymes, 245
microbial enzymes, 245
nontoxigenic strains, 246
proteolytic microorganisms, 245

Pistachio
    components, 395
    extended storage, 396
    fatty acid composition, 396
    tocopherol content, 396
Potato dumpling
    Erdepfel, 147
    napkin, 148
    region, 143–146, 148
    wrapping, 147
Potential industrial applications, 228
Preservation treatments
    drying, 355, 356
    heat, 354, 355
    horchata, 353
    low temperatures, 353, 354
    non thermal technology, 356, 357
    sporulated and fecal microorganisms, 353
    thermal/heat processing, 352
    tigernuts, 352
Prevention of Food Adulteration Act, 103
"Produits du terroir", 14
Prosopis sp., traditional food products
    cultivation and potential uses
        abundance and ecologic behavior, 210
        arid-temperate zone, 210
        attributes, 209
        food properties, 211
        leguminous family, 209
        natural coverage, 210
    pods and flours
        criollos, 212
        crude protein, 213
        fruit varies, 211
        lomento drupáceo, 211
        plants, 212
    regional activities, 213, 214
Protected geographical indication (PGI),
        17–19, 21, 63
    attributes, 20
    Cabrito da Gralheira (see Cabrito da
        Gralheira (PGI))
    European policy (see CAP)
    gastronomic heritage, 19
    Portuguese consumers' quality perceptions,
        21
    taste, 20
Pulses, 189
Pupunha (Bactris gasipaes)
    carbohydrates and lipids, 402
    carotenoids, 404
    composition, 403, 404
    consumption, 402
    feature, 401

    fiber, 404
    gluten and richness, starch, 404
    heart, palm, 401
    lipid fraction, 403
    microbial strains, 403
    nutritional properties, 403
    tocopherols, 404
    water and salt, 402

Q
Quality and consumer aspects
    Argentine Food Code, 132
    Argentine quality protocol, 133–135
    food quality, 132
    requirements, 132, 133
Quality parameters, 382, 383
Queen of forage crops, 202

R
Ram balls (Værballer), 271
Rio de la Plata, 124
Roman Empire, Mead, 331
Romanian, traditional food
    bakery products
        beaten bread crust, 69
        wood sticks, 69
    dairy products
        fermented cheese, 66, 68
        micro units/sheep farms, 66
    MADR, 64
    PDO, 63
    PGI, 63
    TSG, 64
    WTO agreements, 62
Rye bread production, Baltic region
    advantages, 185, 186
    ancient art, 174
    commercial process, 183–185
    European supermarkets, 173
    Lithuanian spontaneous rye, 175–181
    pleasant sweet tone, 174
    recipe and technological parameters,
        183, 184

S
SAGPyA. See Secretariat of Agriculture,
        Livestock, Fisheries and Food of
        Argentina (SAGPyA)
Salted fermented foods
    animal products, 42–44
    category, 32

fishery products (*see* Fishery products)
fruit and vegetable, 44, 45
Secretariat of Agriculture, Livestock,
        Fisheries and Food of Argentina
        (SAGPyA), 124
Seed, Amaranth
    amino acids, 221
    amylopectin, 219
    biopeptides, 220
    components, 219
    coronary heart disease, 222
    dietary fiber, 221
    minor components, 222, 223
    morphological structure, 218, 219
    physicochemical features, 222
    protein fractions, 220
    squalene, 221
    starch granules, 219, 220
    thermal properties, 220
Sheep and Lamb in Norway
    cured lamb thigh (Fenalår), 270
    history and tradition, 267, 268
    Ram balls (Værballer), 271
    Smalahove, 268–269
Shrikhand
    characteristics, 107
    culture, 107
    industrial method, 107, 108
    semi solid mass, 107
    sweetened milk product, 107
    technological developments, 109
    traditional method, 107
    UF *chakka*, 109, 110
Sirene, 120
Slovakian traditional foods
    animal husbandry
        cows, 76
        pigs, 76
        sheep and goats, 76
    cultivated species
        cereals, 75
        fruit plants, 76
        hoed-crop, 75
        legumes, 75
        oil-bearing plants, 75
        tea, 76
        vegetable, 75
    forest fruits, 72
    fungi, 73–74
    herbs, 71–72
    medicinal plants, 74
    nutrition (*see* Nutrition)
    socioeconomic development, 84
    spices and seasoning, 72, 73

Smalahove
    EU directives, 268
    head, preparation, 269
    serving, 269, 270
    slaughtering season, 268
SNF. *See* Solids-not-fat (SNF)
Solids-not-fat (SNF), 103
Soup dumplings
    boil up, milk, 149
    bread crumbs and fry, 150
    Knöderln, 148
    Nockerln, 150
South Portugal and Spain (*Muxama* and
        Estupeta)
    consumption, 276
    fishing and food resource, 273, 274
    Tuna loins, 274, 275
Stockfish
    drying method, 262, 263
    heads and backbones, 263
    Iceland, 262
    Lofoten, 263
    raw material, 263
    whole fish, 263
Stockfish (Bacalao), 265, 266
Sucrose, 126
Sweet dumplings
    apricots, 153
    browned bread crumbs, 153
    honey and poppy seeds, 152
Szamorodni, 325

T
Table olives, Slovenian Istra
    characteristics, 386
    fermentation process, 383
    lactic acid bacteria, 384, 385
    sensory characteristics, 386
    traditional production technology, 384
Tafelspitz (boiled beef)
    history, 257
    ingredients and preparation, 258, 259
Tank fermentation, 159, 160
Tarator, 122
Thailand, traditional fermented foods
    fermentation, 51, 57
    kanom-jeen, 31
    non-salted fermented foods, 45, 46, 49, 51
    safety aspects, 58
    salted fermented foods, 32, 37–39, 41, 42,
        44, 45
    storage and preservation, 31
Tigernuts, 350

Tokaji Aszú
    history, 321, 322
    labelling wines, 321
    official national anthem, 321
    raw materials, grapes, 322, 323
    traditional technology, 323, 324
    types, 325
Traditional Bulgarian dairy foods
    Ayran, 121
    historical and cultural regions, 115
    Kashkaval, 120, 121
    Katyk, 119, 120
    products, 115
    raw materials, 115
    Sirene, 120
    sociological and ethnological studies, 115
    Tarator, 122
    tillers and stock breeders, 115
Traditional food products (TFPs)
    consumer perception, 12–14
    consumers' agreement, 9, 10
    image, 12–14
    origin and locality, 7
    processing and elaboration, 7
    qualitative exploratory research, 9, 11
    sensory properties, 8
    truefood consumer study, 4, 11
    typical traditional food consumer, 11
Traditional speciality guaranteed (TSG)
    EU food law, 64
    MADR, 64
Tuna loins
    fishing and food resource, 273, 274
    traditional processing, 274, 275
Turkish traditional foods
    category, 89
    consumer aspects, 86–88
    culture, 85
    diversity, 89
    factors, 89
    historical influences, 90
    homes/small-scale enterprises, 86
    innovation, 91, 96
    lifestyle and economic conditions, 89
    products, 91
    regions, 91–95
    transmission, 89
Tyrolean bacon dumplings, 151, 152

V
Viennese meat specialities
    Backhendl (fried chicken), 258–260
    minerals, 254
    proteins, 253
    Tafelspitz (boiled beef), 257, 258
    vitamins, 253, 254
    Wiener Schnitzel–Viennese Schnitzel,
        254–256

W
Walnut
    fatty acid composition, 392
    human nutrition, 391
    hydrocarbon fraction, 393
    phenotypic variation, 392
    seed (kernel), 392
    sterol content, 393
    storage and processing, 394
    tocopherol content, 393
    volatile compounds, 393
Wiener Schnitzel–Viennese Schnitzel
    history, 254, 255
    ingredients and preparation
        baking, 256
        coating, 256
        meat, 255
        plating, 255

X
X-ray computerized tomography, 167
Xuanwei ham, 243–247
    crossbreeds, 248
    green ham, 240, 241
    with industrial processing and consumer
        demands, 248
    industry development, 238
    lipids and proteins, 241
    market research, 248
    microflora, 248–249
    microorganisms, 248
    physicochemical and microbial
        changes (see Physicochemical
        and microbial changes,
        Xuanwei ham)
    physicochemical properties
        amino acids and volatile organic
            compounds, 242, 249
        dry-cured ham, 249
    pig production, 239
    salting procedure, 240
    Wujin breed, 238

Y
Yeast beer, 313, 314

Printed in the United States
By Bookmasters